THE
HOMEOWNER'S
ENERGY
HANDBOOK

The mission of Storey Publishing is to serve our customers by
publishing practical information that encourages
personal independence in harmony with the environment.

EDITED BY Deborah Burns and Nancy D. Wood

ART DIRECTION AND BOOK DESIGN BY Carolyn Eckert

TEXT PRODUCTION BY Gary Rosenberg and Theresa Wiscovitch

COVER ILLUSTRATION BY © Michael Austin/Jing and Mike Company

INTERIOR ILLUSTRATIONS BY © James Provost, except for © Michael Austin/Jing and Mike Company, 3, 14, and 104; and Ilona Sherratt, 127

ILLUSTRATION EDITING BY Ilona Sherratt

PHOTOGRAPHS AND GRAPHICS COURTESY OF: Paul Scheckel, 27; National Fenestration Research Council, 85; PowerWise Systems, Inc., 99; www.energystar.gov, 73 top left; Web Energy Logger, 103; U.S. Department of Energy, 127; Kestrel Renewable Energy, kestrelwind.co.za, 141; UN Development Program, 262

INDEXED BY Nancy D. Wood

TECHNICAL EDIT BY Philip Schmidt, with contributions from Hilton Dier III, Bret Hamilton, David House, Chris Kaiser, and Ian Woofenden

The information in this book is true and complete to the best of our knowledge. All recommendations are made without guarantee on the part of the author or Storey Publishing. The author and publisher disclaim any liability in connection with the use of this information.

Storey books are available for special premium and promotional uses and for customized editions. For further information, please call 1-800-793-9396.

Storey Publishing
210 MASS MoCA Way
North Adams, MA 01247
www.storey.com

Printed in the United States by Courier
10 9 8 7 6 5 4 3 2 1

Library of Congress Cataloging-in-Publication Data

Scheckel, Paul.
  The homeowner's energy handbook / by Paul Scheckel.
    pages cm
  Includes index.
  ISBN 978-1-61212-016-4 (pbk. : alk. paper)
  ISBN 978-1-60342-847-7 (e-book)
  1. Renewable energy sources—Handbooks,
      manuals, etc.
  2. Dwellings—Insulation—Handbooks, manuals, etc.
  3. Architecture and energy conservation—Handbooks,
manuals, etc. I. Title.
TJ808.S328 2013
696—dc23
                                    2012032592

THE HOMEOWNER'S ENERGY HANDBOOK

Your Guide to Getting Off the Grid

PAUL SCHECKEL

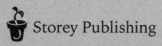

Storey Publishing

For Silas,

who reminds me every day

of the boundless renewable energy

inside us all.

Acknowledgments

THIS BOOK IS not, and could not be, anything close to what I imagined at the conceptual stage. It is much more and much better. For this, I owe a debt of gratitude to all those who offered their time, experience, expertise, encouragement, and support for this project.

It has been nothing short of a pleasure to work with the dedicated professionals at Storey Publishing. I specifically want to thank Deborah Burns for her enthusiasm, followed by much patience with my seemingly endless delays; Nancy Wood, whose organizational and editing skills kept me on track, and who continued to be patient (and even encouraging) with even more delays; Phillip Schmidt for his incredibly thorough technical review and ability to draw clarity from foggy bottom; and James Provost for his excellent illustrations.

There are many more to thank for their various contributions, including:

You, the reader, for taking the time to understand and explore renewable energy and energy efficiency. Without individual motivation and action, nothing will be as you imagine it could be.

My family, for their extended patience during the frenzy of endless deadlines and with my obsession with getting the hands-on projects just right. The porch, garage, yard, garden, and even the closets are littered with the remnants of prototypes.

The following professionals provided advice, guidance, technical review, and personal contributions: Ian Woofenden, *Home Power* magazine; William David House, Vahid Biogas; Bret Hamilton, Shelter Analytics; Hilton Dier, Solar Gain; Gordon Grunder; John Dunham; Jim O'Riordan, O'Riordan Plumbing and Heating; Tina Webber, Swing Green; Paul Gipe, wind energy expert and advocate; *Home Power* magazine, for all things renewable; Tracy Vosloo, Kestrel Wind Turbines; Chris Kaiser and Powell Smith, Mapawatt Blog; Josh Van Houten for help with advanced chemistry; Paul Harris, the University of Adelaide; Lori Barg, Community Hydro; Chris Pratt, Open Sash; United Nations Development Programme; Aprovecho Research Center; The Journey to Forever Project; Solar Energy International; Cornell University, Waste Management Institute; Spenton LLC, wood-gas camp stoves; Powerhouse Dynamics, home energy monitors; Blue Line Innovations, home energy monitors; PowerWise energy monitoring dashboards; Jock Gill, biomass expert and advocate; All Power Labs, wood gasifiers; NASA's Langley Research Center.

Contents

PART TWO: RENEWABLE ENERGY

Introduction

AS I WRITE this, I have just lit the fire on my first batch of homemade biogas. Food scraps and pig poop have been successfully transformed into a gas, similar to propane or natural gas, that we can cook with. Biogas is the combustible result of the decay that happens in nature just as easily as the sun shines or the wind blows. The challenges in harnessing these energetic gifts from nature lie in collecting, controlling, storing, and often transforming the primary energy resource into a form that can be used to meet a particular need. Much of this book is all about exploring the options for meeting your energy needs through natural resources, along with the processes involved in focusing their potential toward some particular need.

For years I've been fascinated with the idea of making biogas, but I was intimidated by what appeared to be complex and exacting science in the recipe requirements for optimum gas production. But experience is the best teacher and, after all, this simple process of biomass decay happens all by itself in nature. So how hard could it be to create the environment for gas to not only happen, but to actually be produced? I found a 55-gallon airtight barrel in the "inventory" (as I like to call it; my wife calls it something else) behind the garage. I dumped in a 5-gallon bucket of compostable food scraps and a smaller bucket of poop from our two pigs, along with a pile of grass clippings. Then I filled it halfway up with water and waited. One week later, combustible gas was bubbling out of the barrel — no exact recipe or scientific calculations required.

The Road to Self-Empowerment

MAKING YOUR OWN energy is, in every sense of the word, empowering. Watching the biogas burn reminded me of the feeling I had when I bought my first solar panel 25 years ago and set up an off-grid room in the rental house I shared — thrilling! This feeling of independence set me down the path of exploring renewable energy and energy efficiency as both a vocation and avocation.

I've spent lots of professional and hobby time exploring the processes involved in meeting my family's energy needs through natural and locally available resources. I do this mostly because it's fun but also because I want to wean myself off the myopic, destructive, and often corrupt global power structure of energy addiction. So, in addition to all the fun I've had, my family also enjoys a certain level of autonomy from the energy supply machine, and that feels pretty good. But that's just me. You may have different reasons for making (and saving) energy.

What's the Alternative?

Renewable energy sources have been dubbed "alternative energy," a phrase that marginalizes the true value of these traditional energy sources. The fossil fuel era will be a tiny blip in human history, and nuclear energy is not only too expensive, it also continues to struggle with its own, self-made image. These modern fuels are really the "alternative" to traditional energy. Humanity will need to use what nature offers, without breaking the budget of the natural capital available to all of us and against which we have borrowed heavily.

Nature has been providing earthly inhabitants with abundance for millennia, and these natural resources are available to all. The processes of harnessing energy work best if you apply a little knowledge and provide a catalyst to get things started, then simply get out of the way and let nature do its thing — and don't ask for more than you need. Making your own energy comes with a new awareness of efficiency and facilitates a change of perception around comfort and convenience.

Cost-Effectiveness

AS AN ENERGY efficiency and renewable energy consultant, I find it challenging to overcome some of the rationale my industry uses to sell efficiency. And I need to be honest: The industry is filled with energy geeks who are passionate about what they do, and this generally is a good thing. However, most of us enjoy talking about details that put the average person to sleep. You want a yes-or-no answer; we reply with building-science theories and numbers.

When it comes to promoting the benefits of energy efficiency improvements, many of us have focused primarily on cost savings, simple payback, and return on investment. Financial payback is not why we buy most things, but it's always the first question when it comes to energy improvements at home. Cars, couches, and music do not offer financial returns and are not usually considered investments, yet we buy them not just because we need them but because they make us feel good. Sitting on a couch listening to music is better than sitting on the floor in silence. But if your couch is next to a drafty window, it's your comfort that's at stake, not your financial situation.

I've avoided lengthy discussion about economic evaluations in this book because if you're reading

---

We are like tenant farmers chopping down the fence around our house for fuel when we should be using Nature's inexhaustible sources of energy — sun, wind, and tide.... I'd put my money on the sun and solar energy. What a source of power! I hope we don't have to wait until oil and coal run out before we tackle that. — THOMAS EDISON, INVENTOR (1847–1931)

---

it, you probably have a number of motivations for wanting to make and save energy. But if you're stuck on the idea of monetary cost-effectiveness, the following are some simple ways to compare costs and return on investment (just keep in mind that simple analysis involves making guesses about certain things and leaving some important considerations out of the equation).

SIMPLE PAYBACK CALCULATION

To get a ballpark idea of the value of energy production from a renewable energy system, first add up your installation costs for the system. Then, multiply the annual maintenance costs by the expected life of the system. Add the two sums to find the lifetime operating cost of the system. Next, figure the energy production of the system over its lifetime. Divide the lifetime operating costs (installation and maintenance) by the lifetime energy production value to find the cost per unit of energy. Here's an example using a solar electric power generating system:

- A 3-kilowatt PV (solar electric) system costs $15,000 to install after all incentives (if you've taken out a loan for the system, don't forget to include the cost of financing). Assume it will produce reliable power for 25 years. PV systems do not require much maintenance if you don't have batteries, but assume that you will need to replace the inverter every ten years at a cost of $2,000 (this is a guess, because who knows what inverter technology and cost will be in 10 years). Your lifetime cost to own and operate the system is $19,000.

- You expect power production to be about 3,600 kilowatt-hours (kWh) per year. Multiply that by 25 years, and you'll have produced 90,000 kWh of solar electricity over the system's life.

- $19,000 divided by 90,000 kWh = $0.21/kWh

- Compare this to current electricity prices, and try to guess what the price might be in 25 years. Estimates of energy escalation (the rate at which energy costs will rise over the rate of general inflation) range from 1 to 10 percent, depending on whom you ask. Recent

history proves that the energy sector is highly volatile in many ways, and economics react quickly to volatility.

UP-FRONT COST AND LIFETIME COST

When replacing older appliances in your home, it pays to spend a little more up front for a more efficient model. Buying the cheapest product often results in the highest lifetime cost, with the cost of energy to operate the equipment often far exceeding the purchase price. The incremental cost of the high-efficiency choice often will more than pay for itself over the lifetime of the appliance.

If you're hoping to supply your home with a renewable energy source, consider the cost of increasing the size of your renewable system to meet the additional needs of lower-efficiency equipment. Buying energy is almost always more expensive than saving it. As you'll see in chapter 5, it's probably worthwhile to evaluate how much energy an appliance currently is using to see if the energy savings alone are worth the cost of upgrading. Don't just assume that because it's old, it must be an energy hog.

More than Money

BEYOND SAVING MONEY, there are so many more reasons to "get efficient" and to make your own energy. Many of the non-energy-related benefits of lowering your energy consumption fall into more subjective or emotional categories (which is how we humans tend to make decisions) and may indeed outweigh the cost savings of making efficiency improvements.

Nice to know, though, that energy efficiency upgrades and renewable energy systems are among the few things you can buy that will actually pay for themselves over time, promising to offset the cost of that kitchen renovation you've been wanting. If you want a more detailed analysis of costs, savings, and carbon footprint, you can find links to energy analysis tools on my website (see Resources). The tools are fairly simple to use, but they require some knowledge of your existing and proposed conditions.

Here's a partial list of benefits or "value propositions" (in addition to saving money) to consider when you're thinking about making a change in your energy situation. You may have your own need or desire to add items to the list.

- Using less energy
- Increasing comfort
- Reducing carbon footprint
- Improving energy security
- Gaining energy independence and autonomy
- Stabilizing energy bills
- Diversifying your energy portfolio
- Increasing resilience to natural disasters and interruptions in energy supply
- Reducing financial risk from exposure to unstable energy markets
- Increasing home market value
- Reducing home maintenance
- Increasing convenience with modern, efficient, and smarter appliances
- Minimizing mechanical systems size and cost
- Eliminating or reducing reliance on air conditioning and humidity control
- Spending money locally to reduce the money you send out of your state or country
- Using less water
- Improving indoor air quality, occupant health, and safety
- Making your house more durable
- Reducing animal intrusion
- Minimizing noise between outdoors and in
- Adding storage space with a dry basement
- Reducing mold, dust, and other allergens
- Eating more fish (mercury emissions from coal power plants poison fish and anything that eats fish)

- Reducing combustion-related particulate-matter emissions (helps reduce chronic asthma)
- Feeling good just by knowing you can do it yourself, to be independent and meet your family's needs

Limitations and Opportunities

EACH CHAPTER IN this book could have been a book in itself. My intention is to offer the basics of specific technologies and present enough information so that you can understand the principles, without getting overly technical. If you want to go deeper, there are references to more information to help increase your expertise in any given area.

Where practical, some chapters offer a do-it-yourself project that can give you a feel for how things work. This hands-on learning experience will help you understand how to make and use your own energy and, with a little trial and error, expand beyond the experimental approach to help meet your energy needs. As with any good hobby, you can spend as much or as little time as you like in the pursuit of homemade energy, with varying degrees of success. Once you get started, you will discover new ways of making things work better in your specific situation.

Industrialized societies like to think big. When building a commercial enterprise, there are economies of scale to consider, but when it's just you and the family to provide for, the economics change drastically. The key ingredient is your own motivation for (at least partially) meeting your resource needs. As for other ingredients, I encourage you to use resources that are available locally, especially waste material. Anytime you can use waste material to generate energy, the "energy profit ratio" of that material greatly increases (see Defining EPR on the following page).

Efficiency of Systems

NOTHING IN THIS world yet discovered will deliver greater than 100-percent efficiency in terms of energy used, produced, converted, or transferred. While new heat and hot water equipment might offer fuel efficiency percentages in the high 90s, these values don't account for the energy required to extract and deliver the fuel to your home.

In most cases, you're doing great if you get 50 percent efficiency from a system. Consider that most energy systems take the raw, or primary, energy resource and convert it to the desired form. Each conversion saps energy from the original source and effectively reduces overall process efficiency.

For example, electric heating elements are 100 percent efficient at converting electricity into heat. No amount of fancy gizmos or slick marketing will ever change that fact. However, the process of generating the electricity and getting it to your home might be only 30 percent efficient. If you want to get your electricity from the wind, you need to convert the kinetic energy in the wind to mechanical energy that turns the generator, which converts it into the electrical power that you want. That entire process might be 25 percent efficient. If the electrical energy produced from the wind generator is stored in a lead-acid battery, itself having an efficiency of about 70 percent, you will need to put 30 percent more energy into the battery than you take out in order to keep the battery charged.

I can't emphasize enough that the efficiency of renewable energy systems is a distraction. Capturing free and abundant renewable resources does cost something, but the power keeps on coming. The measure most of us care about is cost-effectiveness, which means: How much will it cost to get a quantity of energy over a period of time? How you value that energy is not always about money.

SYSTEM EFFICIENCY = 3%

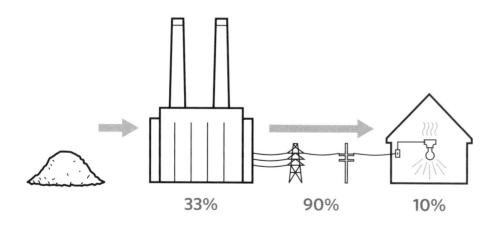

33%   90%   10%

◀ **The efficiency of a system is determined by multiplying the efficiencies of each component of that system together. A coal power plant might be 33% efficient at converting coal to electricity; the power distribution network (power lines, switches, transformers between the plant and the house) can lose up to 10% to inefficiencies; and an incandescent light bulb is only about 10% efficient at converting electricity to light. The overall efficiency of converting coal into light in this case is only about 3%, with the remaining 97% of the energy being lost as heat.**

Have Fun and Be Careful

THE CONCEPTS IN renewable energy are relatively simple, but the details and nuances may take a lifetime to master. This book is intended to engage you in understanding the basics, but know that when you are done you'll have only scratched the surface.

While you can learn the basics in the comfort of your reading chair, combining this knowledge with even a small project will give you the hands-on experience you need to gain the skills and confidence for taking your energy goals further.

Finally, don't be careless. There are no good shortcuts. Shortcuts create problems, cause premature failure, impact components in unforeseen ways, and increase the risk of personal injury. If you're unsure about how to do something, consider all the different and potentially catastrophic ways something can fail, and consult a professional or your favorite curmudgeonly naysayer.

I applaud your willingness to take responsibility for your energy needs. I sincerely hope you will experience the joy of invention in harnessing the opportunities offered by our home planet to meet those needs. I guarantee you will find many rewards if you take it personally, and take action!

Only Humans

Humans are the only inhabitants of the earth to use energy that does not come directly from the sun. All life is in a constant struggle to gain more and more energy in different forms — it seems to be the basis for a species' success. Whether you're on or off the grid or you use renewable energy or not, energy prices and energy impacts will continue to grow. Any efforts to reduce your use will allow you to meet your needs with fewer resources, expand your energy options, and help to put humans on a sustainable energy path.

PART ONE

HOME ENERGY EFFICIENCY

BEFORE INVESTING IN renewables for your home, it's always best to reduce your energy consumption. A big part of doing this is becoming aware of how and when you use energy in your home. Do you really need your living spaces to remain at a constant temperature and humidity level throughout the year, or can you learn (or relearn) to widen your range of comfort? After you've completed your own home energy audit, you can decide where best to focus your efforts for comfort and cost control.

Most of us do not associate homes with performance, as you might when comparing which car to buy. But with the proper understanding, along with a strategic approach and attention to detail, your home can be transformed from an old clunker into an ultra-efficient, high-performance place of comfort and energy autonomy.

Efficiency can become a challenge for you — and even a game for the whole family — to see how low your energy use can go. Both game and challenge can be enhanced through real-time monitoring and tracking with a home energy monitor, offering an immediate sense of reward and consequence. Make a plan, keep track of your savings, and consider the future keepers of your home — and the planet.

Getting Ready for Renewables

BACK IN THE early 1990s, when I was converting gasoline-powered cars to electric operation, my partners and I thought this was obviously the next step in transportation. Technology was indeed ready, and we even found a few customers, but we quickly discovered that it was much easier to convert cars than to convert driver behavior.

Batteries are the limiting factor in the viability of electric vehicles, and efficient driving habits can increase the range of travel by up to 50 percent. Try as we did, it seemed nearly impossible for many to unlearn wasteful driving habits. Stomping on the accelerator from a dead stop does nothing to increase an electric car's speed, but it puts a huge drain on the batteries.

When it comes to energy use, most of us have some of the same bad habits with our houses as with our cars. Getting your home — and yourself — ready for renewables requires understanding what renewable resources you want to use, as well as the practicalities of how to harness that energy. This chapter will help to guide your decision-making, with a broad view of what's required "behind the scenes."

The Big and Little Energy Picture

THERE IS MUCH that you can do to get ready for renewables. You can do it all at once, if resources allow, or you can take baby steps with incremental improvements to make your home "renewable-ready," ultimately resulting in substantial savings. Shifting toward renewables requires that you closely examine your relationship with energy and your expectations around how it serves you — and accept changes in both. Where is your energy coming from now, and how will you get it, or make it, in the future? On the fossil fuel main line, we don't just buy energy, but we also buy the convenience of throwing a switch or turning a dial to meet our needs with very little planning, thinking, or heavy lifting required. We have come to expect energy to be an invisible and seamless part of our lifestyle, keeping us in a very narrow bandwidth of comfort, without being anywhere near the raw materials, processes, or awareness of this huge global infrastructure.

Yet fuel prices are rising fast, as are concerns about all the resources required to sustain increasing levels of imported energy — resources ranging from money to land to human lives. When you think about it, it seems crazy to take fuels like natural gas and oil (which are highly processed, costly, complex, and difficult to obtain), transport them around the world, and burn them for heat. It's crazier still when you realize that the industry

◀ Renewable energy resources are all around us. Sun, wind, water, trees, and even food waste can be converted to heat, hot water, or electricity.

of refining petroleum products is the single biggest energy consumer in the United States.

All of these factors make homegrown options attractive on many levels. Equivalent heat is available from the sun or from "low-grade" (minimally processed, easily obtainable) biomass fuels found much closer to home. Drawing from local sources to heat your home can be as simple as planning at the design stage to take advantage of solar heat gain, or as complex as using a ground-source heat pump to scavenge heat from the earth and redirect it into your home. Somewhere in the middle of the cost and complexity scale might be installing a wood stove in your home,

RECOGNIZING OUR ENERGY BAGGAGE

For perspective: If we each took responsibility for societal energy impacts, every American would take receipt of one-third of a pound of high-level radioactive waste (currently stored at nuclear reactor sites), plus another quarter-ounce produced on our behalf each year. We would also receive a few ounces of airborne mercury that is produced by burning coal and eventually rains down onto our soils and water.

The fish we can't eat due to the high mercury content in their flesh cannot be decontaminated. Much land has been rendered uninhabitable due to oil spills, nuclear accidents, and hydropower reservoir construction. Energy security is a huge expense.

We are all complicit in filling the bags of energy liability because we continue to demand the lifestyle that cheap energy brings us. We have reached the point where we can't force that bag closed even by sitting on it! The zipper has popped and the baggage is spewing out.

or adding an active solar thermal or solar electric system with substantial collection area and sufficient storage capability, coupled with effective and efficient delivery of that energy.

YOUR ENERGY VIEW-SHED

Society has energy choices to make, and so do you. These are not easy decisions because there are economic, environmental, social, political, and personal values associated with any energy source. These values can be viewed from both micro and macro vantage points, and you can bring your own value propositions into the discussion.

The question is: What do you want in your backyard — or in policy-speak, your energy "view-shed"? In terms of energy impact, the view from your house might have both visible and invisible components. You might see electric transmission lines, coal-burning power plants with their smoke-stacks, or wind turbines. Less visually dramatic but equally tangible impacts of energy production include the effects of smog when you breathe. You might not see an entire mountaintop removed to get at the coal or uranium underneath, but you know that the natural world has been affected.

You'll experience fewer impacts, a better view, and easier rest if you and your neighbors are using energy efficiently. If you opt for energy status quo, then you also choose to accept your share of the pollution and other impacts of various generation sources.

THE EFFECTS OF OUR CARBON FOOTPRINT

The term "carbon footprint" refers to the annual amount of greenhouse gases (GHG), or the gaseous emissions of substances that have been shown to contribute to the effects of climate change, for which every human is directly or indirectly accountable. There are many such gases, and they are often expressed in terms of **carbon dioxide equivalency** (CO_{2e}) because CO_2 is the primary GHG.

The average American has a carbon footprint of about 23 tons per year, nearly 80 percent of which comes from burning fossil fuels. That's enough carbon dioxide (CO_2) to fill over 18 average homes full of this potent greenhouse gas. If you took all the oxygen (O_2) out of the 23 tons of CO_2, you would have a pile of carbon weighing almost 6½ tons. This is often called carbon equivalency or C_e.

Where do we store these gases other than the closet of our atmosphere? And the fallout, literally and figuratively, is that we must put up with days where ozone, smog, and unexpectedly high particulate matter limit our ability to breathe while increasing societal health care costs.

Take a Deep Breath

Our nation spends $15 billion each year on asthma medication. The increase in people suffering from asthma is partly due to increases in air pollution from dirty coal-burning power plants around the world. And we all share the same air. In America, many eastern states are downwind of coal power plants in the Midwest. When it's hot in Ohio, for example, locals crank up their air conditioners, and that ultimately raises the level of asthma-triggering particulate matter in the air of New England states.

To continue down the road of energy status quo means accepting continuous environmental degradation from acid rain and adjusting to a shifting global climate. We'll have to accept mining disasters, wars, increasing military budgets, and political destabilization as we wrangle for resources, as well as displaced people, extinguished wildlife, rising costs, and continued warnings to avoid certain foods that absorb various pollutants. And, every once in a while, we'll need to accept radioactive rain and witness productive farmland being reduced to wasteland. The environment in which we live is a closed system in motion. In terms of energy impacts, our backyards have expanded to include the entire planet.

▲ Animals are completely dependent on natural resources for survival. Humans have the capability to capture, harness, and manipulate these resources to serve specific needs. Our only limitations are imagination and desire.

COMPARING ENERGY TRADEOFFS

All energy systems have a downside, and renewable energy sources are not to be excused for their contribution to environmental and social impacts. Land is submerged for huge hydropower operations, and people are displaced from their homes. Substantial wind resources are often found within sensitive ecosystems where roads and construction projects will take their toll. These installations may have a permanent impact on the landscape and local environment, but they do not have the continuous impacts of resource exploitation and pollution production that fossil energy sources do.

Basic Considerations

MAKING YOUR HOME ready for renewables means taking incremental steps toward meeting your own energy needs with minimal impacts. This can be done in a relatively pain-free way while you're doing other renovation work, as long as you plan for the eventuality of integrating renewables into your home and lifestyle.

ON AND OFF THE GRID

The "grid" is the network of electrical generation (power plants), transmission (power lines), and distribution (transformers, controls, and switching stations) required to generate and move electrons to your home. This varied network ties communities, states, and regions together much in the same way superhighways are connected to rural roads. (Imagine cars as the electrons, having been generated at an auto manufacturing plant and pushed out onto the highways.)

Nearly every home in the developed world is connected to the power grid, and eventually this grid will be "smart" enough to move electrons around the country seamlessly, taking them from wherever they're generated (such as the solar panels on the roof of your house) to wherever they're needed (such as a neighboring state where a power plant has just been closed for maintenance).

Remote rural places may be so far from power lines that — if you want electricity — there is no power grid to connect to. Your choices are either to bring power lines to your site and connect to the grid, or build your own off-the-grid power generation system where you keep all the electrons you make. Typically, the term "grid" is applied specifically to the electrical power network, but it can be used metaphorically (as I often do) to refer to the similar network of fossil fuel production and delivery systems. How far "off the grid" can *you* get?

As you make your plans, consider your needs:

- What do you want to use renewables for?
- How much energy will you need after efficiency improvements?
- What equipment is needed and where will it go?

At the same time, consider available resources that can help you meet those needs:

- Do you live in a sunny place?

- Do you have at least an acre of land and plenty of wind?

- Do you own wooded property?

- Are you near falling and/or flowing water?

- Are you a farmer with excess crop waste and manure to manage?

- Are you (or do you know) a restaurant owner with waste vegetable oil and/or food scraps to dispose of?

- How much energy can the available resources yield?

Of course, everything starts with efficiency. You don't want to buy or produce energy only to lose or waste it to inefficiencies. Reduce the energy use of your dwelling and maximize efficiency through weatherization and upgrades to heating and cooling systems and appliances. A small energy footprint allows you to meet a greater portion of your energy needs through a diversity of options.

Reducing your use means that a smaller, less costly renewable energy system can meet a greater portion of your needs. If you're off-grid, it also means less reliance on a fossil fuel generator to keep your system's batteries charged.

While it's certainly possible for renewables to meet all of a home's energy requirements, it's typically most cost-effective to supply up to 80 percent of your annual energy needs with renewables. This is because adding sufficient capacity for that last 20 percent, which you may need only for a short-lived, worst-case scenario, can double the system capacity and cost. In most cases, the technology is relatively manageable. The more challenging aspect to consider is your lifestyle and how much you are willing to personally engage in the process of assembling and managing your own energy systems.

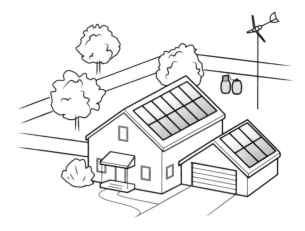

▲ An efficient home, inhabited by a conscientious family, can harness a diversity of natural resources to meet the majority of its energy requirements.

MANAGEMENT AND MAINTENANCE

If you're a hands-on person and are able, willing, and available to address maintenance issues as they arise, you might have the flexibility to experiment with various approaches and systems to see what works best for you. On the other hand, if this level of oversight isn't realistic or desirable, you'll probably need to hire a professional to install and maintain a more conventional system, and you will need to budget for maintenance. Set realistic goals around how much purchased energy can be offset with renewables.

Don't forget to look into state and local zoning restrictions that may affect structure height or the visual appearance of your home. Be ready to work with neighbors and officials to change laws that may prohibit the use of solar collectors, wind towers, micro-hydroelectric turbines, or even clotheslines.

Living with renewable energy will change your awareness of that resource. To make the most use of it, you'll need to be open to changing your habits to live within the requirements of renewables and possible constraints on their availability.

Defining Btus

A British thermal unit (Btu) is the amount of energy it takes to raise the temperature of one pound (one pint) of water by $1\,°F$, or about the same energy released when a wooden match is completely burned. The energy output of many heating, cooling, and cooking appliances is measured in Btus.

RENEWABLE ENERGY OPTIONS

There's a good chance that your home is more ready for renewables than you might think. Sunlight falls everywhere, and proper design (or redesign) of your home can allow you to take advantage of solar heat gain to offset your heating needs while also reducing your cooling loads.

Even if your home isn't ideally oriented, a rooftop or yard with good access to the southern sky offers the potential to harness solar energy for use in water heating and electrical power for the home. A reasonable flow of wind or water offers additional opportunities for home power generation. Wood is widely available in split logs (cordwood) or pellets, and either option can be a good choice for home and water heating (see below).

HEATING AND COOKING WITH WOOD

If cordwood or pellet heat is in your future, you'll need to consider the best equipment and setup for your needs. Decide if you prefer cordwood or pellets, then if you want to burn the fuel in a stove or in a furnace or boiler. Some wood heating systems can be located outdoors. Outdoor systems have the advantage of putting the equipment and mess outside but come with the added expense of moving the heat to the indoors. The chimney is another factor. Inside chimneys perform better and last longer than chimneys on the outside of your house, simply because they are warmer (yielding a better draft) and aren't exposed to the elements.

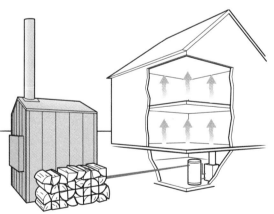

▲ Outdoor wood boilers are becoming a popular choice for home and water heating. Be sure to select a model that meets US EPA guidelines for low particulate emissions. Less smoke is not only an indicator of efficiency, but will surely lead to better relations with your neighbors.

Next comes the question of where to store a ton or two of pellets or a few cords of wood to keep them dry and to minimize handling. Wood pellets are often sold in 40-pound bags, so they can go anywhere. Bulk pellets require a large storage bin, or hopper, near the heat system. Cordwood, on the other hand, has a high moisture content. If you store it indoors, much of that moisture ends up in the air as the wood dries, and this can lead to problems with mold or mildew. Better to keep it outside under a well-ventilated shelter.

Wood can also be used for cooking, and not just in a campfire. The top of a wood stove is a convenient place to keep a kettle of water for hot drinks, or to simmer or fry a meal. A hot bed of coals is a great place to bury and cook potatoes or assorted veggies wrapped in foil. A homemade metal box (with air vents to control the temperature) sitting a couple of inches above the top of a wood stove can be an effective way to bake.

A pound of wood packs a heating value of around 8,000 Btus, roughly equivalent to cooking on a gas range set to a high flame for about 30 minutes. If you're looking for a fuel-free cooking solution, the sun is the obvious option. But solar cooking is a slow process and limited to locations with plenty of sunlight. Even in an ideal location, the sun delivers between 200 and 300 Btus per hour for each square foot of solar collector area. Solar cooking is certainly cleaner than wood cooking, but with good stove design, wood can become much more clean-burning and efficient. See chapter 12 to explore this option.

Systems and Planning

WHETHER YOUR HOME is new or old, you can affordably make renewable energy part of your future by phasing it in to help control costs. For example, a little advance planning goes a long way toward minimizing labor costs of future installations of renewable energy systems. It's a good idea to work with a professional energy consultant or installation contractor to understand all the major components of the technology you're considering and to plan for things that will be hidden in the walls, such as wiring and plumbing.

We collect the sun's heat through the windows in our home and control it by opening and closing shades. The heat is stored in high-mass material and released when the sun no longer provides heat. ▶

COLLECTION, STORAGE, AND CONTROL

Renewable energy technologies have three basic interacting systems: collection, storage, and control. You must provide a place for each of these systems to reside in (or outside) your home, as well as means for them to interact with and connect to one another.

Integrated within these systems is the process of energy conversion. With solar hot water, for example, the solar thermal collectors may live on the roof (or on a ground-mounted rack) on the south side of the house, converting the energy in the sunlight into fluid heat. A solar hot water storage tank may live in the basement alongside a backup water heater, with controls (sensors, an electronic brain, and wires) between the collectors and storage tank to control fluid flow and heat exchange between the sun and the storage tank. Another important element, of course, is the plumbing that connects the collectors to the storage tank.

For solar electric systems, there are collectors (solar panels), possibly storage batteries, power management controls, and wiring between them all. If you're considering wind energy, you'll need to know where the tower will go and how the power cables will get to your batteries or interconnect with utility power lines.

With utility-connected (grid-tied) electrical systems, the grid serves as the storage facility to which electricity is delivered as it's produced, and from which electricity is drawn when needed. Renewable energy sources are seldom consistent, so matching the supply rate to storage capacity is an important consideration. Most energy systems will also have some sort of monitoring facility so that you can see what is happening within each subsystem.

A BUILDING PLAN

As you make home improvements or renovations, think about the master plan you have for all of the systems you're considering, and provide for their future inclusion as you work. One of the best things you can make available to your renewable energy installer is a **chase** (groove) between the roof and the basement that allows easy access and plenty of space for running wires, plumbing, or even a chimney without having to cut into walls.

If you're doing roof work, installing supports for future solar collectors will save time and money down the road when you're ready to put up your panels. This requires knowing the dimensions and mounting requirements of the equipment, so be sure to do your research. As mentioned, it's a good idea to bring in a system designer or experienced installer early in the process, and make sure the roof can handle the additional load of solar panels.

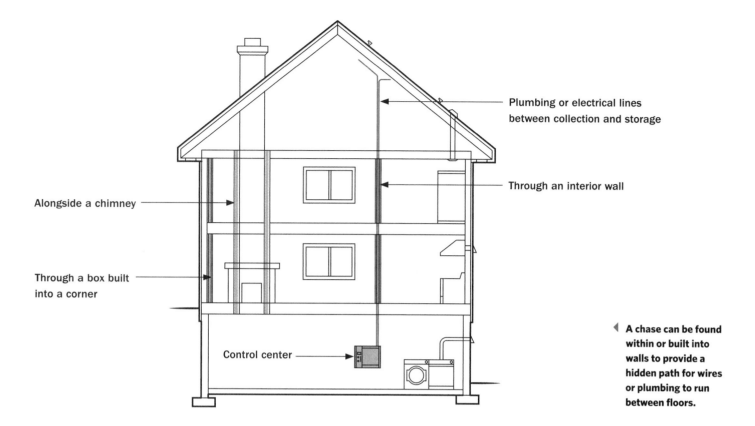

Plumbing or electrical lines between collection and storage

Through an interior wall

Alongside a chimney

Through a box built into a corner

Control center

A chase can be found within or built into walls to provide a hidden path for wires or plumbing to run between floors.

Renewable Habits

PUT TECHNOLOGY TO work for you, but don't expect technology to do it all. Living with renewable energy is about living within the means that nature provides. Adopt renewable habits, such as "one person, one light," and simply be aware of all energy being used in your home.

There are times when the sun doesn't shine or the wind doesn't blow, and those are times for conservation. But when nature gives, take advantage of the opportunity for abundance. For example, save your hot-water clothes washes for when the sun can heat the water.

Becoming aware of your energy habits and applying energy-smart strategies can make a big difference in the size and success of your renewable energy system. You can live in the greenest, most efficient dwelling and still use a ton of energy if you have not adopted efficient habits. Here are some tips to help enhance your renewable acumen.

Readiness tip #1: Increase your energy awareness by understanding what's happening in your house and why.

- Are there lights on that don't need to be?

Piggyback Projects

Knowing your long-term needs allows you to piggyback projects with little or no extra cost. When we did some work on our driveway, I took advantage of having a backhoe and crew already on-site. They dug out trenches for running conduit between my house and a future wind tower site, as well as for piping between rain collection barrels. It took less than an hour of additional backhoe time, and I ended up a step ahead on two projects.

- Do appliances have standby loads that always consume power?

- If you have a private water system, do you know when your well pump is on?

- Is the furnace pilot light on in the summer?

- Are the computer's energy-saving features turned on?

Readiness tip #2: **Assess your energy use** on every level by doing your own energy audit (see chapter 2). For example:

- Look at every outlet; know what's plugged in and why.

- Learn to read your electric and gas meters and understand where every last Btu or kilowatt-hour is going. Examine a year's worth of energy bills, look at monthly and seasonal trends, and think about what happens in your home during those periods.

- Try to determine how many fuel units are used for heating, hot water, air conditioning, and other electrical uses.

- Know something about everything in your home that uses energy — when it's needed and why, how much it uses while operating, and how best to control its operation.

Yes, money (and energy) *do* fall from the sky! ▶

Readiness tip #3: **Research products** when replacing lights and appliances, and use only the most efficient models you can find. The ENERGY STAR website (see Resources) is a good resource that lists thousands of products and their energy consumption. Plan ahead by researching for future appliance purchases so you know what you want. That way, if an appliance breaks down and you need to replace it right away, you'll know what to buy and not end up with an energy hog simply because it's on sale.

Readiness tip #4: **Adopt the most efficient practices,** preferably those that don't use any energy at all. These include:

- Hanging clothes to dry on a passive solar clothes dryer (a clothesline)

- Employing passive heating and cooling strategies

- Using solar-heated water

- Watching the cat or the kids (or the neighbors) instead of TV

- Taking advantage of nighttime air to cool your house with open windows and fans, then closing the windows and shades before the air warms in the morning

Readiness tip #5: **Control what you can.** This might include:

- Keeping the thermostat as low as you can in winter and as high as you can stand it in summer

- Making sure your water heater is set no higher than 120°F

- Installing low-flow showerheads and low-flow aerators on faucets

- Putting appliances with phantom loads (see page 41) on switched or automatically controlled power strips

- Turning off the water heater if you're away from home for more than a few days

- Keeping humidity levels under control by removing moisture at its source; if you must use a dehumidifier, pay attention to the relative humidity and do not over-dry the space or dry more space than necessary

- Using timer controls and occupancy sensors for lighting that tends to get left on

- Using switched power strips that allow you to turn things off (such as an entire entertainment system or office peripherals) with ease

Readiness tip #6: Minimize optional or discretionary uses of energy, such as clothes-drying, outdoor lighting, and use of air conditioning when outdoor temperatures are not life-threatening.

DOING YOUR HOMEWORK

As you explore practical renewable energy options that fit your needs, location, climate, and lifestyle, be sure to research state, local, utility, and federal incentives that may be available for renewable energy and energy efficiency projects. Contact your state energy office and local electric and gas companies about services and incentives, and ask a tax professional about applicable federal tax credits for efficiency upgrades and renewables.

In addition, here are some of my favorite resources that can help you identify incentives and keep you abreast of developments in renewable energy and energy efficiency (see Resources for websites):

- **Database of State Incentives for Renewable Energy** An online resource for incentives and policy information for renewables and energy efficiency improvements, including initiatives sponsored by states, local governments, utilities, and some federal programs.

- **Tax Incentives Assistance Project (TIAP)** Developed as part of the Energy Policy Act of 2005, this online resource helps homeowners and businesses make the most of federal income tax incentives for renewable energy and energy-efficient products and technologies.

- *Home Power* **magazine** An excellent resource for all things renewable. Content ranges from homeowner profiles to highly technical details

on all aspects of creating and maintaining home energy systems.

- *Home Energy* **magazine** Another good periodical and website devoted to all matters of efficiency. Though it's geared primarily to the energy professional, interested homeowners will find lots here to chew on.

The Value of Electricity

I READ SOMEWHERE that a human in decent physical condition can generate only about ¼ horsepower (hp) — or around 200 watts — of energy for any length of time. I got a chance to prove this when the Vermont Energy Education Program asked me to build a bicycle-powered generator for a school energy demonstration project.

Mounted on a modified bicycle trainer, the rear wheel of the bike drives an electrical generator that powers four light bulbs. First, I screwed in four 100-watt bulbs and jumped on the bike. I could barely move the pedals. It was like trying to bike up a vertical incline. I switched off three of the four bulbs and pedaled happily for a couple of minutes before breaking a sweat. Then a second bulb was switched on, and after a minute or two of producing 200 watts, the lights dimmed as my energy was drained and I pedaled more slowly.

To generate one kilowatt-hour (kWh) would require pedaling for five hours with two 100-watt light bulbs switched on (2 x 100 watts x 5 hours = 1 kWh). At a cost of only 10 or 20 cents from the power company, a kWh is a pretty good deal!

The instructors using the bicycle generator sometimes offer a $10 bill to any student who can produce 10 cents worth of electricity. When the 100-watt incandescent bulbs are replaced with high-efficiency, 25-watt fluorescent bulbs, all four light up with the equivalent light output of the four incandescents — and minimal complaint from the rider. Still, no rider has yet been willing to ride for long enough to take home the $10.

POWER EQUIVALENTS

Here are a few comparisons to help you get a feel for what a kilowatt-hour is:

- A gallon of gasoline contains the energy equivalent of over 36 kWh.

- A car battery stores less than 1 kWh of energy.

- It takes 861 food calories (more than ¼ pound of butter) to supply the energy equivalent of 1-kWh — about the amount of energy you'd burn up during two solid hours of high-impact aerobics.

If you like math, you'll like what James Watt did back in the 1700s. He determined that an average horse could lift a 550-pound weight one vertical foot in one second. This rate of work is now known as horsepower (hp), and can also be expressed as 550 foot-pounds per second. A horse can perform work, and so can electricity. Horsepower can therefore be expressed in terms of electrical power. Electric motors are often rated in horsepower.

In an ideal world, one horsepower is equal to 746 watts or 0.746 kilowatts. However, in reality no activity or process is 100 percent efficient, so a 1 hp motor will demand about 1,000 watts, more or less, depending on the motor's efficiency and how hard it is working. It's helpful to understand this relationship when determining the power consumption of the various motors around your home or when estimating the size of a motor you might need to do some work for you.

Energy Action in Cuba

BEFORE 1990, CUBA enjoyed robust trade with the former Soviet Union. Major components of the trade relationship were Cuban sugar and Russian oil. That partnership came to an abrupt end with the fall of the Soviet economy, which subsequently led to the crash of the Cuban economy. This started what the Cubans euphemistically call the "Special Period," when dramatic reforms and austerity measures became necessary for the nation's survival.

Due to the demise of their major trade ally, the Cuban people suddenly found themselves without jobs, money, or resources, and they suffered lengthy daily power outages. Cuban leaders understood that people needed basic services, but that sacrifices would have to be made.

The government immediately invested in public transportation, purchased one million bicycles, mandated energy-efficient lighting and refrigeration, upgraded its power grid, expanded the use of renewable energy, and developed electric rate structures that provided affordable electricity to meet basic needs while discouraging overuse.

Out of economic and practical necessity Cuba reduced its energy consumption by half over a period of four years. It has now become a global leader in practical, innovative approaches to energy efficiency, renewable energy, and community energy solutions — all on a very tight budget. Cuba also looked to increase international cooperation. It now exports technical expertise in health care and has its own solar electric panel assembly

Hydroelectric station ▶

facility, which meets its growing needs with some left over for export.

I was impressed by the small hydroelectric power station (above) that used 30-year-old Russian technology to provide power for 57 households. The same size system might provide enough power for four average American homes. Each family takes pride in some level of "ownership" of the station and understands the limitations of a finite resource. If one family is being an energy hog, the whole neighborhood feels it. The local school takes power priority and has a solar electric system as a backup.

You might think that all this frugality makes Cubans grumpy, but the society has worked such circumstances to its advantage. For example, the high price of chemical fertilizers (manufactured from fossil fuels) has facilitated advances in organic farming using locally produced compost as fertilizer. This saves money while growing healthy, local food in a closed-loop system.

Coffee was once an import, but the connection between agriculture and economy is very strong: Why pay someone else for something you have the resources to do yourself? The result, I'm happy to report, is quite delicious. Farming in that industry

has become a well-paid and highly sought-after vocation. One local grower offers benefits that exceed the standards of even progressive U.S. employers.

Throughout these struggles, every citizen has been provided with health care, a home, and education. But I'm not trying to put a happy face on all of this. Change is always a struggle, and there was substantial change on many levels. Not everything tried has worked. Many of those bicycles are now rusting away; despite good intentions, there was no infrastructure for repairing or even riding bicycles in many places. Also, the bicycles chosen were frumpy, single-speed, utilitarian clunkers rather than something one would actually look forward to riding.

Energy advances in Cuba were the result of dire circumstances that led to a quantum shift in awareness, policy, behavior, and community-level action. Island people generally have an innate sense of finite resources, and we can learn a lot from the Cuban response to hardship. For more information on Cuba's energy situation, I recommend a video called *The Power of Community — How Cuba Survived Peak Oil*, produced by Community Solutions (see Resources).

MY EXPERIENCE

Every once in a while, an experience can completely crumble some long-standing preconceived notion. Such was the case when I returned from a weeklong trip to Cuba with a group of 20 other energy professionals, sponsored by Solar Energy International and Global Exchange (see Resources for websites). The purpose of the trip was to explore Cuban solutions for curbing energy consumption and increasing the use of renewable energy in the face of dire circumstances.

Out of economic and practical necessity Cuba reduced its energy consumption by half over a period of four years. It has now become a global leader in practical, innovative approaches to energy efficiency — and on a very tight budget.

Bicycle Power

A bicycle is an extremely efficient mode of transportation. With a multi-gear drive mechanism, it can be modified to efficiently deliver mechanical power that can be used for anything from grinding grain to spinning an electrical generator.

There are several approaches to using a bicycle to provide electrical or nonmotive mechanical power. Options range from chain or gear drive to direct mechanical drive. The design featured in the project on page 29 uses a bicycle trainer that lifts the rear wheel off the ground, and replaces the trainer's resistance unit with an electrical generator. Mechanical power generated at the roller shaft is coupled directly to the generator shaft, and output voltage depends upon how fast you pedal.

To produce electrical power with a bicycle, you need to generate low-voltage direct current (DC), not alternating current (AC). This is because AC appliances require a steady voltage (120 or 240 volts) and frequency (60 or 50 hertz) for proper operation. That would require careful control of the generator's speed, expressed in rotations per minute (rpm).

When mechanical energy input, such as the spinning of a bicycle wheel or wind turbine, is used to create electricity, the speed (and therefore the frequency in the case of AC) will vary according to the force applied. In order to achieve greater versatility with power management, using DC power provides the necessary flexibility. Generating 12-volt DC power allows you to charge batteries, and battery power can be converted to AC power using an electronic inverter, or it can directly supply DC appliances, such as those made for camping and recreational vehicles.

Build a Bicycle-Powered Battery Charger

This project shows you how to build a 12-volt DC battery charger with a DC-to-AC inverter, so you can use the power generated to supply conventional (AC) electrical devices. I've built seven generators with this basic design, but no two are identical. Bicycle trainer designs are constantly changing, and this necessitates modification of the design. Since the trainer you use is very likely to differ from mine, there's no point in offering detailed templates, and the parts may vary. My goal is to provide enough guidance to get you on the right track using your own parts and materials.

The Parts

The most costly part required for this project is the DC generator, if you buy it new. However, you may be able to find a suitable used unit for a fraction of the cost of a new one. Check online for used and surplus electronic equipment suppliers.

The type of trainer I use has a friction roller that's driven by the bike's rear wheel. Prices for trainers vary widely, but magnetic and air-resistance types typically are the least expensive. You will remove the resistance unit and connect the generator to the shaft of the roller. Therefore, look for a simple trainer with a flat-bottomed roller mount and a roller assembly that can be removed.

The Wiring

The electrical wiring must provide safe and efficient means to transfer power from the generator to the battery. This requires the proper gauge of wire, solid electrical connections, a sturdy connector between the generator and battery and between the battery and inverter, and fuses to protect against over-current. You will also need a diode, which is an electronic "check valve" that allows current to flow in only one direction. Without the diode, electricity would flow from the battery and spin the generator as a motor, and you would have an electric bike.

This charger design uses clips on the wires that connect to the battery, allowing for easy disassembly and transportation. If you have a more permanent location for the bike and charger, consider using wire terminals that offer a more solid connection to the battery. Most of these parts can be found at auto or electrical supply stores.

The Generator

The generator essentially is a motor that can be operated in reverse to produce electricity rather than consume it: You supply the mechanical energy to spin the shaft, and the generator turns it into electricity. Choose a DC permanent-magnet motor (brush or brushless) rated between one-sixth and one-third horsepower (125 to 250 watts), capable of generating up to 20 amps of current and about 14 volts when spinning somewhere in the vicinity of 2,000 rpm (see Figuring RPM on the next page). The output voltage will vary with the generator speed, which depends on how fast the bike wheel is spinning.

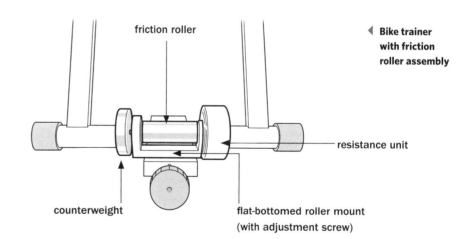

friction roller

◀ **Bike trainer with friction roller assembly**

resistance unit

counterweight

flat-bottomed roller mount (with adjustment screw)

Warning!

The electricity produced by this generator can be potentially lethal. Also, batteries contain harmful acid and can deliver incredible amounts of energy if the terminals are shorted (connected together). If you are uncertain about any electrical or mechanical aspects of this project, consult an expert or qualified professional who can help.

FIGURING RPM

The power specifications for this battery charger are based on using a 26" O.D. bicycle tire. I recommend using a smooth tire with little or no tread (to reduce noise) that is about 1" smaller in diameter and won't appreciably affect performance. Likewise, using a 27" or 700 mm tire will provide similar output.

The friction roller (on the trainer) has a diameter of 1.2". This gives you a drive ratio of about 22 to 1, meaning that for every one revolution of the bicycle wheel, the roller (and therefore the generator shaft) spins 22 times. If you ride the bike at a speed of 7.5 mph, the wheel spins at about 100 rpm, and the roller spins at around 2,200 rpm. Experience suggests that this is a reasonable compromise for effective power production by a wide range of riders.

MATERIALS
Trainer Assembly

Bicycle trainer
One 6" x 14" piece ¼" aluminum plate
12-volt DC motor/generator (Dayton 3XE20)
Two ¼" x 1½" #20 bolts
Two ¼" #20 nylon locking nuts
Four ¼" x 2" #20 bolts with eight nuts
One jaw coupling with I.D. (inner diameter) to match generator shaft O.D. (outer diameter)
One jaw coupling with I.D. to match friction roller shaft O.D. (the *body* size of both jaw couplings must be the same)
One jaw coupling insert sized to match both jaw couplings

Electrical: Generator to Battery

One ¼" spade-type crimp connector for 12 AWG wire (for diode anode lug)
Two 2" lengths of ¼"-diameter heat-shrink insulation tubing
One ring-type crimp connector for 12 AWG wire with ¼" hole (for diode cathode stud)

One 1N1190A diode
One 16" length 12 AWG wire
One wire nut (sized for two 12 AWG wires)
One automotive-type fuse holder
20-amp automotive fuse
Two battery clips to attach 12 AWG wire to battery (one clip should be red, the other black)
One 8" length of 1"-diameter protective wire loom
Two wire ties

Electrical: Battery to Inverter
(see Facts below)

Two 12" lengths and one 24" length of 4 AWG wire
Six ring-type crimp connectors with ¼" hole (for 4 AWG wire)
Eight 2" lengths of ½" heat-shrink insulation tubing
DC-rated fuse or circuit breaker (sized per inverter manufacturer specs)
Two battery clips (for 4 AWG wire; one clip should be red, the other black)
One 12-volt to 120-volt inverter
Battery
DC voltmeter

A FEW ESSENTIAL FACTS
Notes on the electrical parts used to connect the battery to the inverter:

- **Connectors:** The type and size of connectors and cables you use, as well as the kinds of connections you make, will be determined by your specific setup and size of fuses, cables, battery clips, and inverter connection. The wiring harness described here is sized for a 600-watt inverter and incorporates a resettable circuit breaker. Use smaller wire and connectors for a smaller inverter. See the table on page 179.

- **Heavy-gauge wire:** For greater flexibility, use fine-strand welding cable instead of standard battery cable.

- **Inverter:** Size the inverter according to the needs of the appliance you hope to operate. Anywhere from 200 to 600 watts would be an appropriate size for this project. True sine wave inverters will offer better performance than "modified" sine wave models, but the former are more expensive.

- **DC-rated fuse or circuit breaker:** Size this according to the inverter manufacturer's specifications. I used an 80-amp resettable DC-rated circuit breaker for a 600-watt inverter. The DC rating is critical, since DC-rated devices are electrically different from AC devices.

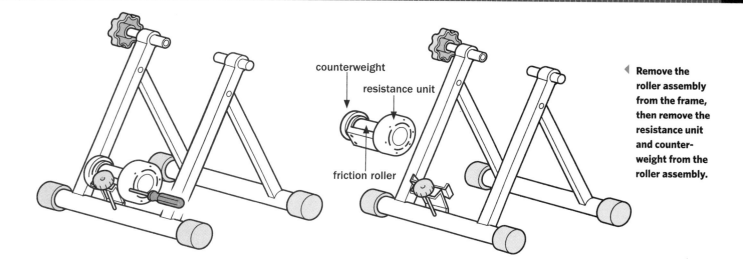

counterweight

resistance unit

friction roller

◀ Remove the roller assembly from the frame, then remove the resistance unit and counter-weight from the roller assembly.

1. Disassemble the trainer roller assembly.

The friction roller shaft will likely have a housing for the resistance mechanism on one end and a counterweight on the other end. Remove any screws as needed to get inside the housing, then disconnect the resistance mechanism from the end of the shaft. Remove the counterweight from the other end of the shaft; the weight is likely to be threaded onto the shaft. Remove any bolts or screws as needed to separate the roller assembly from the mount on the trainer frame.

2. Modify the friction roller.

You will connect the generator to the end of the roller shaft that had the counterbalance (which may have a screw thread) using a jaw-type coupling. The coupling won't hold well on threads, so you have to cut off the threaded portion, using a grinder with a metal cut-off disc or a hacksaw or reciprocating saw. Secure the roller shaft in a bench vise to hold it while you make the cut.

3. Cut and install the generator mounting plate.

The generator mounting plate supports the weight of the generator and connects it to the roller mount on the trainer, while the entire assembly remains adjustable to fit various sizes of bike wheels.

Using a 6" x 14" piece of ¼" aluminum, cut a notch to create a tab that slides underneath the roller mount. For my trainer, the notch measured 3½" x 2¾". Cut the plate with a jigsaw and fine metal blade.

Drill two ¼" holes in the roller mount for attaching the generator plate. Slide the aluminum plate underneath the mount, and align the plate so it's square with the roller mount, then clamp it in place. Poke a center punch through the holes in the mount to transfer their locations to the aluminum plate. Remove the plate and drill two ¼" holes through the plate. Clean up the holes and metal shavings, then attach the generator plate to the roller mount with two ¼" x 1½" #20 bolts and locking nuts. Note: You may need to cut a notch in the plate to provide clearance for the trainer's adjusting screw.

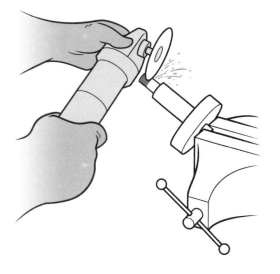

◀ Cutting off the threaded section of the roller shaft

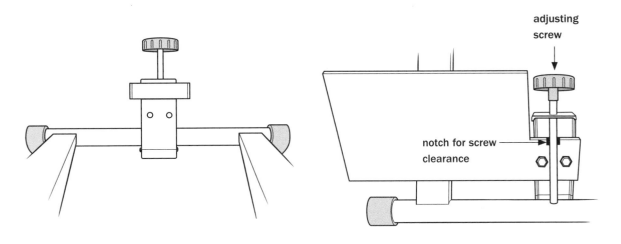

▲ **Top view: ¼″ holes drilled through roller bearing mount**

▲ **Bottom view: Generator mounting plate installed with bolts**

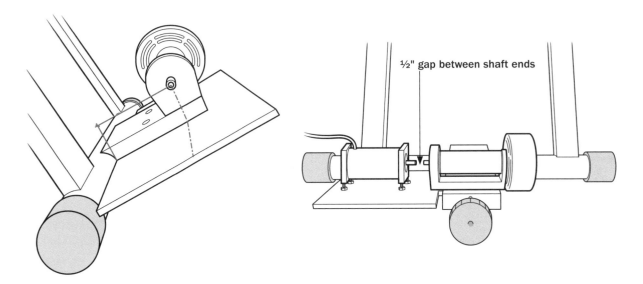

▲ **Transferring the centerline of the roller shaft to the generator mounting plate**

▲ **Aligning roller and generator shafts**

4. Fit the generator to its mounting plate.

Reinstall the friction roller onto its mount. Carefully measure where the centerline of the roller shaft extends over the generator mounting plate. Use a straightedge or small square to mark the centerline across the mounting plate; this line will serve as a guide for locating the holes for the generator mounting bolts.

Place the generator on the mounting plate and align the center of the generator shaft with the roller shaft, keeping their ends precisely ½″ apart (this space is required for the shaft couplings). Using the centerline on the mounting plate as a reference, carefully measure and mark the locations for the generator mounting holes. Note: It may be easier to complete the assembly by removing the mounting plate.

Drill four ¼″ holes through the plate for the ¼″ x 2″ #20 generator mounting bolts. The bolts will thread directly into the generator housing. To adjust the height between the generator and its

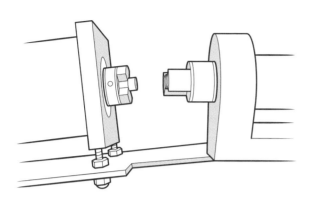

▲ Fitting the generator and plate into position, with the couplings loosely fitted onto the generator and roller shafts

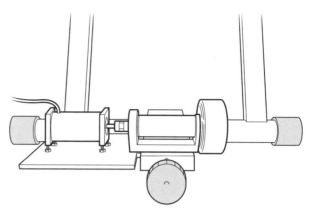

▲ Completed generator and roller assembly

mounting plate (for aligning with the roller shaft), add a nut on each side of the mounting plate, and thread the bolts into the generator housing. Later, you can adjust the nuts as needed to move the generator up or down, then tighten them to secure the generator in the correct position. Leave the nuts loose for now so you'll be able to move the generator for fitting the couplings.

5. Complete the mechanical assembly.

Slip a jaw coupling onto the end of the generator shaft, using the appropriate size coupling; do the same with the roller shaft. Install the rubber jaw coupling insert. Reinstall the mounting plate with the generator onto the roller mount while aligning the shaft couplings.

Use the double-nut system to adjust the generator height to align the two couplings.

Once the shafts are perfectly aligned, slide the jaw couplings together and tighten all hardware, including the setscrews, to secure the couplings to the shafts. Test-fit the bike on the trainer. **NOTE:** The mounting plate may need additional support so the generator's weight does not bend the plate and cause misalignment; a simple wood block custom-cut to size and tucked under the free end of the mounting plate will suffice.

6. Prepare the generator wiring.

Strip 2" of sheathing from the loose end of the generator output cable to expose the three insulated conductors inside. The green wire is intended for equipment grounding, but it is not required in this case; you will be working with only the black and white wires, both of which will carry electrical current.

Operate the generator with the bicycle mounted on the trainer by pushing a pedal down with one hand, causing the back tire to spin. Use a voltmeter to identify which of the conductors is positive and which is negative. Polarity depends upon the direction of rotation. Mark the positive conductor with a piece of red tape. Strip ¼" of insulation from the positive wire, and strip about ½" from the negative wire.

Attach a spade terminal to the positive wire of the generator and crimp it using a crimping tool. Slide a piece of heat-shrink insulation tubing over the terminal and apply heat with a hair dryer so that the insulating tube shrinks tightly over the wire and connector.

7. Attach the diode.

The diode has two ends: the anode and the cathode. The anode is the end marked + (plus) and connects to the positive conductor of the generator. The cathode is the end marked – (minus)

Diode with schematic diagram ▶

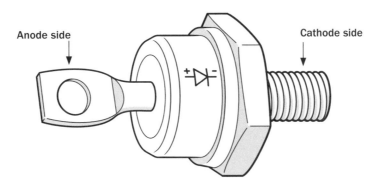

Anode side

Cathode side

and connects to the battery's positive terminal. Look for this represented by a schematic diagram printed on the body of the diode.

Attach the spade terminal of the positive generator wire to the anode of the diode. Slide a piece of heat-shrink tubing over one end of the fuse holder wire, attach the ring terminal to the fuse holder wire and crimp it. Apply heat to the heat-shrink tubing. Bend the terminal at a 90-degree angle and attach it to the threaded stud on the cathode side of the diode, using the nut supplied with the diode.

8. Complete the generator output wiring.

Strip about ½" from each end of the 12 AWG wire. Use a wire nut to connect the negative side of the generator to one end of the 12 AWG wire.

Prepare the other end of the fuse holder wire by stripping back about ½"of insulation. Attach each of these two wires to its corresponding battery connector (the fuse holder wire is the positive wire, and so is connected to the red battery clip) by crimping and soldering the connections, or use a terminal connector and screw. The type of connection you make depends on the type of battery connector you have, but avoid wrapping bare wire around a screw, as this is not a secure connection.

Make sure all exposed electrical connections are covered and protected to prevent exposure and physical stress.

Slide the wire loom over the diode and its connectors, and secure it in place with wire ties. This will help to protect the connections and parts both physically and electrically. You can dress

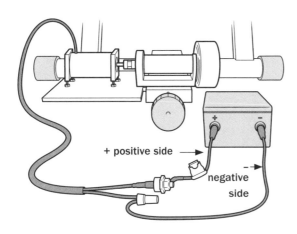

+ positive side

– negative side

▲ **Completed wiring between generator and battery. Cover the connections with wire loom for protection.**

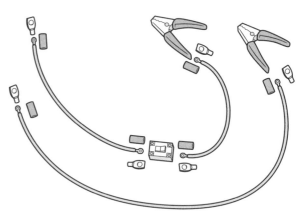

▲ **Parts for battery-to-inverter wiring**

up the entire wiring harness by covering all of it with wire loom. If the bike generator is likely to be moved around a lot or otherwise abused, you can increase the durability of the wiring harness by using a piece of PVC tubing in place of the loom. In any case, be sure to leave access to replace the fuse if needed.

9. Connect the battery to the inverter.

Be sure to read A Few Essential Facts on page 30. To complete the wiring connections, strip about ½" from each end of all three 4 AWG wires. Slide a crimp connector onto each end and crimp the lug securely onto the wire. Slide a piece of ½" heat-shrink tubing over each lug, leaving the hole exposed, and apply heat.

Connect one end of each of the two short wires to the fuse or circuit breaker. This is the "positive" side cable. Attach a red battery clip to the cable coming from the fuse or circuit breaker connection point marked "line." Attach the free end of the positive cable connected to the other side of the fuse or circuit breaker marked "load" to the inverter's positive terminal.

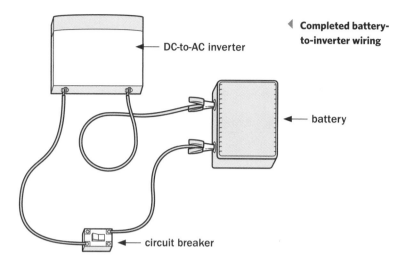

◄ Completed battery-to-inverter wiring

DC-to-AC inverter

battery

circuit breaker

Connect one end of the long wire to the negative battery clip, and connect the other end to the negative side of the inverter.

Follow the manufacturer's instructions to connect the inverter to the battery. Typically, you would attach the negative side to the battery first, then make the positive connection. Observe polarity carefully, as electronic equipment can be destroyed if hooked up backward.

USING YOUR BATTERY CHARGER

Once you're all hooked up, you can start pedaling to charge the battery. But you won't know when the battery is full, when to stop pedaling, or how hard you need to pedal to keep the battery full of juice. The only way to know is to hook up a voltmeter to the battery and watch the numbers. You can find inexpensive voltmeters at electronic supply stores, or devise a scheme to hook up a panel-mounted automotive-type meter to the battery.

Try not to let the battery charge level drop below 11.5 volts or rise above about 14.5 volts. Pedal faster or slower to vary the charging rate. Higher and lower voltages can damage batteries. Higher voltages may damage the inverter. There's no need to have everything connected all the time; the inverter needs to be connected only when power is being used, and the charger must be connected only when you're pedaling to charge.

Please be aware that incredible amounts of energy are stored in a battery, and its acid electrolyte will burn through skin, clothing, and lots of other things. If a tool or other piece of metal creates a short circuit across the battery terminals, the result could be a melted tool and an exploded battery with splattered acid. Insulate the battery terminals and consider enclosing the battery in a protective box to keep the young and uninitiated away.

2

Do Your Own Energy Audit

PERFORMING AN ENERGY audit in your home is a great way to learn about your house, assess your habits, reduce your energy use, save money, fix problems, and find lots of things you thought you'd never see again. An audit implies investigation, and not necessarily action. But why bother with the investigation if you don't intend to do anything with the information you gather?

Therefore, this chapter covers a number of improvements you can make to reduce your energy use, along with some discussion to give you a basic understanding of the issues. Your goal is to identify the amount of energy used by the myriad products and appliances in your home, so you know where to begin to look for savings.

While some auditing tasks are best left to a professional, there are plenty of things you can learn to do on your own. You may not find one silver bullet that saves you barrels of energy and money, but if you tackle the manageable items — taking care of the biggest things first, then working on the smaller changes over time — your savings can really add up.

Getting Started

IMAGE THAT YOU'RE entering your house for the first time. When visiting any new place, we look at all the things the locals take for granted: architectural details, sounds, smells, cracks in the sidewalk, the unique view — even the main attraction that you may have traveled thousands of miles to see. Approach your energy audit as if you were going on vacation to some exotic place.

As a professional energy auditor, I use my skills, experience, and tools to find obvious and hidden energy users and wasters throughout a home. It might take two to four hours to test and investigate, and I won't catch everything, due to time and other constraints. As a homeowner, you're not subject to these constraints, so you're in the best position to spend the time it takes to investigate the energy use and savings potential in your home.

Your energy audit starts with an inventory of everything in your home that uses energy, followed by a review of your energy consumption. Gather a year's worth of utility bills from all of your energy suppliers. How much of each fuel do you use in a year? What are the seasonal variations? Find the daily energy consumption of each fuel during each month of the year by dividing the total monthly usage by the number of days in that month.

You may use more electricity in the summertime when the air conditioner is running, then less in the fall, and more again in the winter when the furnace is operating. While you may find that your energy use varies seasonally, you probably have a consistent **base load**. This is energy-geek lingo for the lights and appliances that are used consistently year-round. Gas base load might include things like water heating and cooking, while your electric base load might be made up of refrigeration, lighting, and clothes washing.

Base load may be reflected in the entirety of your electric bill during the few months out of the year that do not include use of heating or cooling energy systems, or other season-specific energy users. Energy use that changes with the seasons, such as heating and air conditioning, is not base load. Like a good detective, don't take anything for granted, but gather the information and the evidence, then look objectively at those findings.

"AAA" APPROACH TO ENERGY SAVINGS

My professional audit of any home starts as I pull into the driveway, taking in a broad view of the house. On one job I saw a garage with two 300-watt halogen floodlights burning brightly in the middle of the day. When I asked why, the answer was "the switch broke." Those lights were costing the homeowner $48 per month and the fix was a simple $2 switch. Even if the homeowner paid an electrician $75 for the repair, the simple payback (see page 42) would be less than two months.

With the above example in mind, let me introduce my "AAA" approach to energy efficiency:

Toolbox

No special tools are required to understand and accomplish many of the basics — your primary inspection tools are your eyes and ears — but a few basic tools will make your job easier:

- Flashlight
- Screwdrivers
- Tongue-and-groove pliers
- Tape measure
- Wooden BBQ skewer
- Dust mask or (better) respirator
- Notepad and pen
- Camera
- Fearless curiosity

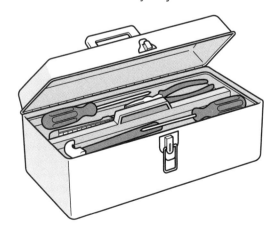

- **Awareness** of the equipment and conditions you have in your house, along with your habits (notice that you have lights, that the lights are on, and ask yourself, "Why?")

- **Assessment** of the energy use of the equipment in your home and how your habits affect energy consumption and costs (calculate the energy used and cost of the lights being on)

- **Action** to reduce your use (replace the switch)

Electricity

LET'S START BY taking a look at the usage of electricity in your home, using the AAA approach. I've been in many homes where so much work has been done to them over the years that it's hard to figure out what's going where and why. One home had mysteriously high electric bills that we tracked to an electrically heated radiant floor in the kitchen that was unknown to the current owners. It was found only with the aid of an infrared camera.

AWARENESS

Take an inventory of electrical appliances and anything that's plugged into an outlet. Crawl around, look under and behind things to find those dusty power strips hidden by the entertainment center, then find and list everything that's plugged in. Remember that some electrical users, such as water heaters and furnaces, are wired directly to a house circuit (which leads to the circuit breaker box) rather than having a cord and plug that connects to an outlet. Go to the circuit breaker box and identify what every breaker is for. If there are mysteries, get to the bottom of them by turning breakers on and off and identifying what's connected to each circuit.

The Sample of Household Electricity Usage chart on the next page shows a billing history of consumption both numerically and graphically. Remember that electrical energy is expressed in kilowatt-hours (kWh), which is the electrical quantity unit for which the power company bills you. The electric company may not read your meter on the same day of every month, and this may cause confusion when one read period is longer or shorter than another. For this reason, knowing how many kWh are used in a day is a more useful way to understand electrical consumption.

ASSESSMENT

To assess your electrical use means to understand how much electrical energy each appliance uses. Power consumption of appliances can be discovered in several ways, although some are more accurate than others.

Disaggregation

Looking down the table of numbers in the billing history, we see a base load of around 28 kWh per day or 840 kWh per month in May and October. This monthly consumption is higher than that shown in the table because the read periods for those 2 months are less than the number of days in the month. Multiplying kWh per day used by the number of days in the month serves to "true-up" the monthly consumption.

If summer vacation in July lasted for the whole month, we could see what the house itself uses in a month (all the appliances left on but no occupancy use, such as lights, TV, electronic games, and so forth). But the actual 2-week break does

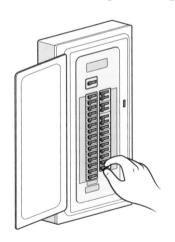

▲ Breaker boxes should contain electrician's notes indicating what's on each circuit (but these are often incomplete). Switch the breaker to the OFF position to cut power to the circuit.

▲ Older electrical service panels have fuses. Unscrew the fuse completely (and set it aside) to shut down the individual circuit.

SAMPLE OF HOUSEHOLD ELECTRICITY USAGE

Read date	# days between meter readings	kWh/ read period	kWh/ day	Comments
January 15	30	1950	65	furnace blower, electric heat, lots of lights
February 12	28	1274	46	more furnace use, less electric heat
March 9	25	1045	42	same as February
April 15	37	1150	31	close to base load, still a little heat
May 14	29	785	27	base load use; no heat, no air conditioning
June 15	32	1505	47	opened the pool June 1st, air conditioning
July 19	34	1231	36	two-week vacation, left pool pump on
August 15	27	1810	67	pool, air conditioning, dehumidifier
September 16	32	1152	36	less of all the above
October 14	28	804	29	base load
November 14	31	1220	39	start of heating season
December 15	31	1739	56	same as January, less electric heat

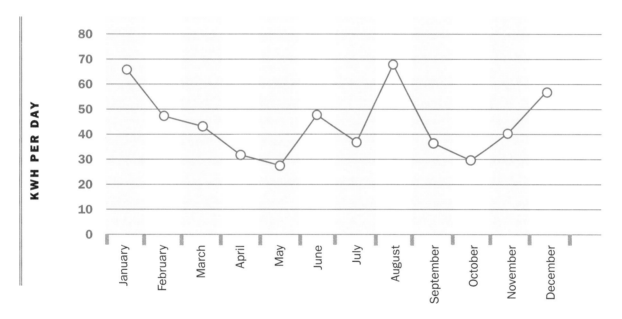

offer a clue to consumption habits. The 1-horsepower (HP) pool pump runs continuously and uses about 24 kWh per day (one HP is roughly equivalent to 1 kilowatt). Subtracting 24 kWh from the actual July usage of 36 kWh per day, the resulting 12 kWh per day is the lowest monthly reading — even lower than the base load months. This tells us that occupant behavior contributes quite a lot to the home's power consumption.

Knowing the base load and the pool pump use can help tease out the usage of the air conditioner.

For example, the base load is 840 kWh per month, the pool pump uses 730 kWh per month (1 kilowatt x 730 hours in a month), and we'll estimate the dehumidifier at 50 kWh per month, for a total of 1620 kWh. This family used 1810 kWh in August, so the remaining 190 kWh is air conditioning, plus any additional occupancy-related items that occurred during that month. This kind of disaggregation is not precise, but it works well to gain perspective on the bigger energy-use picture.

Appliance Power Rating

An easy but not always accurate way to assess electrical use is to look at the tag on the appliance indicating the operating voltage and the power consumption in watts. If it shows amps, multiply amps by volts to get watts. Most appliances and anything that plugs into a standard outlet require 120 volts; electric dryers, ranges, and many heaters require 240 volts.

1. Add up how many hours each appliance is turned on during an average month.

2. Multiply watts by hours of on-time and your answer is in watt-hours.

3. Next, divide by 1,000 to arrive at kilowatt-hours (kWh), then multiply kWh consumed by your cost per kWh and you know how much the appliance is costing you each month.

Here's an example of a television that draws 5 amps and is on for three hours every day:

5 amps x 120 volts = 600 watts x 3 hours
= 1,800 watt-hours
1,800 watt-hours ÷ 1,000 = 1.8 kWh

How much is that TV usage costing you? Look at your electric bill to find how much you pay for each kWh and multiply that by the number of kWh the appliance uses. For this example let's say your electricity rate is $0.12/kWh:

1.8kWh x 0.12 = $0.216 per day x 365 days
= $78.84 per year

A simple wattmeter plugs in between the wall outlet and appliance and tells you exactly how much electricity the appliance uses. ▶

To get the most accurate understanding of the energy use and cost of individual appliances, use a wattmeter (such as Watts Up? or Kill A Watt; see Resources) to measure the actual power used by the appliance. This is the only option for determining the actual consumption of appliances that frequently cycle on and off, like refrigerators and freezers.

ACTION

After assessing your electrical use, it's time to start exploring some options and make a plan for action. Let's say you've added up all your lighting use over a month, and it totals 100 kWh. If you pay 12 cents a kWh, that's $12 per month, or $144 per year. If you replaced all your incandescent light bulbs with LEDs or compact fluorescent bulbs, and maybe installed occupancy sensor switches where needed, you would use about one-quarter to one-third of the power to provide the same amount of light.

As you replace those bulbs, take a close look at where the light falls. Does it light up the countertops or the top of the cabinet? Put light where you need it by changing the lighting layout to make best use of those lumens.

Refrigerators, air conditioners, ceiling fans, computers, televisions and set-top boxes, clothes washers, and dishwashers are all candidates for efficiency improvements. Explore the many options and compare the energy consumption of many different appliances at the *ENERGY STAR* website (see Resources).

Planning Your Energy Investments

Knowing the existing appliances' power use before researching new ones gives you the knowledge you need to calculate savings potential and make an informed replacement decision. Don't replace appliances just because they're old. Assess their energy use first and compare their energy consumption with new equipment. See chapter 5 for more information about energy use monitoring.

Once you have an understanding of what's happening in terms of base and seasonal loads, and exactly where that energy is going, you can begin to explore improvement options and savings scenarios. When you consider making an investment, you want to know what rewards that investment might offer you over time. Often, that reward is quantified in terms of how many dollars you may save over a period of time. Investments in energy efficiency have value beyond financial returns, such as the immediate payback of increased comfort. Efficiency is a good excuse to upgrade, since there is almost always a financial gain in terms of energy cost savings.

◀ Most home electronic equipment is never really "off" unless unplugged, and it can eat up a surprising amount of electricity.

Changing Habits

Awareness and habits go hand in hand. Changing habits starts with identifying what you currently do and how you do it. This means washing full loads of dishes and laundry, turning off lights that are not needed, and simply being aware of what is in use in your home. For example, if you have a private water system, do you know when the well pump is on? A leaky toilet valve can lead to a mysterious increase in electrical use due to the increased run-time of the well pump.

Don't forget about gas standby loads (see box below), such as pilot lights. A water heater or furnace pilot light might use 1,000 Btus each hour, while gas range pilot lights can use 250 Btus. If your heating season is six months long and you leave the furnace pilot light on during the summer months, you'll have paid for nearly 4,400 cubic feet (or 44 therms; one therm is equal to 100,000 Btus or 100 cubic feet of natural gas that was ultimately wasted. Many modern gas appliances have electronic ignition so no pilot is needed.

WHAT ARE STANDBY LOADS?

Many appliances appear to be off but are really using a small amount of power. Eliminating these standby or "phantom" loads represents savings that really add up. When it comes to controlling standby power, a few inexpensive, well placed, and intentionally used switchable or automatically controlled power strips are well worth the investment. A switched power strip turns everything plugged into it on or off with a single switch.

A controlled power strip (see the Smart Strip in Resources) uses the on/off action of one component plugged into it to switch other components plugged into it on or off. For example, when you turn on your computer monitor, the power strip automatically turns on the printer, modem, and anything else plugged into it. When the monitor is powered off, the power to all peripherals is automatically switched off. Some power strips also have built-in power meters so you can monitor how much electricity the plugged-in devices are using.

Although a few watts of standby energy use per appliance may sound like small potatoes, the combined energy use of these loads adds up fast. Phantom loads in a typical American household amount to over a kilowatt-hour per day. You might find some surprises as you meter the many gadgets plugged into our modern homes.

My microwave oven uses 5 watts of power when it is "off." That is, it presents a constant power drain of 5 watts. That works out to 120 watt-hours per day, or 44 kWh per year. Since our total household usage is only 7 kWh per day, that's equivalent to about six days of power consumption over the course of a year. The nuker is now plugged into a $7 switchable power strip that's easily accessible. This power strip has a simple payback of a little over a year.

Hot Water

CONTINUING THE AAA APPROACH, let's take a look at your hot water usage. Hot water is probably the second largest energy user in your home, after space heating (if you live in a heating-dominated climate).

AWARENESS

Clear a path to your water heater and take a good look at it. Answering a few questions will get you started down the AAA path, offering clues about the condition of your water heater, how much energy it might be using, and whether or not it's time to repair or replace.

- What is its energy source? Gas, electric, or something else?

- Is the water heater rusty or leaky? How old is it?

- If it's a gas-powered unit, examine the flue. Is it connected to the chimney and does it slope upward? Is there any rust flaking from the flue pipe?

- Have you ever performed any maintenance such as draining, flushing, or changing the anode rod?

- Do you run out of hot water frequently?

Open a hot water faucet and listen carefully while standing quietly next to the water heater. Do you hear anything? If you hear hissing or popping, it could indicate a leak or perhaps the sound of the heat being applied to sediment buildup inside the tank. Be sure that electrical connections and thermostats are safely covered.

Check your water faucets. Do they have aerators installed to restrict the flow? Aerators are available that deliver various flows, but the maximum flow rate of 2.5 gallons per minute (gpm) is a federal regulation initiated to save both water and energy. Showerheads and faucet aerators that deliver less than 1.5 gpm are considered "low flow." Look for flow rate printed on the side of the aerator or showerhead.

Hot water habits are hard to break. One study performed by Lawrence Berkeley National Laboratory concluded that as much as 20 percent of water-heating energy is wasted by waiting for hot water to arrive at the faucet. If you run hot water and it barely gets hot by the time you turn off the water, all that heat is lost in the pipes, and you'll save by using cold water instead.

ASSESSMENT

The average American uses about 17 gallons of hot water each day. To assess your hot water consumption you'll need a few additional tools for measuring temperature and flow rates:

- Thermometer

- 1-quart measuring jar

- Stopwatch or watch with second hand

To measure the hot water temperature, put the thermometer into the jar and run hot water into it until it's good and hot. Check the temperature — it should be 120°F or less. You may have a mixing valve on the outlet of the water heater, which helps prevent scalding temperatures at faucets by mixing cold water with the hot. The presence of a mixing valve can give you a false reading about

A low-flow showerhead with on-off control lets you stop the water flow while you shampoo and soap up. The water stays at your set temperature for when it's time to rinse. ▶

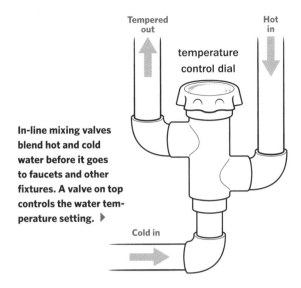

Tempered out

Hot in

temperature control dial

In-line mixing valves blend hot and cold water before it goes to faucets and other fixtures. A valve on top controls the water temperature setting. ▶

Cold in

To check the flow rates of your showerhead, turn on the water and adjust the flow to where you normally set it when you shower. Hold the jar under the showerhead and time how many seconds it takes to fill. Flow rate is measured in gallons per minute (gpm), so you'll need to make the conversion from a 1-quart jar, using the Flow Rate chart below or the following calculation:

60 (seconds per minute) ÷ 4 (quarts in a gallon) ÷ number of seconds required to fill the jar = gpm

As an example, if it takes five seconds to fill your 1-quart jar, then:

60 ÷ 4 ÷ 5 = 3 gpm

If your showerhead flow rate is more than 2.5 gpm, it's probably worthwhile to get a new one with a lower flow.

the actual tank temperature. If you can adjust the mixing valve, turn it to the highest setting before taking your temperature reading at the faucet, but this still may not give you an accurate reading of the tank temperature.

Look at the temperature and pressure relief valve (TPRV or TPR valve) located on the top or side of the water heater. It should not be dripping or have any evidence of rust or corrosion. The purpose of this safety valve is to blow off steam in the event of high temperature or pressure inside the water heater. Sometimes these valves can stick and need to be replaced, but you can do a simple check. With a bucket under the outlet of the TPRV, operate the lever to allow water to escape. *Be careful*: It will be very hot and come out forcefully. Measuring the temperature of this water is an accurate way to find the tank temperature setting.

FLOW RATE
(IN GALLONS PER MINUTE)

# of seconds to fill container	Quart Jar	Gallon Jug
3	5.00	20.0
4	3.75	15.0
5	3.00	12.0
6	2.50	10.0
7	2.14	8.6
8	1.88	7.5
9	1.67	6.7
10	1.50	6.0
12	1.25	5.0
15	1.00	4.0
20	0.75	3.0
30	0.50	2.0
40	0.38	1.5
50	0.30	1.2
60	0.25	1.0

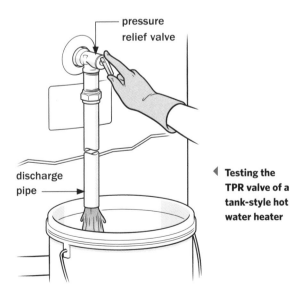

pressure relief valve

discharge pipe

◀ Testing the TPR valve of a tank-style hot water heater

Multiply the showerhead's gpm by the number and length of showers each week, and by the percentage of hot water in the mix. See Hot Water Used in a Shower (below) for a way to determine this. Then add in the hot water used for laundry, dishes, and washing. These can be tricky values to uncover since older clothes washers can use up to 40 gallons a load, while modern front-loading machines use 10 to 20 gallons. You probably do laundry loads at various temperatures, so you'll need to do some estimating.

Water Heater Efficiency

The efficiency of a water heater is reflected in its **Energy Factor** (EF) rating. Electric water heaters have an EF of around 0.90, meaning that given a certain hot water usage (based on a standardized test), the heaters convert 90 percent of the electrical energy supplied to them into hot water. The remaining 10 percent is heat lost through the walls of the storage tank.

Gas water heaters generally have a lower EF, due in part to the additional heat lost out the flue. On-demand water heaters have a higher EF because there is no hot water storage tank through which to lose heat. The EF for on-demand gas water heaters can be as high as the upper 90s for high performance models.

When I need to estimate hot water usage for a household, I use a shortcut that will get me in the ballpark of expected annual hot water energy use. It's based on an average daily hot water use of 17 gallons per person per day. Here's the formula, followed by an example using a four-person household and a gas water heater with an EF of 0.65:

3.8 million Btus x number of people
in the household ÷ EF
3.8 x 4 ÷ 0.65 = 23.4 million Btus (MMBtu)

The answer doesn't mean much unless you know the energy content (in Btus) of the fuel you

---

HOT WATER USED IN A SHOWER

The water coming out of the showerhead is a mix of hot and cold. You can easily measure the temperature of the shower water when you measure the flow rate by putting a thermometer into the measuring container. To determine how many gpm of your shower flow is hot water requires a bit of math, or you can just use the table below.

PERCENT OF HOT WATER MIX FOR 104°F (40°C) SHOWER

Hot water temp*	% mix of hot water
104° (40°C)	100%
110° (43°C)	89%
115° (46°C)	81%
120° (49°C)	75%
125° (52°C)	70%
130° (54°C)	65%
135° (57°C)	61%
140° (60°C)	57%
145° (63°C)	54%

*Assumes cold water temperature of 55°F (13°C)

You can use the following formula to calculate the percentage of hot and cold water in your warm water mix if you know three things:

1   Hot water temperature

2   Cold water temperature

3   Warm water shower temperature

The formula compares the difference between the hot- and warm-water temperatures to the difference between the hot- and cold-water temperatures.

H = Hot water temperature

C = Cold water temperature

W = Warm water temperature

The percentage of cold water in the warm water stream is equal to:

(H − W) ÷ (H − C) x 100 = percent of C
(130 − 104) ÷ (130 − 55) x 100 = 35% Cold, or 65% Hot

If your shower flow rate is 3 gpm, and 35 percent is cold, then 65 percent is hot. That works out to about 2 gpm of hot water flow in the shower.

use to heat water. The Fuel Energy Content chart on page 57 supplies this information. If you have a natural gas water heater, 23.4 MMBtu translates to 234 CCF (or therms) of gas each year. For an electric water heater, divide MMBtus by 3413 to express the value in kilowatt-hours (23,400,000 ÷ 3413 = 6,856 kWh).

ACTION

Reducing water use and the energy used to heat water takes both technology fixes and behavior adjustments.

- Repair dripping faucets.

- Use cold water whenever possible.

- If you find yourself running the hot water for less time than it takes to get hot, use cold water instead.

- Insulate the tank and pipes (materials are commonly available at hardware stores).

- Install low-flow showerheads and faucet aerators.

- Turn the water heater temperature down to just over your favorite shower temperature. If you're concerned about *Legionella* or other bacteria, one effective solution is to shock the tank each month by turning the temperature up to 140°F for a day and flushing the hot water pipes at this high temperature.

- Wash laundry in cold water. If you feel you need to wash in hot water to control allergens, try Allersearch or De-Mite cold-water detergents (see Resources).

- Upgrade hot-water appliances, such as dishwashers and clothes washers, to more efficient models that use less water.

- Turn off the water heater when you're away from home for more than a few days.

- Plan ahead. If the water heater is more than 12 years old and hasn't been maintained, shop now and pick out the water heater you want so that when the time comes, you don't have to shop while you mop. Decisions made in such an emergency tend to ignore efficiency.

- When it's time to replace the water heater, consider a high-efficiency on-demand unit. Be aware that with on-demand water heaters, the best thing about them is that you can get hot water all day long. The worst thing about them is that you can get hot water all day long — meaning that you will not realize any savings if you're taking longer showers, because the hot water never runs out. You will also find yourself waiting longer for hot water at the tap because the water needs to heat up first. As with all energy-using appliances, the incremental cost of a more efficient model will be recouped over its lifetime when it is properly used to take advantage of its efficiency.

Adjusting the Thermostat

To adjust the thermostat on an electric water heater, first disconnect the power at the circuit breaker. *Failure to disconnect the power before working on an electric water heater can result in death!* It is very easy to slip with a tool or finger and hit a live wire.

1. Remove the heating element access cover(s) located on the side of the water heater; there are usually two screws holding each cover on.

2. Pull back the insulation that may be behind the access panel, filling the space around the

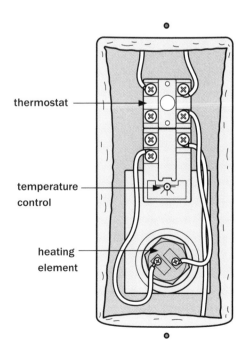

thermostat

temperature control

heating element

◀ **Behind an electric water heater element access cover. Assume that all wires are electrically live unless the circuit breaker is off.**

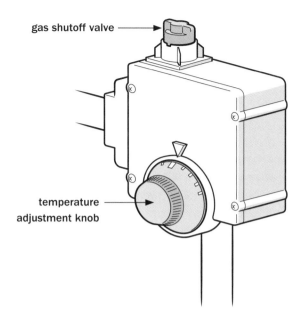

Thermostat on a gas water heater ▶

gas shutoff valve

temperature adjustment knob

thermostat. You may also find a plastic shield covering the wiring with an access opening to adjust the thermostat.

3. You should now be able to see the thermostat. It should look like the illustration on page 45. The thermostat is the rectangular piece toward the top; the heating element is below the thermostat and is screwed into the tank. While you're there, look for leaks or rust around where the element disappears into the tank.

4. Use a screwdriver to turn the dial to a lower temperature.

Some thermostats have no temperatures printed on the thermostat, just the vague words "warm" and "hot." Or maybe it is not even adjustable at all, as some thermostats are preset to a specific temperature. If there are two heating elements, set both the top and bottom thermostats to the same temperature. If you have a new thermostat installed, be sure the plumber sets the temperature where you want it— not at the factory preset.

The thermostat on a gas water heater is usually mounted outside on the bottom half of the unit and is fairly obvious. You don't need to turn off the gas to make the adjustment. Simply turn the thermostat knob to the desired water temperature.

Finding and Adjusting an Aquastat

Oil and indirect-fired water heaters (those that use a hot water boiler to heat domestic hot water) use a not-so-obvious aquastat. An aquastat is a temperature-sensitive switch designed for use in water systems. It senses water temperature by means of a probe that's in contact with the water to be monitored and connected to a switch that activates a hot water circulator. When the water cools off, the switch closes, and the heat source is activated, sending hot water to the storage tank. The aquastat is often inside a small gray box attached to the water heater. The temperature adjustment is inside its housing. Sometimes it is accessible, sometimes not.

To adjust the temperature, turn off the power to the boiler or water heater at the circuit breaker (the wiring you'll be working near is typically low-voltage, but turn off the power to be safe). If you don't see the temperature adjustment, remove the control box cover (there are usually one or two screws holding the cover on), and you will see a well-camouflaged temperature dial inside. Look closely at the dial for temperature markings, and turn it with your finger to make the adjustment.

Inside an aquastat. The notched dial on the left adjusts the temperature. ▶

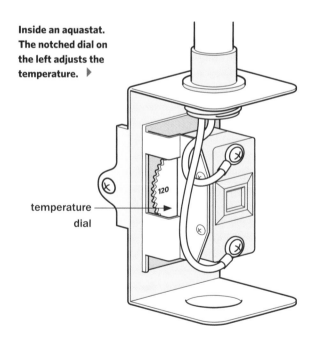

temperature dial

120

Flushing the Water Heater

The average lifetime of a water heater is 10 to 15 years, but simply flushing once a year to remove sediments and mineral scale buildup can increase efficiency, reduce maintenance, and extend its lifetime.

Look for a descaling or de-liming product (such as Mag-Erad or Un-Lime) at hardware stores. Alternatively, you can use white vinegar or muriatic acid (mixed with 10 parts water to 1 part acid). Be careful: Too strong an acid solution can eat through the walls of the tank, especially if it is old or damaged. I've had success with pouring a quart or two of vinegar into the tank through a plumbing union at the tank inlet, then adding a gallon or two of water and letting it sit for a few hours.

A union is a threaded plumbing connection that allows you to easily disconnect the water supply pipes. Adding a union will only cost a few extra dollars when installing the water heater and is well worth the convenience. If your tank is more than 15 years old and you've never flushed it, or if there is evidence of rust or leaks anywhere on the tank, you can flush the tank, but *do not* use the acid solution.

Here's how to flush your water heater:

1. Turn off the power or fuel supply to the water heater, and make sure you know how and where to relight the gas pilot when you're finished.

2. Turn off the cold water supply to the water heater.

3. Open a hot water faucet to allow air into the water heater or the water won't drain completely.

4. Attach a hose to the drain valve of the tank, run the other end into a utility sink, bucket, or outdoors, and open the valve to drain the tank.

5. Drain the first few gallons of water into a bucket and take note of the water condition as it comes out of the tank. Look for discoloration, rust, sand, or flaky mineral deposits.

6. Once the tank is empty, and with the drain valve still open, pulse the cold water by quickly turning the cold water inlet valve on and off. This will help loosen scale.

7. Close the water supply valve and the drain valve.

8. Loosen the plumbing union on the hot or cold pipe. If no union exists, you will need to have a plumber install one. If you have an electric water heater, you can access the tank via the hole that exists after you remove the heating element (you do this with a socket wrench after taking off the electrical wires).

9. Mix the flushing solution.

10. Using a funnel, pour the mixture into the water heater and let it sit for as long as the instructions indicate.

11. Pulse the cold water, and drain the tank again.

12. Repeat steps 10 and 11 until the water runs clear.

13. Close the drain, remove the hose, attach the union, and open the cold water inlet valve to fill the water heater. Leave the hot water tap open until the air is cleared from the line and water flows.

14. While you're at it, check the TPRV for proper operation by manually activating its lever (see page 43). Sometimes this valve can stick due to rust, debris, mineral deposits, or age. Replace the valve if the operation is not smooth or water trickles out instead of gushing.

15. Turn on the circuit breaker or the gas valve, and relight the pilot.

Anode Rod Inspection

To get even more life out of your water heater, remove and inspect the anode rod for excessive deterioration every few years, and replace it if necessary. The mysterious anode rod lives and dies silently inside your water heater, sacrificing itself so that your water heater may live. It works to prevent rust inside the water heater by attracting the electrochemical activity that would otherwise corrode the steel tank.

Anode rods are about ½" in diameter, 3 or 4 feet long, and are usually made of aluminum, magnesium, or zinc. The rod is sometimes attached to the hot water outlet side of the water heater but is often separate. It should be removed and inspected every few years — more often if you have hard water. If you have a water softener, you may want to check the anode rod every year. With a water softener, hard minerals in the water are exchanged for salt, and this salt can consume an anode rod up to three times faster than calcium carbonate (the typical anode-consuming mineral in hard water). Once the anode rod is depleted, your water heater's days are numbered.

New anode rod (top) and spent, used rod (bottom) ▶

Install a new water heater to allow enough room for maintenance. ▶

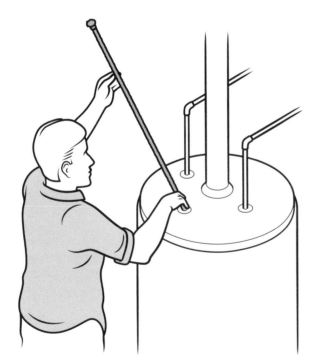

Heating and Air Conditioning

HEATING AND AIR conditioning are collectively called **space conditioning**. Together they represent the largest piece of the energy-use pie in most homes. High-efficiency heating and cooling appliances are available that will help reduce energy consumption. But as important as efficient equipment is, improvements in insulation, reductions in air leakage, and an increase in your tolerance for variation in temperature and humidity will all work together to exponentially reduce space conditioning energy use.

AWARENESS

Get to know your heating and air conditioning systems by taking a close look and asking some basic questions.

- How old are they? Average lifetime is 15 to 20 years.

- Do you have a furnace (blows hot air), a boiler (circulates hot water or steam), or a space heater, such as a wall-mounted gas heater or wood stove?

- What is the heating fuel?

- Do you have the heating system professionally serviced at least once every two years?

- For central air conditioners, is the cooling coil kept clean? This is accessed inside the ductwork and may require professional service.

- For room air conditioners, is there a gap around where the unit goes through the window, providing a direct path for losing cooled air to the outdoors?

- Listen to your heating and cooling equipment while they operate. Pay attention to any odd noises and whether they turn on and off fairly frequently (called **short-cycling**), which could indicate a functional problem but also may indicate that the system is oversized for the home.

Maximum efficiency with heating and AC systems requires periodic, professional cleaning, tune-ups, and efficiency testing. Heating system inspections include, among other tasks, combustion analysis and a flue draft check. Inspect the flue for solid connections and an upward slope toward the chimney. Examine the chimney for loose mortar (if it's a traditional masonry structure), and have a chimney sweep inspect the inside of the chimney. Any rust or disconnected flue pipes need immediate attention.

Next, check that the space conditioning distribution system is in good shape. Air ducts for a forced-air (furnace and/or central air conditioning) system is should be solidly connected, sealed at the seams and connections, and insulated. On hot water (boiler) systems, the water pipes should be insulated. Make sure to have working carbon monoxide alarms on each floor (see Resources). For central air conditioning, a professional will need to check that the refrigerant charge level is correct, along with other routine inspection items such as checking for proper temperature delivery and dehumidification control.

Efficiency and Expense

The energy required to condition your home for comfort is based primarily on the ability of the home's **thermal envelope** (the boundary between indoors and out) to retain the energy delivered to that space. It also depends on the efficiency of the heating and cooling system, as well as the habits of the occupants. A more efficient thermal envelope allows you to use smaller, less expensive space conditioning systems. Many heating and cooling systems are oversized, meaning they can supply more heating or cooling energy than what's required by the home. This may not sound like such a bad thing,

but it's a situation that can lead to inefficient operation and discomfort.

Once you've made efficiency improvements to the thermal envelope, an oversized heating and/or cooling system may effectively become even more oversized than before, exacerbating the problems. You should still complete any necessary insulation and air leakage improvements (see chapter 3), but follow these by having a professional heating contractor or energy auditor perform a heat load analysis. This will give you accurate data for sizing new heating or air conditioning equipment for your more efficient home.

ASSESSMENT

How much fuel do you use to heat your home in an average heating season? How much electricity do you use to keep cool? If your heat, hot water, and/or cooking appliances share the same fuel source, you will need to review a year's worth of energy bills so that you can tease out the base load from the seasonal loads, much as you did for electrical use (see page 37).

Assessing the condition of a "heating plant" (the furnace or boiler) and its associated fuel usage is a matter of the heating plant's combustion efficiency and heat distribution system's (air ducts' or water pipes') efficiency. There are two measures of efficiency for heating equipment. The first, called instantaneous or **steady state efficiency** (SSE), is analogous to the highway fuel efficiency of a car. The SSE can be measured and adjusted by a technician using a combustion analyzer and the unit's efficiency at converting the fuel energy into heat energy is reported in terms of a percentage.

The second measure of efficiency is a seasonal value called the **annual fuel utilization efficiency** (AFUE). It is based on a standardized factory test, and can be compared to the combined city and

---

Hot and Cold

A furnace that is too big may short-cycle (operate with rapid on-off cycles), causing wide temperature swings. An oversized air conditioner may cool off the home quickly but will not be effective at controlling humidity, a significant factor in summertime comfort.

highway fuel efficiency of your car. AFUE will be lower than the SSE because it accounts for warm-up and cool-down losses, along with the heat lost up the chimney.

Every new heating plant will have the AFUE listed on the energy use guide that is fixed to its cabinet. Older heating equipment may have an AFUE of anywhere between 75 and 85 percent, while modern high-efficiency gas equipment can range from 90 to 98 percent.

To evaluate a central air conditioning system, special equipment and skills are required to test for the proper refrigerant charge. The steady state efficiency rating of an air conditioner is the **energy efficiency ratio** (EER), and the seasonal efficiency rating is the **seasonal energy efficiency ratio** (SEER). Older central air conditioners (when properly installed and adjusted) have a SEER of 10 or less, while newer models can achieve SEER 16 or higher. SEER and EER are measures of how many Btus can be delivered for each unit of electricity consumed. Higher EER and SEER values mean greater efficiency, though requirements should be tailored to your climate.

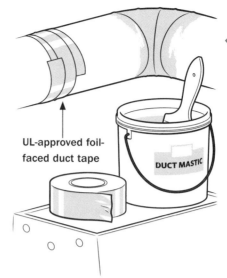

UL-approved foil-faced duct tape

Duct joint sealed with mastic applied "thick as a nickel" to all seams on supply and return ducts

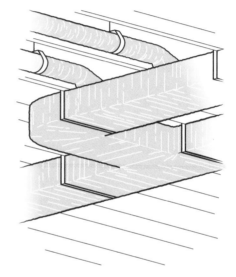

After the duct joints are sealed, wrap all ducts completely with insulation. Air supply ducts are a higher priority for insulation than return ducts.

ACTION

Once you've evaluated your heating and cooling system components, you can take the steps needed to increase their effectiveness at doing their respective jobs.

- Check the furnace filter monthly and replace as needed.

- Clean the furnace fan blades if they're accessible (usually behind the filter), making sure that the power is off first. If you have a furnace and central air conditioning, the same fan is used to move both heated and cooled air.

- Seal ductwork joints with mastic or foil tape (not standard plastic/cloth "duct" tape), and wrap ducts with insulation.

- Clean dust off the air conditioner evaporator coil located inside the ductwork, if accessible. This task may require a professional.

- Keep forced-air registers free and clear.

- Insulate hot water or steam boiler pipes.

- Save up to three percent of your heating energy for each degree you turn your thermostat down throughout the heating season. Use a programmable thermostat so you don't have to remember to turn it up or down.

- Control sources of interior heat and moisture; keep them in or let them out based on seasonal needs.

- Have a professional perform an efficiency test on the heating system and perform a "clean and tune" service. Have the technician check the heating plant's steady state efficiency and, if it's below 80 percent, consider an upgrade or replacement to the highest efficiency equipment available.

- Have a technician check for proper air flow through the ductwork and evaluate the duct system for proper sizing and layout of supplies and returns.

- Ask your heating contractor about upgrading to a modern, efficient, electronically controlled furnace fan motor.

- Have a professional check the refrigerant charge of your central air conditioner.

- Add more heating zones to improve the temperature control of different areas in the home.

- If you need a new furnace or boiler, gas heating systems are available with efficiency ratings over 95 percent, while better oil systems are around 85 percent efficient.

Thermal Envelope

THE THERMAL ENVELOPE of your home includes everything between (and including) the paint on the inside to the paint on the outside of the walls. Framing materials, wallboard, insulation, siding, windows, doors, roof, and foundation walls — anything that separates the inside from the outside — make up a building's envelope.

The envelope of a home needs to keep you comfortable, protect you from the elements, and be durable and aesthetically appealing. All of the envelope components, including the materials they are made of and how they are put together, affect the energy consumption and durability of the building.

AWARENESS

Begin by taking a broad view of your home. Stand on the sidewalk across from your house and just look at it.

Start at the roof:

- Do you see any broken or flaking shingles?

- Is there moss growing on the shingles?

- Are the gutters and downspouts clean and in good repair?

- Are there any rotten or missing boards on the soffits or evidence of animal damage such as holes or bits of insulation poking out?

- If it's cold and snowy, do you see icicles or ice dams indicating excessive heat loss?

- Does the snow melt off certain parts of the roof faster than other parts, indicating insulation defects?

Move down to the walls:

- What is the condition of the siding? Do you see peeling paint or warped clapboards that may indicate moisture damage?

- Are there unsealed holes for utility cables where air, water, and bugs can get in?

- Check the condition of the windows. Are they double- or single-pane? Are they all straight and tight or are there broken panes and rotten sashes and sills?

- Inspect behind the siding by first looking carefully for hazards like nails and wasp nests, then putting your fingers behind the first course of clapboards just above the foundation; does the wood feel wet, soft, or rotten, indicating water intrusion?

Look at the foundation:

- Sight down the side of the foundation along ground level: Is it straight? Any variations could indicate areas of unwanted exchange between indoor and outdoor environments.

- Are there cracks or holes that might allow water in or indicate structural problems?

- How does the foundation connect to the wooden framing of the house? Are there any gaps in that junction causing air and/or water leaks? This detail may be best seen from inside the basement, looking at the juncture between the top of the foundation wall and wood framing.

- Is the ground wet around your foundation? Gutters and leaders, along with ground that slopes away from the house, will allow rainwater to drain away from the house rather than into the basement.

• Where are the electric, water, and gas meters? Do you know how to read them, so you can gauge your own daily consumption?

• How does electrical service get to your house — overhead or underground? Are there trees growing into the power lines, and do you know where underground electric, gas, and water lines are?

Once you've had a good look around and poked and probed for signs of weakness, you can move indoors and pick apart all the various thermal envelope systems and components in your home. It's important to recognize that all of these systems work together; if you change one thing, it will affect something else. For instance, back in the early days of weatherization efforts during the 1970s, builders and weatherization professionals found that sealing up air leaks in old homes created moisture problems that led to mold and decay in wood framing.

The building needs to be put together so that moisture stays out of walls and insulation, but the design needs to be resilient enough so that when water *does* get into the walls (and it will), there is a way for it to drain and dry. Building science has come a long way in understanding how systems interact with one another and in developing successful strategies for durable, efficient building upgrades.

ASSESSMENT

Assessing the envelope involves evaluating insulation levels and airtightness of all the various assemblies in the house, and this is discussed in greater detail in chapter 3. One critical aspect of the envelope, in terms of energy use and comfort, is how much air leaks between indoors and out.

Air leakage is driven primarily by wind but also by temperature and pressure differences within the house, and between the house and the outside. For every cubic foot of air that enters a building, a cubic foot of conditioned air escapes. You can use an incense stick on a windy day to locate air leaks by observing when and where the smoke moves erratically. Professional energy auditors measure air leakage and identify leakage paths with a "blower door," and they reveal insulation defects with an infrared camera that "sees" heat.

Using a blower door in conjunction with an infrared camera can be invaluable in identifying air leakage and insulation problems. With specific knowledge about the shortcomings of the building envelope, homeowners can focus their air leakage reduction efforts for maximum benefit.

Leaky House Syndrome

Even if your house doesn't feel drafty, air leakage can account for 20 to 50 percent of your home's heating and cooling energy loss. Reducing air leakage is generally a low-cost improvement that can yield substantial energy savings.

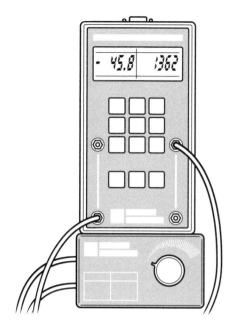

◀ A blower door (far left) seals over an exterior door opening and has a large fan that exhausts air from the house.

◀ Gauges used with the blower door (left) measure air pressure and flow to quantify the air leakage of the building.

◀ An infrared camera measures surface temperatures to identify air leakage and insulation problems. This is what the camera "sees" when focused on the finished room. The darker areas in the infrared image represent cooler temperatures, revealing thermal bridging (a path of relatively rapid heat conduction; see page 63) created by the roof rafters above the sloped ceiling. The camera also shows evidence of a poorly insulated section of the end wall.

Air-sealing a house requires great attention to detail. Air carries moisture, and because heat and moisture naturally move from areas with more heat and moisture to areas with less of the same, the building envelope is the front-line defense in this battle between conditioned indoor air and the outdoor environment.

Ventilation and Humidity

You always want to stop air leakage before adding insulation, as leakiness reduces the effectiveness of insulation. At the same time, too little fresh air can lead to poor indoor air quality and moisture-related problems within the building envelope. The solution to this balancing act is summed up by this phrase from the building science industry: "Build tight, ventilate right." This means that there is no such thing as a house that's too tight, but rather that it may be underventilated.

It is important to regulate the amount of fresh, outside air allowed into the house by using a mechanical ventilation system, such as an exhaust fan or air-to-air heat exchange ventilation system (see chapter 4). Having and using a mechanical ventilation system does not mean you can't open windows when it's nice outside, but it does ensure a supply of fresh air even when the windows are closed.

Keeping **relative humidity** (RH) levels under control can keep you comfortable and keep your house dry. Humans have a wide comfort range in terms of humidity. If your windows regularly develop condensation on the inside during cold weather, it may be an indication of high indoor humidity levels, which in turn may indicate a moisture source problem or insufficient air exchange. Keep track of RH using a hygrometer and operate exhaust fans as needed to remove moisture from the home.

Windows

CONTRARY TO POPULAR BELIEF, replacing your old windows often is *not* the most cost-effective solution to high heating bills. There are several simple, inexpensive repairs and improvements you can make to increase the thermal performance of your existing windows.

When assessing your windows, look at the number of layers of glass (also called glazing) and how tightly the windows close. If the sashes rattle against each other when closed and locked, they are good candidates for air-sealing improvement measures such as weatherstripping and side-mounted sash locks. If the sashes and frames are rotten (wood) or cracking (vinyl), it may be time to think about replacement.

Upgrading old single-pane windows to new double- or triple-glazed units can save energy when they are properly installed to include air-leakage control around the frame. However, air-sealing, insulating, and adding storms are more cost-effective options, especially when these improvements are combined. Replacing windows is worthwhile if they are damaged or if they are causing damage to another part of the building (for example, leaking window sills can cause wooden framing to rot).

If you do replace your windows, it's best to pay the extra cost for highly efficient units. Over the long run, the incremental cost of upgrading to a triple-pane window will pay for itself in efficiency gains and reduced energy use. Since you will probably replace the windows in your home only once, this is a decision with long-term consequences.

Air Infiltration

In most cases, any discomfort you feel when standing next to the window is due to air infiltration around the perimeter of the window frame. If you were to pull off the trim from around the inside of a window, you might be able to see from inside the exterior sheathing or even daylight. Sometimes this gap, or shim space, is filled with fiberglass insulation; however, like a sweater, fiberglass does not eliminate air movement. The fiberglass should be removed and replaced with non-expanding foam or caulk as an air barrier.

Windows occupy 12 to 25 percent of the average home's wall area. Because of their lower insulating value (relative to the framed, insulated wall), windows lose heat much more rapidly than do walls, so it pays to install good, energy-efficient windows to reduce that loss. More windows in your walls increase the potential for both heat loss to the outdoors and heat gain from the sun.

Choosing windows based on their location, their insulating value (U-factor), and their ability to accept or reject the sun's heat (solar heat gain coefficient, or SHGC) will result in greater energy savings and increased comfort (see chapter 4 for

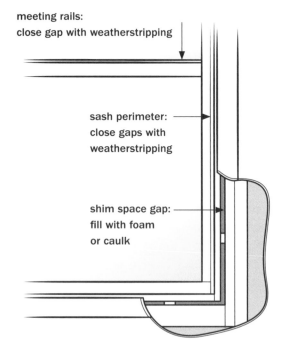

meeting rails:
close gap with weatherstripping

sash perimeter:
close gaps with
weatherstripping

shim space gap:
fill with foam
or caulk

▲ **Where windows leak air**

more information on choosing and locating windows). Whether you live in a hot or cold climate, the right window in the right place offers a good balance of efficiency, light, and solar heat gain.

ACTION

Here are some important steps for improving the performance of your home's thermal envelope:

- Let the sun in when it shines during the winter.

- Hire an energy auditor to measure air leakage with a blower door and perform an infrared scan to direct comprehensive air-sealing efforts and identify thermal defects.

- Add window treatments, such as awnings and insulating curtains or cellular shades, to provide window insulation and solar heat gain control.

- Add insulation to the foundation, walls, and ceilings where appropriate.

- Reduce space-conditioning loads before replacing any heating or air-conditioning equipment, so you can properly size the systems to match the new loads.

Prioritizing Your Improvements

EFFICIENCY IMPROVEMENTS CAN be broken down into things you can do now and things you may need to plan (and budget) for doing later. Some are cheap and easy, some are expensive and difficult, but all should ideally be part of a comprehensive plan that considers the entire house and how the various systems will interact with one another.

These considerations will help you prioritize improvements:

- Low- or no-cost improvements often involve awareness of (and perhaps changing) habits. These can be more difficult than technology fixes but can offer substantial savings.

- Fast and easy improvements that offer instant savings include: shorter showers,

low-flow showerheads, hot-water pipe insulation, efficient lighting, eliminating standby loads, turning down the heat and hot water temperature, washing full loads, weatherstripping windows and doors, repairing leaky faucets and other fixtures, and using a clothesline instead of a clothes dryer.

- Envelope improvements, such as air-sealing and insulation, are very cost-effective and allow you to reduce the size of your heating system when it comes time for replacement. Start by air-sealing the top floor between the ceiling and the attic, then add insulation. Repeat in the basement, then the side walls.

- Heating system distribution improvements can increase your comfort. Add supply and return ducts where heat distribution is poor, and balance, seal, and insulate ductwork. If you have hydronic (hot-water) space heating, make sure the system is zoned for optimal comfort and efficiency, and insulate all the pipes.

- New electrical appliances can offer quick returns on your investment if the old appliance was an energy hog and the new one is exceptionally efficient. Meter the energy use of the old appliance first, so you can assess savings potential.

- Change your lighting plan to put light where you need it.

- Heating plant and water heater replacement may not be cost-effective if the system is moderately efficient and still has some life left in it. When it's time to buy new equipment, have it sized by a professional to match the heating requirements of the home. Pay a bit more up front for comprehensive sizing analysis and the most efficient equipment to save energy and money over the lifetime of the equipment.

- Add storm windows to single- or even double-pane windows. These can be permanent, operable units or removable.

- Improving existing windows and adding window treatments and storm windows are often more cost-effective than replacing windows.

- New windows should be considered only if you're renovating or if existing windows are damaged and need replacing anyway. Choose new windows with energy performance and orientation in mind.

- Add south-facing windows to increase solar heat gain into your home.

- Add thermal mass (tile, masonry, dense materials) to floors and walls to absorb and store solar heat.

- Install renewable energy systems only after making all of your efficiency improvements.

HOW CAN AN ENERGY AUDITOR HELP?

You don't need to call a doctor to put on a bandage, but if you break a bone or need serious attention, self-medication is not the best solution. Don't be afraid to call in an expert energy auditor to help you identify and prioritize where to spend your energy improvement dollars. Often, after I perform a whole-house audit, the homeowner will say to me, "I never knew you could find out so much about a house." Imagine touring a hospital for the first time and being awed by all the diagnostic gizmos used to peek into and probe a human body. Building energy specialists have the tools and skills to diagnose and fix almost any problem your home might have.

An energy auditor will walk through your home from the attic to the basement, examining every room as if on an exotic vacation, taking in all the details. Common diagnostic functions in an audit include a blower door test to measure and pinpoint air leakage within the home, an infrared camera scan to check for insulation voids inside a wall or ceiling, and a duct leakage test to determine the efficiency of the heating/cooling distribution system. An auditor might also perform

---

UNDERSTANDING THE STACK EFFECT

When air leaves a building by way of exhaust fans or air leaks, replacement air must come from somewhere else to take its place; this is called **make-up air**. All buildings leak air. Warm air in your home rises and finds its way out through airflow paths in the ceiling as cool make-up air enters through the lower part of the house.

This air movement sets up a convective heat flow within the house, called the "stack effect" (as in a chimney), which also creates a pressure differential within the house. The upper levels are under higher pressure than the lower levels, creating a natural draft within the house. This is not to say that all air and moisture moving within the house will follow only this path, but the stack effect is an important consideration when addressing air leakage issues.

To reduce the stack effect in your home and its energy-robbing heat loss, seal all uncontrolled air leakage paths in the envelope. Start at the ceiling where the warm, moist air is escaping (causing potential moisture-related damage in the attic), then break the entire air leakage "circuit" by sealing up air entry paths in the basement. Walls and windows can come later (unless they're extremely drafty and your comfort dictates more immediate attention).

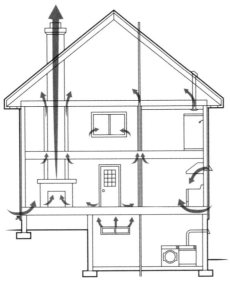

▲ **Air leakage and stack effect**

a combustion test of your heating system and check the refrigerant charge on your air conditioner, as well as make a metered test of your refrigerator to determine its energy consumption.

Additionally, many auditors now incorporate health and safety checks into their work scope, including carbon monoxide tests on all combustion equipment and keeping an eye out for any moisture or mold problems that may exist. Bear in mind that an auditor may not be a building inspector and therefore may not be looking into detailed structural or code-related deficiencies of the home. For a list of contractors with an eye toward comprehensive home improvements, visit the Building Performance Institute website (see Resources).

Hiring an Energy Auditor

Auditors can be trained and certified by energy auditing organizations, and while there are plenty of talented and certified auditors out there, nothing is as important as experience. Before hiring an auditor, ask for references and actually follow

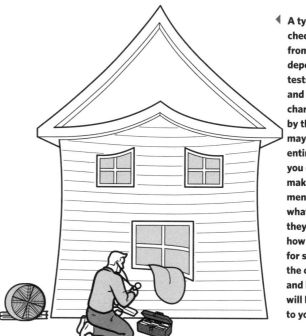

◀ A typical home checkup can take from 2 to 4 hours, depending upon the tests performed, and auditors may charge a flat rate or by the hour. Some may refund the entire audit cost if you choose them to make the improvements. Always ask what specific tests they will perform, how they charge for services, what the cost will be, and how the results will be presented to you.

up by contacting some of the references. Two relevant certifications require auditors to have a minimum level of knowledge and experience (see Resources for websites):

- Residential Energy Services Network certifies contractors through the national *ENERGY STAR for New Homes* program.

- Building Performance Institute certifies contractors through the national *Home Performance with ENERGY STAR* program.

You can find certified energy auditors in your area through both of the programs above or through your state's energy office.

FUEL ENERGY CONTENT

Fuel	BTU/unit*	Unit
Home Heating Oil	138,700	gallon
Natural Gas	100,000	therm or CCF
Liquid Petroleum Gas (LPG)	91,700	gallon
Gasoline	125,000	gallon
Kerosene	135,000	gallon
Coal	21 million	ton
Wood	20 million	cord
Electricity	3,413	kWh
Hydrogen	52,000	pound
Hydrogen	333	cubic foot
Enriched Uranium	33 billion	pound
Solar Home Storage Battery	60	pound

*Note: Energy content per unit of fuel may vary due to additives, impurities, and source.

HOW BIG IS MY CARBON FOOTPRINT?

Carbon (C) is present in fossil fuels. When we burn these fuels, carbon is oxidized (forms a chemical bond with oxygen, O_2) to form the potent greenhouse gas carbon dioxide (CO_2). You can calculate your household's CO_2 production, or carbon footprint, by quantifying how much of various fuels you use over the course of a year, then determining how much CO_2 is released when each unit of that fuel is burned.

When discussing carbon footprints, carbon dioxide is the most common way to quantify greenhouse gas emissions. Greenhouse gases (methane, chlorofluorocarbons, and the like) are often expressed in terms of their "CO_2 equivalent," where their global warming potential (GWP) is compared to that of CO_2, which is assumed to have a GWP of 1. This is often represented as CO_2e, where e means "equivalent". Sometimes you will see a reference to carbon equivalent, or Ce. Carbon makes up 27 percent of the weight of carbon dioxide (the rest is oxygen). To convert a quantity of carbon dioxide to carbon, multiply by 0.27.

It's easy to confuse carbon, carbon dioxide, and their equivalencies, and any one of these metrics can be used to represent carbon footprint. Look closely at the fine print when you're making assessments, comparisons, or calculations.

You can easily determine your *direct carbon footprint,* the carbon or carbon dioxide resulting from the energy you actually buy. It is more difficult to quantify the indirect components of energy use, such as the amount of energy required to produce, transport, and dispose of the food and goods we buy. To determine your direct carbon footprint, simply add up all the energy you buy for your home and travel, and use the chart at right to calculate the pounds of carbon dioxide produced.

For example, if you drive 12,000 miles per year and your car gets 20 miles per gallon (mpg), you'll burn through 600 gallons of gasoline and produce 11,760 pounds of CO_2, or 3,175 pounds of carbon (11,760 x 0.27).

The concentration of carbon dioxide in the atmosphere has been rising since the beginning of the industrial revolution and, according to many climate scientists, is nearing the point of climate crisis. The U. S. Environmental Protection Agency estimates that the average carbon footprint for an American household of two people is about 41,500 pounds (almost 21 tons) of CO_2 per year.

Climate science suggests that in order to achieve climate stability, each individual on the planet has an annual *carbon* (not carbon *dioxide*) "budget" of around one ton — that's about 3.7 tons of CO_2. For perspective, that's the amount of CO_2 released by burning 370 gallons of gasoline. How would you spend your ton of carbon? If carbon becomes taxed or otherwise valued in the marketplace, it will be important to understand these details. For more on climate change and greenhouse gas emissions, visit the Environmental Protection Agency's (EPA) web page on climate change (see Resources).

CO$_2$ EMISSIONS OF VARIOUS ENERGY SOURCES

Type of Fuel	Pounds CO_2 per unit	Unit
Heating oil, diesel fuel	22.4	gal
Natural gas	12.1	therm
Liquid propane gas	12.7	gal
Kerosene	21.5	gal
Gasoline	19.6	gal
Electricity*	1.58	kWh

*National average shown, value varies by state.

Note: Wood is generally considered to be carbon neutral, given its relatively short regrowth period and the CO_2-absorbing properties of trees.

Insulating Your Home

I N AN EFFORT to maintain universal law, order, and equilibrium, heat moves from areas of more energy (hotter) to areas with less energy (cooler). And the greater the temperature difference, the more rapid the energy transfer. Unfortunately, this physical law is tied directly to our bank accounts and the comfort level in our homes, giving you a vested interest in not squandering the labor and cost associated with meeting your energy needs.

Whether it's wrapped around our bodies or stuffed inside our buildings, we use insulation to slow the flow of heat energy transfer. This chapter is not an exhaustive guide on how to insulate your home, nor does it cover all the pros and cons of various insulation choices. Rather, it provides some general guidance and references to available resources so that you can get the understanding and information you need before starting on your insulation project.

How Heat Moves

HEAT MOVES IN three distinct ways, and a good insulation product (or combination of products), along with careful installation, will help slow all three avenues of movement to keep heat loss and energy costs under control.

CONDUCTION

Heat conducts through materials in contact with each other. Your warm hand wrapped around a glass of cold water conducts body heat to the glass. Your hand feels cool as it loses heat and warms the glass along with its contents. More dense materials (such as metal) conduct heat better than less dense materials (such as air). Insulation works by trapping air in tiny pockets, reducing conductive heat loss.

the air at the top of your house (and the water at the top of your water heater) being warmer than at the bottom. This is due to density differences in the air or water. Hot air is less dense than cool air, so it rests on top of the cooler, denser material — thus, the conventional understanding that "hot air rises."

Convective heat loss in a building occurs in several ways. It is primarily the result of wind acting on the outside of the home, carrying away heat radiated and conducted from the house and leading to greater air leakage within the home. Temperature differences between materials, such as a cold window and warm air, create air movement called convection currents. Cold air migrating into a warm, insulated wall cavity creates convection currents within the cavity, reducing the effectiveness of the insulation.

CONVECTION

Heat transfer due to movement within or between liquids and gases is called convection. When you blow on a hot drink, convective heat transfer occurs between the drink and the air as the passing air extracts heat from the liquid. Another example is wind causing convective heat loss from your skin. Convection is also responsible for

RADIATION

Heat passing through space (across the room or across the universe) from one object to another is said to radiate. A hot wood stove and the sun both radiate heat that you can feel on your skin. When you stand next to a cold window, your body radiates heat toward the relatively cooler window, causing you to lose heat and feel cold. Dropping

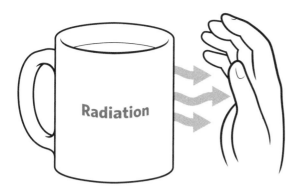

Radiation

an insulated shade over the window makes you feel warmer because it reduces the flow of radiant heat transfer from your warm body to the cold window surface. Reducing radiant heat flow away from your body makes you feel warmer and is often more effective than turning up the thermostat.

In a similar way, **low-E** (E means emissivity) coatings on windows also help to reduce radiation heat loss. Emissivity describes a material's ability to emit, or radiate, energy, and is the key to understanding heat radiation. In general, more reflective materials have a lower emissivity, while materials that absorb heat have a higher emissivity. Radiational heat transfer will occur as long as there is a temperature difference between two objects, and the objects are separated by space (a fraction of an inch or 100 million miles).

Aluminum foil is a good example of a radiant barrier. Hold a piece of aluminum foil between you and a heat source (or between you and the cold window), and almost all of the radiant heat energy transfer is eliminated, having been reflected away. But radiant barriers may also be good *conductors* of energy. If your piece of aluminum foil is in direct contact with a hot surface, it gets hot as heat is readily conducted through it.

A foil-faced radiant barrier under the roof will reflect heat away from the home in hot weather, provided there is an air space between the radiant barrier and roofing material. In combination with good attic insulation, proper air leakage control, and reflective roof colors, a radiant barrier is a valuable component in a system that will help increase comfort and reduce cooling loads

by keeping the attic cool and slowing heat movement to the living area. Some insulation products have integrated foil-faced radiant barriers, and other products are available with a radiant barrier fixed to sheet material that can be used for a roof deck (see Resources).

A radiant barrier is not, in and of itself, insulation. For a radiant barrier to be effective, there must be a calm air space between it and any adjacent materials; otherwise, the radiant benefit can be short-circuited through conduction or convection. Some roofing materials are designed to decrease cooling loads in hot climates. You can learn more about these from the Cool Roof Rating Council (see Resources).

Heat transmitted through the building envelope is driven by the temperature difference between indoors and out, along with wind loads acting on the outside of the house. Conduction, convection, and radiation, along with air leakage, all contribute to heat loss in your home. In addition to insulation, addressing air leakage between indoors and out and eliminating air movement within the insulation reduce heat loss by all three transmission paths.

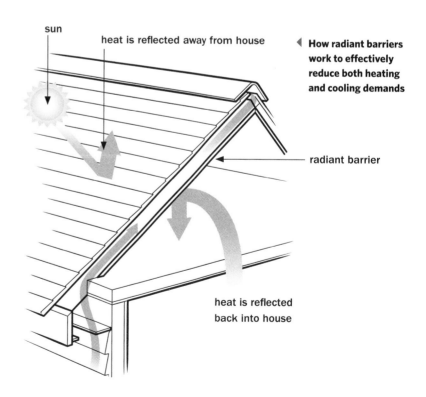

sun

heat is reflected away from house

How radiant barriers work to effectively reduce both heating and cooling demands

radiant barrier

heat is reflected back into house

Measuring Insulation Value

IN THE UNITED States, all insulation products must comply with Federal Trade Commission testing and labeling requirements. This makes it easy to compare insulation ratings and know that, for example, R-19 is R-19 regardless of product or manufacturer. However, the product must be installed properly to perform at or close to its rated R-value.

U-FACTOR AND R-VALUE

The performance of insulation products is stated as **R-value**, or resistance to heat flow. R-value has no dimension, meaning that there is no engineering value associated with it, but higher R-values indicate higher levels of insulation, and thus slower heat movement through the insulation.

Some products, notably windows, are described by a different heat transfer value, called **U-factor**. U-factor is a measure of the thermal conductance of a material. The U-factor expresses how much heat, in Btus (see page 20), is transmitted through one square foot of material in one hour when the temperature difference between opposite surfaces is 1°F (.56°C). A higher U-factor indicates greater conductivity of heat and corresponds to a lower R-value.

U-factor and R-value are the inverse of each other. Mathematically, that means:

$$1 \div \text{R-value} = \text{U-factor or}$$
$$1 \div \text{U-factor} = \text{R-value}$$

Therefore, a wall with an insulating value of R-19 has a U-factor of 0.0526, while a window with a U-factor of 0.40 has an insulating value of R-2.5.

WHERE YOU NEED INSULATION

The question of where to install insulation may appear to have an obvious answer: walls, ceilings, floors, and foundation. Yet one basic concept is consistently overlooked: The building's "thermal boundary" must be well defined in order to develop an effective insulation plan. The thermal boundary is comprised of the building assemblies that delineate indoors from out.

To identify your home's thermal boundary, draw a cross section of the building onto paper. You may need several cross-sectional diagrams depending upon how complex the building is. Now, with a different color pen, add a line that indicates where there's insulation. You should be able to trace the insulation line of the cross section with your finger all the way around the house — from the roof, down the walls, across the foundation, and back up to the roof on the other side — without lifting your finger off the paper.

If you need to lift your finger to jump to the next place that is insulated, you have a break in the thermal boundary. This means you have uncontrolled heat loss to (or gain from) the outdoors. You may find that you have two parallel but unconnected thermal boundaries, such as an attic floor and ceiling that are both insulated, or insulation in both the back side of kneewalls and the roof

above. This "doubling up" is simply a waste of material. Understanding where insulation is, where it isn't, and whether it's effectively installed is the first step in developing an insulation plan.

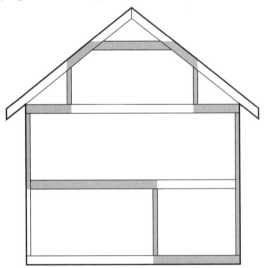

▲ **Thermal boundary. Create a simple cross section of your house to note where insulation exists — or not. The entire thermal boundary should have a continuous line of insulation around the living space.**

THERMAL BRIDGE

Insulation is not, however, the only material in the building envelope. Wood or metal framing also affects the overall R-value of a building assembly. Each stud, rafter, or joist presents a "thermal bridge" — in effect, a short circuit — within the insulated assembly. Wood has an insulating value of about R-1 per inch of thickness. For example, a 2x4, which actually measures about 1½" x 3½", has an R-value of about 1.5 when laid flat or 3.5 when on edge.

In terms of U-factor, a 1"-thick piece of wood conducts heat at the rate of 1 Btu per square foot per hour per degree of temperature difference. A wooden 2x6 stud will conduct heat faster than the insulation-filled wall cavity between the studs but a little slower than an empty cavity with no insulation. What this means is that when it's warmer indoors compared to outside, the studs in a wall will be cooler (on the inside of the wall) than the insulated wall cavity. On the outside, the studs will be warmer than the insulated part of the wall because the studs conduct heat away from the warm indoors faster than insulation does.

A thermal bridge can be fixed by using a "thermal break," which is simply a less conductive material (such as a piece of insulation) between two thermally conductive materials. For example, in addition to insulating the stud cavities in a wall, a well-insulated building will have a continuous insulation layer, such as rigid foam board, over the studs so that the thermal connection, or conductive channel, presented by the stud frame is minimized. Many energy auditors use an infrared camera to locate thermal bridges within building assemblies.

In theory, we can calculate the overall R-value of each component in a building and arrive at an average R-value for each assembly or for the building as a whole. In practice, however, the actual thermal performance of a building depends a great deal upon how well the insulation was installed.

Insulation Inspection

TO DEVELOP AN insulation plan, first identify where it currently exists, then determine where it needs to be and how much of it is needed. I like to start at the top and work down.

ATTACKING THE ATTIC

Some attics have floors, some are open and insulated in between rafters, sometimes instead of an attic there will be a cathedralized ceiling covered with finish material, and many homes have some combination of these. If you have an attic with access, it's relatively easy to see what's up there. A cathedralized ceiling, however, requires a removal of the finish material or an infrared inspection. Put on some old work clothes, gloves, and a respirator, get a tape measure and a flashlight, and go look! Be careful to step only on wooden framing members. Once you're in the attic, ask yourself:

- What is the condition of the insulation?
- Is the insulation a consistent depth?
- Is there evidence of damage from animals?
- Are there places that are matted down or trampled?
- Any evidence of water damage?
- Can you see the top of the ceiling surface below (the back side of the drywall or plaster)?
- Are there any visible framing materials? (This would present a thermal bridge between the house and the unconditioned attic.)

While you're up there, look for vents from bathroom exhaust fans, dryers, or other appliances that may be terminated in the attic. This situation can ultimately lead to moisture, durability, and health issues such as mold and wood rot. All exhaust ventilation systems should terminate outside the building. Terminating an exhaust in front of an attic vent or louver is not reliable because airflows could bring the exhaust right back into the attic.

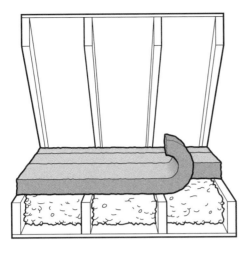

▲ **Properly installed fiberglass insulation in an attic. There are two layers of insulation: the first lies between the framing members on the attic floor; the layer on top is perpendicular to the one on the bottom, eliminating the thermal bridge of wood framing.**

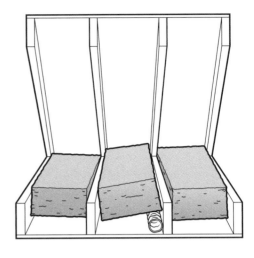

▲ **How not to install insulation. Insulation must be in contact with the surface it is intended to insulate, or it will be less effective. There is also the potential to crush the flexible vent duct under the insulation.**

Fan boxes or lighting fixtures mounted in the ceiling below may be visible and often are sources of significant air leakage between indoors and out, if we consider the attic to be essentially outdoors. Recessed lighting fixtures should be properly specified for insulation contact, a designation known as "IC-rated." This means it's safe for them to contact insulation without concern about overheating. When installing IC-rated fixtures, caulk between the fixture housing and ceiling to air-seal it, then cover the top of the fixture in the attic with insulation. Do this only with IC-rated fixtures; covering standard recessed fixtures with insulation may create a fire hazard.

Chimneys

If chimneys or plumbing pipes are poking through the attic floor, there is a good chance that a "bypass" condition exists. A bypass is a place where heat and air are permitted to move around an insulated assembly, breaking the thermal boundary and reducing the effectiveness of the insulation. All bypasses must be sealed against air leakage with appropriate materials, such as rigid foam, caulk, wood, or sheet metal. Fibrous insulation will not prevent air movement, so simply stuffing fiberglass into a hole is not effective. Any materials used on or

near a chimney or other combustion appliance vent must be approved for use on hot surfaces. If you are in doubt, check with your local code inspector or the product manufacturer before proceeding.

WALL INSULATION

Checking wall insulation is a bit more challenging, but it's easy to find enough clues to draw some conclusions. First, find out how thick your walls are. Measure the depth of a window or door frame, and subtract the thickness of the sheathing, siding, wallboard, and the amount the trim stands out on each side; what's left is the wall cavity depth. Most standard wall framing is either 2x4 or 2x6, yielding a 3½" or 5½" cavity depth, respectively.

Another way to determine the depth of the wall cavity and, if you're lucky, also see what kind of insulation exists (if any), is to find or make a small hole in a hidden area, such as a closet. Poke a wooden (**not** metal, in case there are exposed electrical wires) barbecue skewer into the hole, mark where it stops with your finger, and measure how far it went in. There should be a bit of resistance caused by the insulation, and you might find some material caught on the skewer or twisted around the drill bit you used to make the

exterior siding   wall sheathing

interior wall board

trim

stud

insulated
wall cavity
thickness

▲ **Cutaway of a framed, insulated wall**

hole. Be sure to spackle over the hole you made or otherwise seal up that air leak.

Alternatively, you can usually find a gap in the wallboard around electrical boxes for light switches and outlets. Turn off the power at the circuit breaker, remove the cover plate from the switch or outlet, and poke your wooden skewer into the gap between the electrical box and the wallboard. Wiggle it around and try to pull out a bit of insulation to see what type it is. A professional energy auditor may use a borescope, a flexible fiber-optic viewing tool, to look into wall cavities.

WHAT LIES BENEATH

Basements and foundations may be insulated on either the inside or the outside, and covered with finish material. Rigid foam board is commonly used for insulating the exterior of concrete foundations. This may be covered with stucco or other durable material on the exposed portion of the wall. If you don't see exposed cinder blocks or concrete on the outside of the foundation, knock on the surface. If it sounds hollow, there's

probably rigid foam-board insulation under the covering.

In some cases, the basement ceiling may be insulated instead of the walls, but there is no need to insulate both. Decide where you want your thermal boundary to be and install your insulation there. Note that it is quite difficult to separate the basement from the rest of the house in terms of air leakage, so best practice often leans toward insulating and air-sealing the perimeter of the foundation rather than the basement ceiling. This has the added advantage of bringing the basement into the conditioned area of the house, which can begin to make it a more useful space. Also, if there are ducts or hot water pipes running through the basement, it's best to insulate the walls, not the ceiling.

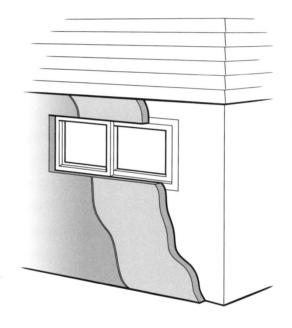

◀ **Rigid foam board insulation around the perimeter of a concrete foundation. This insulation should be covered to protect it from sunlight and physical damage.**

Most basements are partially above, and mostly below, grade (the ground level around the building). The most important place to insulate is above grade, the area of greatest heat loss. Since the average ground temperature for any given location is approximately the same as the annual average outdoor air temperature, the temperature difference between the basement and the ground is usually much less than the difference between indoor and outdoor air temperatures. However, if you plan on heating your basement, it is important to insulate the full height

THE EFFECTS OF CONDENSATION

Basements are often cool and damp — a potentially nasty combination. Condensation occurs when warm, moist air comes into contact with a cool surface. This means that opening windows in a cool basement during a warm, humid summer will only increase the dampness of the basement, as moisture from the air condenses on the relatively cool foundation walls.

Condensation can also occur inside above-grade walls, but it's usually the result of warm, humid air migrating out into the cooler wall cavity. If the dew point (the temperature at which moisture condenses out of the air) is reached within or behind the wallboard or insulation,

moisture-related damage can occur, and there is potential for mold to grow on these hidden materials.

Therefore, it's important to keep moisture-laden air out of places where dew points are reached, especially where the materials cannot readily dry out. The best approach to solving this issue is to eliminate the movement of humid air into these spaces by paying close attention to air-sealing details (see page 72). Sometimes this can be difficult or impossible to achieve, in which case the materials must be resistant to moisture damage, and there must be a means for draining and drying the space where condensation occurs.

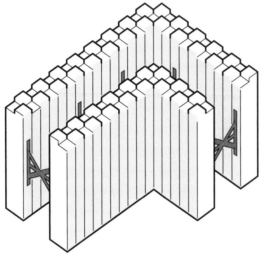

Insulated concrete form. The insulation provides the form into which the concrete foundation is poured. ▶

of the entire foundation perimeter for maximum efficiency and effectiveness of the insulation.

Insulated concrete forms (ICF) can be used when pouring new basement walls. They serve as a pouring form and also provide insulation on both the inside and outside of the wall.

How Much Insulation Do You Need?

UNLESS YOU LIVE in a climate that requires no heating or cooling, the more insulation you have, the less space-conditioning energy you'll need to stay comfortable. There may be a point of maximum cost-effectiveness and return on investment, but with today's volatile energy market, that is a moving — and often subjective — target.

With our current awareness of global energy issues, it's increasingly evident that we need to build durable, long-lasting homes, using minimal materials with low embodied energy (the energy required to produce the material), creating little waste, and requiring a minimum amount of energy to condition for comfort. Minimal home energy use makes it possible to achieve a fair amount

Need New Siding or Roofing?

When it's time to replace the siding or roofing on your house, look into the value of increasing the wall-cavity insulation and adding a few inches of sheet insulation, such as rigid foam board, to the outside of the wall sheathing or roof decking before putting the new siding or roof on. (Note: Do this on the roof only if the rafters are insulated; don't do it if only the attic floor is insulated.) See chapter 4 for more information.

of autonomy and offers some protection from the risks of a volatile energy market, while making it more practical to meet your energy needs with renewable energy.

Understanding Heat Loss

Heat loss math shows that 90 percent of conductive heat loss is eliminated with an insulation level of only R-11 as compared to R-1. This doesn't mean that it's not cost-effective to install much more insulation; there are huge savings to be earned with higher R-values. Increasing the level from R-11 to R-100 reduces energy use by an additional 90 percent.

Over time, the extra investment you make to achieve more substantial energy reductions will pay for itself in many ways, including lower energy costs, increased comfort, smaller carbon footprint, and increased energy security. Many small improvements made on an individual level add up to huge global impacts.

Upgrade When You Can

Given that the majority of the cost of insulation is usually the labor of having it installed, it is often prudent to take advantage of the work crew's being on site to install as much insulation as possible. The incremental cost of upgrading from some minimal level to the highest practical level typically is not unreasonable. If you have 3 feet of height in an unused attic, why not fill it entirely with insulation? Take advantage of other work being done on your home to add insulation and fix air leaks; the savings can help to pay for the remodeling work.

Building science researchers recommend target levels of insulation for both new construction and existing home retrofits in heating climates (see Recommended Insulation R-Values, above). These are based on cost, practicality, and durability. Additional analysis can take into account the tradeoff of the global-warming potential of manufacturing and installing these products vs. the amount of energy they might save over their lifetimes.

Recommended Insulation R-Values

- Under concrete floor slab/basement slab: R-10

- Foundation walls: R-20

- Above-ground walls: R-40

- Ceilings: R-60

- Windows: R-5 (triple pane)

Choosing and Installing Insulation

THERE ARE MANY different insulation products on the market today that can greatly enhance the thermal performance of your home. It can be difficult to choose which product, or combination of products, will work best without some clear guidance from an expert who lives in your climate.

The expert you need might be an energy auditor or a builder who specializes in high-performance home design or energy retrofits. He or she should understand what your house needs, your long-term goals, and how your house will behave when improvements are completed. Choosing the right product and how much of it to use depends on many factors:

- Climate

- Availability

- Practicality

- Cost

- Suitability for the specific assembly that needs insulating

- Building science issues, such as air and moisture permeability properties

- Durability

- Fire resistance

- Need for sound control

- Embodied energy

- Recycled content

- Recyclable value at end of life
- Global-warming potential
- Ozone-depletion potential
- Health risks due to off-gassing of blowing agents
- Local code requirements

To further explore these issues, two unbiased references to begin with are BuildingGreen Inc., (publishers of *Environmental Building News*) and Green Building Advisor (see Resources for websites).

INSULATION OPTIONS

If you understand the basic principles of how insulation works and what it needs to do in a specific situation, you'll have a good start toward understanding which product(s) to use. Some cutting-edge products hold great promise but are too costly to use at this point except in high-end or critical situations. These include products made from translucent silica, and vacuum-insulated

INSULATION R-VALUE COMPARISON

Product	Average R-value per inch
Fiberglass batts, standard	3.4
Fiberglass batts, high-density	3.8
Fiberglass, spray-applied	4.0
Cotton batts	3.5
Cellulose; loose fill, dense pack, damp spray	3.2–3.8
Extruded polystyrene (blue/pink/grey board)	5
Expanded polystyrene (bead board)	4
Polyisocyanurate (foil-faced)	6–8
Spray foam, open-cell	3.7
Spray foam, closed-cell	6
Mineral wool	4.0
Foamglass, cellular glass	3.4
Perlite	3.7

panels with an insulating value of R-30 or more per inch (see Resources).

INSTALLING INSULATION BY TYPE

Many insulation products can be installed by the average homeowner with little experience — but always remember that proper installation is just as important as using the right product in the right place. Consider your health and safety first when removing or installing insulation products (also see Vermiculite Warning on page 74). Read all manufacturers' precautions and always use a respirator when working with foams and fibers, in addition to providing adequate ventilation.

Batts and Blankets

Batt or blanket insulation, made from fiberglass or cotton, can be rolled out and cut to fit into place. The R-value will be stamped on the package and/or on the batts. Look for high-density batts offering greater R-value per inch. Installation is fairly simple with the right tools and attention to detail. Long-bladed scissors or a long, sharp knife and a straightedge make cutting easier. The insulation should be fully lofted and cut to fit snugly inside and around anything inside the cavity. Don't forget to seal up air leaks between indoors and out, or between the house and attic (where plumbing and wiring is often run through drilled holes), with caulk or spray foam before adding insulation.

Loose-Fill Insulation

Three common types of loose-fill insulation are cellulose, fiberglass, and mineral wool. All of these materials are available in bags of manageable size and are made from some percentage of recycled raw material. The R-value of each depends in part upon the density with which it is installed. Overly dense materials conduct heat more readily, while too little density allows excessive air movement within the material, increasing convection.

Look at the label on the bag to determine the depth required to reach the desired R-value, as well as the coverage per bag at the desired

▲ Poorly installed wall insulation with compression, gaps, and voids

▲ Insulation properly cut to fit around wires and fit snugly into wall framing

depth. Keep in mind that loose-fill insulation will settle (up to 20 percent), and it's the settled depth and R-value that's important. The material can be poured out onto an attic floor and pushed into place, but some home supply stores rent or loan a blowing machine with the purchase of the insulation.

Loose-fill fiberglass is essentially the same material found in fiberglass blankets or batts but in a different form. It has an insulating value of about 2.7 per inch when installed at a density of about 1 pound per cubic foot.

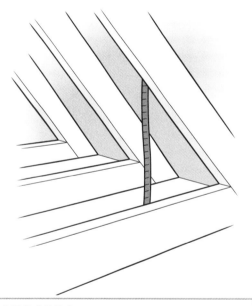

◀ Prior to installing loose-fill insulation, install paper measuring tapes at various locations around the attic to help gauge the insulation depth (and thus the expected R-value).

USING VAPOR BARRIERS

Some insulation products include a vapor retarder, or vapor barrier (such as kraft paper or foil), which may or may not be appropriate to use in your climate or for the specific construction type or assembly you're working with. If you're in doubt about whether or where to use a vapor retarder, consult with a local building expert; it may mean the difference between a lifetime of maintenance-free durability and comfort and a future expensive repair due to moisture damage within the wall. As mentioned earlier, the more important part of moisture control is to prevent moisture-laden air from entering a cold building cavity through air infiltration. It has been found that far more moisture is transported into a building cavity via air movement than through water vapor migration through building materials. Only vapor movement *through* materials is controlled with a vapor-retarding barrier.

▲ Loose-fill cellulose or fiberglass blown into an attic is an excellent choice for insulating attics with complex framing, conforming easily around lumber, vents, and other inconsistencies.

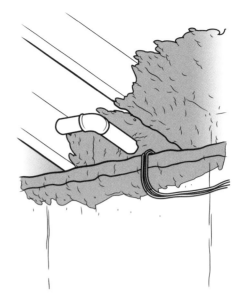

▲ Closed-cell spray foam insulating a basement band joist. Note how the foam seals around penetrations through the joist to prevent air leakage to the outdoors.

Mineral wool (sometimes called rock wool) can be made from minerals, ceramic, or the stone slag left over from processing iron ore. It is naturally fire resistant and has an insulating value of about 3.3 per inch when installed at a density of about 1.7 pounds per cubic foot.

Cellulose is often made from recycled paper products that are treated for fire and insect resistance. It has an insulating value of about 3.7 per inch when installed at a density of about 2 pounds per cubic foot.

Dense-Pack Insulation

Dense-packing insulation is a way to optimize the insulating value of insulation within an enclosed cavity. Any of the loose-fill insulation materials may also be dense-packed, but the technique requires a knowledgeable, experienced contractor with the right equipment. The density of the material must be correct to achieve the rated performance: 3.5 pounds per cubic foot for cellulose; 1.5 pounds per cubic foot for fiberglass and mineral wool. If the blowing machine pressure is too high or the walls are not properly braced, it's possible to blow out the drywall on the interior. Pressure that's too low can lead to voids in the material and lower R-values.

Damp-Sprayed Insulation

Cellulose and fiberglass can be sprayed into an open wall cavity or other building assembly. When mixed with the right amount of water, the material clumps together and adheres to the wall cavity. It is ready for wall covering after a day or so of drying. Damp-spraying insulation requires the right equipment used with careful control and attention to detail to prevent both material and building failure. For this reason, it should be considered a professional job. It may be preferred when there are open walls to be insulated because it is faster to apply than dense-pack, and the insulating value is about the same.

Spray Foam

Spray foam insulation is usually synthetic polyurethane or polyicynene, though there are some soy-based products on the market. Material is fed in two parts, combining at a nozzle from which the mixture sprays, and adheres to wood, stone, concrete, or other building material. The foam cures and expands, releasing heat as it does so.

It is available in both closed-cell and open-cell forms; the best type to use depends on the insulating and moisture permeability requirements of the specific job (consult with a building energy

▲ Insulating walls with cellulose installed from the outside requires using a blowing machine that packs the cellulose into the wall cavity to a predetermined density.

▲ Shooting spray-applied cellulose into an open wall. Binders in the material keep it in place.

▲ After the excess material is screeded off, the insulation is allowed to air-dry, then the wall is ready for finishing.

professional for on-site guidance on this subject). At about R-6 per inch for closed-cell and R-3.6 per inch for open-cell, spray foam is a good choice for irregular surfaces or where air leakage control is required in addition to insulation.

Expanding spray foam is available from hardware stores in small cans for small jobs. For mid-size jobs, such as insulating the perimeter of a basement band joist, two-part closed-cell foam can be purchased in a kit that will cover up to 600 board feet. Board footage is a measure of coverage area and thickness of the insulation. With spray foam, you can apply a 1" to 4" layer, depending upon whether you want to air-seal only or to also add lots of insulating value. With a little practice, you can achieve good results.

Kits are available at efficiency specialty stores or on the Internet from The Energy Federation, Tiger Foam, and Foam It Green, to name a few (see Resources). For larger jobs or for working in colder conditions, it's best to consult a professional.

Rigid Foam

Rigid foam boards made of polyisocyanurate or expanded or extruded polystyrene can be fastened to wood or metal framing, typically on the exterior, using construction adhesive or screws. Basements and crawl spaces also can be insulated with rigid foam on the inside or outside. Extruded polystyrene foam boards are resistant to moisture and can be in contact with the ground, but they must be protected from physical damage, sunlight, rodents, and bugs, such as ants and termites.

◀ Spray foam kits for DIY application include the foam, hoses, spray gun, and spray tips.

Rigid foam boards are commonly used on the exterior of wall and roof structures to prevent thermal bridging. Different types are effective for various applications, including insulating basement walls, band joist cavities, and other new construction or retrofit applications. ▶

Stopping Air and Moisture

INSULATION IS NEARLY useless if air is allowed to move through it, and fibrous insulation (such as all loose-fill materials) itself will not stop air movement. Therefore, it is extremely important to eliminate any air leakage paths within and through the assembly before insulating.

Leakage paths commonly exist around the boundaries between spaces, such as:

- the top-floor ceiling into the attic
- interior or exterior walls where the top of the wall extends into the attic

IMPORTANT INSTALLATION DETAILS

Regardless of the type of insulation you choose, proper installation is key to realizing maximum thermal performance. This means full and even coverage within the entire cavity, without voids or compression. Compression reduces the insulating value, so don't be tempted to stuff an R-19 batt (usually 6" thick) into a 2x4 wall cavity.

Be sure to cut the insulation so that it fits snugly around and behind anything inside the framing cavity, such as plumbing, wiring, and structural bridging or blocking. The insulation must be in full and direct contact with all six sides (including the interior wall finish) of the cavity it is intended to insulate. When installing fiberglass batts with foil or kraft paper backing, always staple the tabs onto the front, not the inside, of the stud.

Avoid gaps in the insulation. Leaving just 5 percent of an assembly uninsulated (50 square feet in a 1,000-square-foot attic) can increase heat loss through that assembly by up to 40 percent! Keep in mind that insulation gaps usually aren't in the form of one big void but rather many small gaps throughout the area, all of which contribute to the reduction of insulating performance.

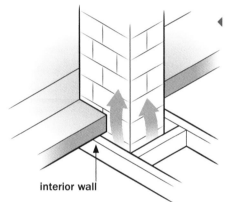

interior wall

◀ **Pulling back the insulation around the chimney, you will often see a gap between the chimney and the attic floor framing. This provides an unwanted path for air movement between the basement, the attic, and all floors in between.**

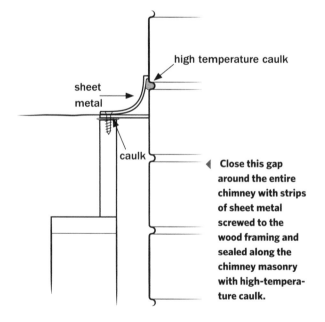

high temperature caulk

sheet metal

caulk

◀ **Close this gap around the entire chimney with strips of sheet metal screwed to the wood framing and sealed along the chimney masonry with high-temperature caulk.**

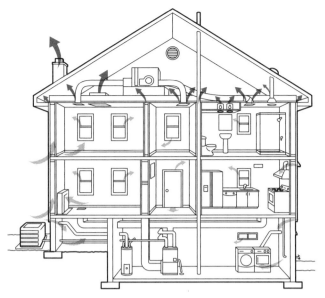

Common paths for air leakage

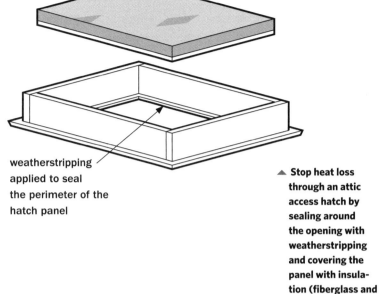
insulation covering hatch panel

weatherstripping applied to seal the perimeter of the hatch panel

▲ **Stop heat loss through an attic access hatch by sealing around the opening with weatherstripping and covering the panel with insulation (fiberglass and rigid foam board insulation are two practical options).**

- plumbing, electrical, and duct runs or chases

- plumbing vents and chimneys

- attic access hatches

- any place where two different materials meet, such as where the wooden sill plate meets the top of the concrete foundation wall, or where the bottom plate of a stud wall meets the plywood floor decking.

These areas should be caulked, foamed, or — if it's a large hole — sealed with an appropriate sheet material such as rigid foam board insulation, drywall, or sheet metal.

In order to perform at its rated R-value, insulation must be used in conjunction with an air barrier to prevent **wind washing**, the effect of air forcing its way into the wall and through the insulation, removing heat and reducing the insulation's effectiveness. Imagine wearing a sweater on a cool day: You'll be warm until the wind blows, but then you have to put on a windbreaker.

The entire thermal boundary of your home should be in contact with an air barrier. Ideally this includes an airtight interior wall surface as well as an exterior air barrier (such as plywood sheathing and building wrap) with airtight surfaces around all the edges to prevent air movement into the insulation from either direction.

◀ **Insulation with interior air barrier. Note that all seams and connections to framing are taped.**

Common places where an air barrier might be missing are spaces behind kneewalls, built-in shelves, fireplace surrounds, bathtubs, shower enclosures, dropped interior soffits, and walls adjoining porch roofs.

DEALING WITH MOISTURE

Wet insulation has almost no insulating value. That's why it's extremely important to keep water and water vapor out of insulation products. This is best accomplished with a durable, weather-resistant exterior finish, along with an airtight interior finish to slow or eliminate moisture-laden air moving from the living spaces outward to the insulation.

Note the lack of the term *waterproof*, which is, in a practical sense, not possible with traditional building materials. The best approach is to allow moisture to drain away from, and dry out of, the wall or other assembly. Because water vapor molecules are smaller than air molecules, it is possible for materials to allow water vapor through while blocking air movement. This is the basis for products such as Gore-Tex and some building wraps. A vapor retarder on the inside of the wall frame is different from an air barrier (see Using Vapor Barriers on page 69).

It's also important to note that there is no requirement for a house to "breathe." This is an outdated concept based on early attempts to weatherize older homes without our current understanding of building science, particularly in regard to the dynamics of air and moisture movement. As with those older homes, your house behaves in certain ways that involve all of its systems interacting in a balance. This balance may or may not produce the results you want, but the systems are at least familiar to you. As you make additional improvements, expect to rethink how your home behaves.

You face many choices when it comes to building or remodeling your home. For long-lasting energy savings, lower maintenance, and increased durability, proper insulation and air-sealing techniques are at least as important as the finish materials you choose. What good is a shiny new paint job on a car that doesn't run well or uses so much fuel you can't afford to drive it?

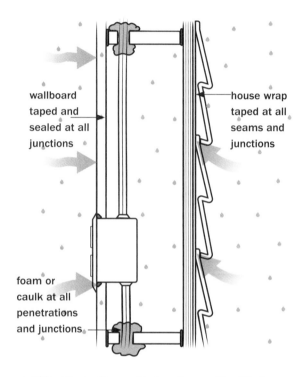

wallboard taped and sealed at all junctions

house wrap taped at all seams and junctions

foam or caulk at all penetrations and junctions

▲ This cutaway shows a simple wall assembly with elements to stop airflow while allowing for the movement of water vapor. A well-sealed drywall finish on the inside stops airflow and slows water vapor. On the exterior, sheathing and building wrap keep air out of the wall cavity and let water vapor drain away to keep the cavity dry.

VERMICULITE WARNING

If you are renovating your home and you find vermiculite insulation (commonly used between the 1950s and 1970s), be aware that many samples of vermiculite have been found to contain the carcinogen asbestos. Vermiculite looks a bit like a small, broken wood chip — a fibrous-looking chunk of compressed material about ½" across. The best thing to do is to leave this insulation alone and not stir up any dust. If you are concerned about its presence, you can have it professionally removed, but it will likely be costly and disruptive. In some cases it's possible to blow in new insulation on top of the old while causing minimal disturbance, but this should be done only by licensed contractors and in conformance with local code requirements. Many contractors are not taking on this liability as health and safety codes specifically address the issue of asbestos.

Roof Venting

ROOF VENTILATION REMOVES heat and moisture from within the roof framing and attic, while keeping the roof itself "cold" (meaning closer to the outside temperature than the indoor temperature). This helps to prevent snowmelt and ice damming (see Anatomy of an Ice Dam on page 76).

A properly vented roof will move outdoor air into a vented soffit at the eaves, up along the roof slope through a vent channel between the roof deck and the insulation, and out through a ridge vent or gable-end vent. The vent channel is commonly created with a rigid, corrugated foam sheet designed for the purpose, but you can also use rigid foam insulation installed 2" away from the roof sheathing if you want to add R-value.

In addition to a vent channel that creates air space underneath the roof deck, the air coming into that air space by way of the soffit vent needs to be shunted away from the insulation. Install a baffle at the eave-end of the insulation, inside the rafter or joist bay, to prevent air movement into and within the insulation. The baffle also prevents insulation from blocking the soffit vent.

The illustration at right shows properly insulated, vented, and baffled roof insulation. Note also that the ceiling insulation extends over the top plate of the wall, eliminating a potential path for losing heat. It is widely believed that increased attic venting will prolong the life of asphalt roofing shingles by keeping them cool. But research shows that venting has very little, if any, effect on shingle temperature. The most important issue in shingle temperature appears to be the color of the shingles. Light-colored shingles reflect sunlight and don't get as hot as dark shingles. The same insulation, venting, and baffling schemes apply to other roofing materials, such as metal and slate.

AIRFLOW PATH

Like all air movement, the airflow path from soffit vent to ridge vent requires an imbalance in pressure. Typically, heat loss from the house (or from solar heat gain into the attic) drives this air movement. As heat passes through the attic insulation, the air in the attic is warmed, causing it to rise and exit via the ridge vent, drawing in make-up air through the soffit vents. This airflow also removes any moisture that may be in the attic area, reducing the potential for condensation buildup within the attic.

Given this information, you may be tempted to increase the ventilation in your attic by adding an attic exhaust fan. Be very careful with this approach and don't exceed the venting area's ability to move air. The fan may create a negative pressure condition in the attic, with respect to the inside of the house. If this happens, air can be pulled from inside the house into the attic. Moving conditioned air into the attic wastes energy and increases the potential for mold growth in the attic.

Roof venting was born long ago as a moisture control strategy for cold climates. Builders knew that heat and moisture would escape from the house into the attic and cause the problems we've just discussed. The old-time fix was to increase air movement through the attic space to quickly remove that heat and moisture. With today's building techniques, it is not much of a stretch to build a ceiling assembly that is well air-sealed and contains enough insulation to prevent problematic levels of moisture and heat flows. While ventilated roofs are not necessarily a bad thing, they can be avoided with proper design and building.

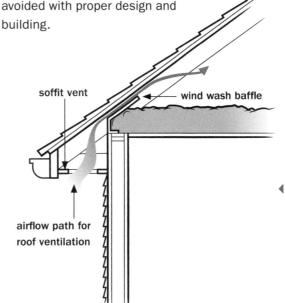

soffit vent

wind wash baffle

airflow path for roof ventilation

◄ **Insulation extends all the way over the top plate of the wall, with the end of the insulation shielded from wind wash by a baffle.**

ANATOMY OF AN ICE DAM

If you live in a cold region, take a look at the roofs in your neighborhood the next time it snows. If you can see vertical stripes, you are looking at the thermal bridging effect of the rafters, warm areas that melt snow faster than the insulated cavities between the rafters. If you see ice dams and icicles, or places on the roof where all the snow has melted, you know that a serious insulation or air-leakage problem exists.

Icicles and ice dams are the result of poor insulation and/or air leakage paths between heated spaces and the roof. This may be due to poor roof drainage and sun melting the snow, but more often the problem is due to heat moving rapidly away from inside the house — an indication of wasted energy and ineffectual thermal and/or air boundary control.

TROUBLE SPOTS

Icicles often lead to ice dams, especially around dormers, roof valleys, or other hard-to-drain areas. When snow melts on a roof, runs down in the form of water, then refreezes near the eave in an ongoing process, the result is an ice dam. Ice can push its way under roofing material, causing water leaks and damage to the home. Understanding how ice forms on a roof is to understand much about the nature of insulation and air leakage.

To state the obvious, snow melts when it's warm. Your roof deck should not be warm enough to melt snow. If it is, you're losing heat from indoors. The source of that heat may be poorly insulated duct-work in the attic, poor ceiling or wall insulation, or air leakage from indoors into the attic or cathedral ceiling space. The failures listed below result in thermally connecting the indoor heated space with the roof, which, as you now know, warms the roof and melts the snow:

- Poorly insulated or air-sealed spaces behind kneewalls

- Interior walls that connect to the roof or attic space without proper insulation or air barrier detailing

- Insulation that does not cover the entire ceiling due to voids, gaps, or the failure to extend it all the way over the top plate of the wall below

- Outdoor air moving through the soffit vent and into the insulation where it rapidly removes heat (wind washing)

- Thermal bridging caused by rafters and complex roof framing details

Good planning while framing the roof, and attention to insulation and air-sealing details, will help reduce heat-loss into the attic, keeping the roof cool and preventing snowmelt.

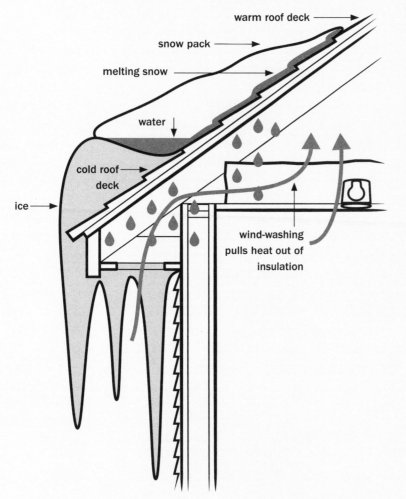

▲ How an ice dam forms. Heat escapes from the house, warms the roof, and melts the snow. Melted snow turns to water, and when the water runs down the roof and hits the cold eaves (where there is no source of heat to warm this area), it freezes and turns to ice. When water turns to ice, it expands; this expansion is more powerful than most roofing materials can withstand. The ice can literally migrate into the soffit or attic, damaging anything in its way.

When "High Performance" Doesn't Perform

A HOMEOWNER ASKED ME to take a look at his 6-year-old home in northern Vermont. Frost was forming in the attic on the inside of the roof sheathing, and he was concerned that the moisture would eventually rot the roof. I agreed to visit the home, although it was now June and the frost was long gone. Walking up to the two-story house, I noticed that the windowsills were unusually deep for typical new construction.

BRET HAMILTON is an energy efficiency and building science consultant. He is also my business partner in Shelter Analytics, LLC, based in central Vermont (see Resources). Here he tells how a missed insulation detail in a high-performance home caused significant energy loss and created a building durability concern. We have come to expect modern, efficient homes to perform in an effective way to deliver maximum efficiency and long-term durability. In order to meet this expectation, apparently small details become extremely important.

I inquired about this, and the homeowner told me that he had had the builder create a double-studded wall to accommodate 12" of insulation, a little more than double the usual amount. He went on to say that he wanted the house to be affordable for the long term and didn't want high energy prices to dictate his lifestyle when he eventually retired.

The attic was similarly well insulated, with 12" fiberglass batts carefully laid out over, and perpendicular to, 6" batts in the attic joist cavities, making a fairly impressive nominal R-60 attic. Looking closely at the underside of the roof sheathing from inside the attic, I did see the telltale signs of staining from previous "moisture events." These moisture stains usually appear fairly random in their positioning, a function of hidden holes under the insulation, eddies in the air currents, location of attic vents, and so on.

In this attic, the stains were right around the outside edges of the attic, just above the exterior walls. Peeling back the layers of insulation on the attic floor, I saw that the exterior walls were nicely filled with big, fluffy batts of fiberglass insulation, but they were completely open at the tops. If I could have removed the insulation, I would have been able to drop a quarter two stories down the wall into the first floor. This type of framing is called "balloon framing" and dates from the late nineteenth to early twentieth centuries, when wood timbers were both long and plentiful. But I had never seen a new home framed in this style.

The good news was the problem was both apparent and easy to fix. All winter long, the warm, humid air from the home seeped into the wall cavity and rose straight up as if the wall were a chimney (fluffy insulation does not stop air movement), and exited the wall at the attic floor. It then flowed up to the cold roof, where the moisture condensed out of the air and created the frost the homeowner saw on the underside of the roof.

We asked a local insulation contractor to spray foam insulation over the openings at the tops of the walls to seal them off and stop the flow of air. Then we installed a timer on the bathroom exhaust fan to run 30 minutes every hour, around the clock, to reduce humidity levels in the house. The fix worked. Frost no longer forms on the roof deck, and the homeowner reports that the home is more comfortable than ever.

If I could have removed the insulation, I would have been able to drop a quarter two stories down the wall into the first floor. This type of framing is called "balloon framing" and dates from the late nineteenth to early twentieth centuries.

4

Deep Energy Retrofits

HOME ENERGY IMPROVEMENTS generally consist of a wide variety of measures that encompass electrical and thermal upgrades. Small improvements make small impacts. But if you go "wide" enough in your approach, lots of small changes will add up to larger savings: efficient lighting and appliances, low-flow showerheads, weatherstripping around windows and doors . . . you get the idea. Energy savings can range from minimal to 20 or even 30 percent if you're diligent and work with a good contractor.

You can also choose to go "deep" with your improvements; rather than just adding insulation to your attic, for example, you first prepare the attic thoroughly by sealing up air leaks between the house and the attic, repair any damage, address the attic and roof from a "renewable ready" standpoint, and then (finally) install as much insulation as possible. While working on the attic and roof, you might also make preparations for a future, similar treatment to the walls. With all of this done, neither you nor a future owner will ever need to address efficiency or durability issues in the attic space again.

Looking at the Big Picture

PIECEMEAL EFFICIENCY IMPROVEMENTS do reduce energy use, but consider the synergies of a holistic approach to go both wide and deep. Taking a systems view of your home's energy-consuming appliances — and how they interact with the envelope assemblies and occupants — allows you to multiply your savings along with improving comfort. As you can imagine, this whole-house **deep energy retrofit** (DER) can be a costly and intrusive process, but short-term pain will yield long-term gain.

One study by the New York State Energy Research and Development Authority (NYSERDA; see Resources) suggests that a comprehensive DER in a cold climate costs about $18 per square foot of shell area (surface area of all six sides of a house), adding up to over $75,000 for the average modern home. This includes many non-energy–related improvements that will likely be encountered along the way.

Admittedly, we are in the "barnstorming" years of product development and installation procedures for such comprehensive energy savings projects. Forward-thinking entities such as NYSERDA, among others, are helping to drive down costs by investing in research projects and utility program developments that will offer national guidance and promote market transformation in this area. For the homeowner, costs and scheduling are made more manageable by developing a plan that allows you to stage, or phase, improvements over time. Addressing immediate needs first allows you to fix the biggest holes in your home's energy bucket while incorporating those improvements into the building's long-term plan without a huge up-front cost.

WHAT A DER INVOLVES

Most homes are designed for appearance and functionality. Unfortunately, they are not always designed around the fundamental idea of matching the mechanical systems (heating, cooling, and ventilation) with the envelope assemblies, and

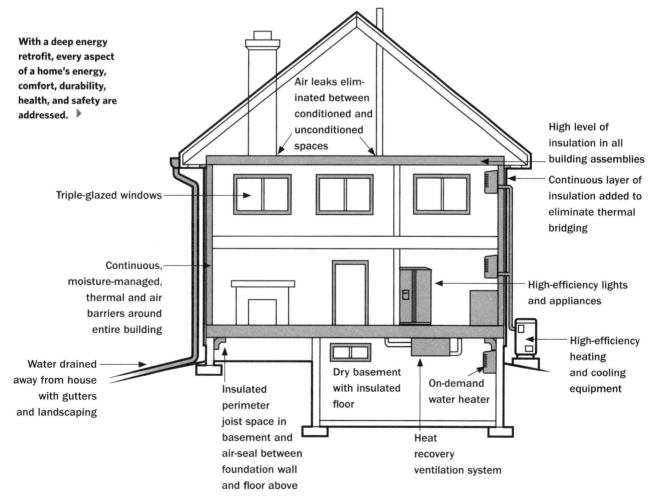

With a deep energy retrofit, every aspect of a home's energy, comfort, durability, health, and safety are addressed. ▶

Air leaks eliminated between conditioned and unconditioned spaces

High level of insulation in all building assemblies

Continuous layer of insulation added to eliminate thermal bridging

Triple-glazed windows

Continuous, moisture-managed, thermal and air barriers around entire building

High-efficiency lights and appliances

High-efficiency heating and cooling equipment

Water drained away from house with gutters and landscaping

Insulated perimeter joist space in basement and air-seal between foundation wall and floor above

Dry basement with insulated floor

On-demand water heater

Heat recovery ventilation system

Performance Levels

Depending upon your climate and the current condition of your home, the requirements of a DER can vary. A typical deep energy retrofit in cold climates, such as those in the northern half of the United States, involves upgrading to the following minimum performance levels:

- R-10 basement or ground-floor slab

- R-20 basement walls with continuous insulation, water drainage, and air and moisture control

- R-30 to R-40 walls with air, vapor, and water control layers

- R-60 roof with air, vapor, and water control layers

- Air leakage reduced to an absolute minimum

- Ventilation system to provide healthy indoor air quality

- Upgraded windows — by adding storm windows or replacing old windows with highly insulating (triple-pane) units with orientation-tuned glazing

- Renewable energy systems where practical

how these relate to the home's performance in terms of energy use, occupant health and safety, indoor air quality, and structural durability. The goal of a deep energy retrofit is to overhaul the way an existing home works with the above points as guiding principles. It often includes many of the following improvements:

- Insulating and air-sealing to substantially improve the thermal envelope

- Upgrading lights and appliances to meet the highest performance specifications

- Adding a whole-house ventilation system, such as heat recovery ventilation

- Upgrading to high-efficiency and properly sized heating, cooling, and hot water systems

- Improving distribution of conditioned air

- Increasing window area on south-facing walls to capture heat from the sun

- Adding thermal mass to floors to capture and store solar heat

- Incorporating renewable energy systems

- Reusing, repurposing, or recycling all building materials removed from the home

Great attention to all detail is required, as are skills, experience, creativity, and a commitment to doing things right, not to mention the tenacity to redo things if necessary until they *are* right. If the costs of improvements are financed at market rates, there can be a fairly long financial payback period. Also (and this is important), the homeowner must be able and willing to accept several months of substantial disruption in the home if all the work is done at once. In short, a DER demands a rare combination of contractor skills and homeowner motivations.

A Team with a Plan

A DEEP ENERGY retrofit plan spells out the methods and approaches for upgrading all the various and interconnected energy-related pieces of the building. The plan is an important piece that should not be ignored; without it, you may later find yourself redoing things that did not go deep enough. Or worse, you may make mistakes that can't be undone without great cost and effort.

The plan starts with a full energy audit and building inspection by a qualified efficiency consultant

The Best Performer on the Block

Together with the homeowner, a DER team can transform an energy-hog building into a high-performance twenty-first-century home that is comfortable and durable — and uses 80 percent less energy than the average home. Your house has the potential to be far more than a container for all your stuff. Substantial investment in well-planned improvements will lead to meaningful savings over the life of the home.

Endgame: Renewables

Efficiency makes renewables more affordable, as less demand from the home means smaller, less expensive energy generation systems. The DER level of energy reduction allows you to meet most or all of your home energy needs with renewables. Think of a DER as endgame planning for the energy use, cost, comfort, and durability of your home.

who can develop and lead an integrated design team. This team is critical to achieving comprehensive and technical excellence with the project. The consultant must be able to communicate well and coordinate the work of an architect, a general contractor, all representatives of trade subcontractors, and the homeowner.

The purpose of the team model is to bring together the decision makers and contractors to engage in goal-setting, planning, road-mapping, accountability, and team-building. Everyone must know what they're getting themselves into and what is expected of them. It's important to find experienced team members who communicate openly and honestly and are willing to take risks in an intelligent and informed manner.

A good plan makes it possible to stage improvements so that you can make upgrades separately and incrementally, if desired, while making sure to accommodate the next improvement phase. The plan can also help you take advantage of natural replacement cycles. For example, roofs need to connect with walls, and walls, of course, have windows in them. When it's time for a new roof or new siding, use this as an opportunity for deep efficiency improvements. Once your thermal envelope is improved, you can optimize the heating and cooling systems to meet the reduced needs afforded by the better envelope. Now it's time to consider on-site renewable energy systems.

UNPLEASANT SURPRISES

In the process of making thermal performance improvements and other efficiency upgrades, you may encounter health, safety, durability, and deferred-maintenance issues that need to be addressed. For example:

- Lead paint
- Asbestos insulation or siding materials
- Failing roof
- Wet basement
- Substandard wiring
- Radon
- Structural issues

These and other problems can present practical challenges to builders and budget challenges for the homeowner. In any case, performing such work on your home will be very specific to its particularities and your climate.

EVERY JOB IS CUSTOM

There are many products on the market today, and many different approaches to the various situations that can arise. When it comes to a deep energy retrofit, it's important to apply building science to your specific project and climate. There are no plug-and-play, off-the-shelf, one-size-fits-all widgets available.

With that in mind, the remainder of this chapter describes some very general approaches to deep energy savings without going into specific products or details, any of which may or may not be appropriate for your situation. A comprehensive DER is not a suitable DIY project for the average homeowner, but there are plenty of opportunities for you to step in during the work phases you are comfortable with.

The methods described here are not intended to represent the "best" approach. There is no such thing (yet) as a *generic* deep energy retrofit, and a comprehensive whole-house treatment involves almost excruciating detail. Therefore, while the following information covers the basic approaches and rationale behind DER improvements, it is by no means a compendium of all the various details involved.

The Basement

BASEMENT FLOORS IN most homes are either concrete slabs, dirt, or gravel. Regardless, controlling water and dampness in the basement is an important part of a DER project. This is because moisture from a wet basement will increase the humidity in the house and can quickly and easily compromise your health (and building materials) as you tighten up and insulate the floors over the basement.

Addressing moisture may mean installing gutters, downspouts, and downspout extensions on the outside of the house, and possibly making landscaping changes, so that water drains away from the building. It may also mean installing a perimeter drain inside the basement, allowing water to drain to a sump where it can be pumped out.

The basic approach for basement slabs is to provide drainage (as needed); insulate the slab with continuous, durable insulation (such as rigid foam board or foam-glass); cover the insulation with a vapor barrier (such as polyethylene); and top it all off with new concrete or another type of flooring.

Basement walls typically are cool, so condensation can easily form on walls under the right conditions. Older basements may also leak bulk ground water after it rains or due to high water tables. You have two basic options for dealing with leaks: redirecting the water before it gets in, or accepting that it will come in and direct it to a place where it can be managed.

Drainage

Walls will be covered with insulation and possibly a finish material, but you do not want these materials to come into contact with a damp foundation wall. One solution is to cover the foundation wall with a waterproof drainage mat (such as a dimpled polyethylene sheet) that directs water to a drain and keeps any moisture away from finish materials. Products (such as Perimate; see Resources) are being developed that can serve as both insulation and drainage plane. Once the water is managed, the wall and the perimeter band joist area can be insulated, air-sealed, and finished.

Drying out a basement and providing means for removing water that does get in not only helps to control potentially damaging humidity levels throughout the house, it also makes the space more livable and versatile.

Above-Grade Walls

THE MOST COST-EFFECTIVE time to upgrade walls is when you're replacing the siding, but there may be good reasons to do the work now rather than wait. Of course, you'll want to investigate what your walls are made of and what's inside them before you start tearing things apart.

The main goal with wall improvements is to prevent air and moisture from getting into the wall while increasing its durability and insulating value. Typically this involves removing the siding and working from the outside. Following are the basic elements and techniques for creating a maximum-efficiency wall.

INSULATING THE INTERIOR WALL

Before insulating, seal up all air leaks in the wall, including all penetrations in the wallboard on the inside, through the siding on the outside, and in the top and bottom plates. The conventional approach is to air-seal the wallboard from the inside, using caulk, spray foam, and other materials. With a few courses of siding removed from the outside of the wall, you can install

Cross section of a basement wall and slab floor with perimeter drainage, sump pump, water barrier membrane, and insulation ▼

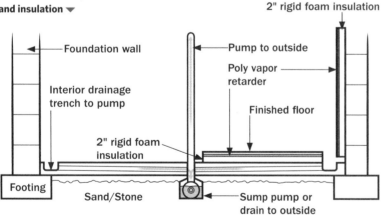

Building Wrap

In some cases it's appropriate to install an air and moisture control barrier, such as building wrap, over the sheathing. This is the common approach in standard wall construction, but may be integrated into other products used in some high R-value wall designs. Building wrap must be taped at the seams and lapped over window and door flashing to eliminate air leakage and to allow water to drain consistently downward.

dense-pack cellulose insulation in the wall-stud cavities, leaving the sheathing intact.

For a more complete approach, you would remove the siding *and* sheathing, pull out all the insulation, and use spray foam to seal leaks from the back side of the wallboard and, as an option, to insulate the wall cavities.

SHEATHING OPTIONS

On most exterior walls, the studs are covered with some kind of structural sheathing, typically plywood or oriented strand board (OSB). This layer must be air-sealed at all joints and at the top and bottom so that no air can move behind the sheathing.

This can be accomplished in part with building wrap and/or rigid insulation taped at the seams, along with spray-on urethane or latex foam or caulk to seal the top and bottom of the wall assembly to top and bottom plates. Peel-and-stick roof underlayment material can be used to seal the lower few inches of the sheathing to the top few inches of the foundation. The goal is to integrate a durable air and water barrier into the wall.

Do the same thing at the eaves by air-sealing the top of the sheathing (or building wrap) to the top plate. This effectively ties the wall's air barrier to both the foundation and the roof air barrier to create a continuous air barrier around the entire building. (Note: If you're working according to a staged plan, be sure to make provisions for this detail so you can easily seal the walls to the roof when you get to your roof work.)

EXTERIOR FOAM BOARD INSULATION

Over the sheathing goes a continuous layer of rigid foam board insulation. Many installations use two layers of 2" foil-faced foam panels, with

the joints staggered so that they overlap, and with the outer panels taped at the seams to seal out air and water. The continuous insulation layer helps to control the thermal bridging effect of the studs and should extend down over the foundation, if possible.

Depending on the materials and approach you use, an additional waterproof membrane may be needed to cover the insulation to drain away water that migrates through the siding. There are several such drainable building wraps (such as HydroGap; see Resources) on the market designed for use as a "drainage plane." Window and doorjambs must be extended to accommodate the added thickness of insulation on the outside of the wall.

MAKING CONNECTIONS

Vertical furring strips are nailed or screwed through the foam insulation and into the studs. These are used as a nailing base for the siding. The windows (below) and doors are installed and

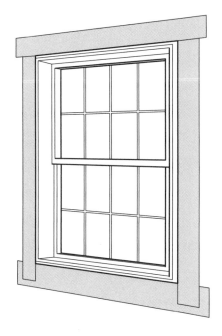

◀ **Windows are sealed over the wall sheathing with a waterproof membrane, such as peel-and-stick tape, that seals the window flashing to the wall (typically overlapping the air and moisture barriers), preventing air and water infiltration into the wall-to-window perimeter junction. This is a critical detail for all window installations.**

CURTAIN WALLS

An alternative to tearing into and improving existing walls is to build new "curtain walls" over the old walls. The advantage to this approach is that it can be less costly and intrusive than a demolition of the existing exterior walls. Curtain wall construction starts with dimensional lumber (2x3 or 2x4) attached vertically to the existing siding to create cavities for new insulation and a nailing base for new siding. The stud cavities are filled with expanding spray foam, then a continuous layer of air-sealed insulation is added over the studs, followed by a water control layer or space, and, finally, the siding.

Exterior curtain wall retrofit. Walls are thickened, windows and doors trimmed as needed, insulation is added, and new siding is installed. ▶

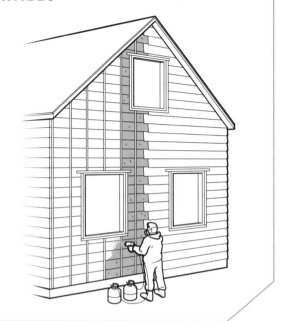

flashed, with attention to both air leakage control and water drainage. After the siding is installed, there will be a gap at the bottom of the wall (and perhaps the top, depending on the construction)

created by the furring strips. This gap must be covered with a screen that keeps out bugs and rodents but allows water to drain.

The drawing at left illustrates how a roof and wall retrofit connect the thermal and air boundaries of those assemblies. The basic approach shown here uses an air barrier over the existing sheathing, followed by two layers of overlapping rigid insulation that is sealed at the seams. The wall assembly uses furring strips as a siding nailer, and the air space created by the furring strips allows water to drain away from the wall. On the roof, additional sheathing is required as both a nailing base for shingles and for additional durability. A weather-resistant barrier covers the exterior roof sheathing, followed by the finished roof. Rigid foam board sealed with spray foam or caulk can be used to block the end of the rafter bays so that insulation can be dense-packed into the cavity.

Wall and roof assemblies, each with continuous exterior insulation applied in a deep energy retrofit. Foam, caulk, insulation, air and moisture barriers, tape and sealants tie together both the thermal and air boundaries of each assembly. Note the small break in the thermal boundary where the wall insulation meets the roof sheathing. This can be considered a compromise to keep the roof overhang intact. The better energy performance solution would be to cut off the rafter tails, allowing for a connection between the roof and wall thermal boundaries. ▶

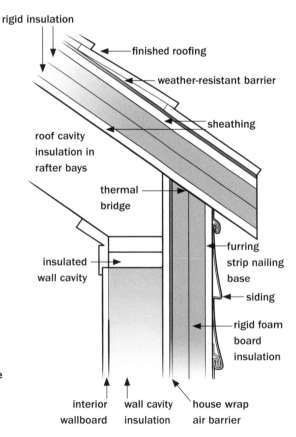

rigid insulation

finished roofing

weather-resistant barrier

sheathing

roof cavity insulation in rafter bays

thermal bridge

furring strip nailing base

siding

rigid foam board insulation

insulated wall cavity

interior wallboard

wall cavity insulation

house wrap air barrier

Windows

WHEN DEALING WITH windows in a deep energy retrofit, there are two common options: 1) You may decide to keep the old windows if they're double-pane units and in good shape; or

Storm window installed over existing window. The entire perimeter of the frame is caulked before installing to provide a good air seal. ▶

2) if the windows are older single-pane units, or they're damaged and in need of replacement, a wall upgrade is the best time to replace them with new high-efficiency, triple-pane windows.

With the first option, you might do well to simply add a storm window so that the windows are now effectively triple-pane (having 3 layers of glass, or glazing). Look for a high-performance, Low-E coated window, and install for airtightness to obtain maximum efficiency (learn more from the Alliance for Low-E Storm Windows, and the Efficient Windows Collaborative; see Resources). In either case, installation details are extremely important. The window must be flashed and sealed to keep air and water out of the wall, while allowing for rainwater to be drained away.

One decision you will need to make when replacing windows in an upgraded wall is between "innies," where the windows are installed close to the inside finished wall (providing a bit more protection), and "outies," where the window is installed flush with the exterior siding (providing a deep windowsill). This is primarily an aesthetic decision, but in terms of energy efficiency, the windows are best placed in the middle of the thermal boundary layer.

With the second option, it's important to choose windows based on performance ratings that vary according to where the window is, which direction it's facing, and what you want the window to do. This is called an "orientation-tuned" glazing approach, and it means that you choose different performance characteristics for windows on different sides of the house. If you're in a heating-dominated climate, you want to maximize solar heat gain during the winter months, but perhaps minimize it during the summer. For cooling-dominated climates, you will want to reduce the solar heat gain. In both cases, you'll want a high R-value (low U-factor) to minimize heat transfer between indoors and out.

It can sometimes take a fair amount of persistence from you (or your builder) to get what you want, as it may not be a standard option in your area. Talking directly with manufacturers can help, since they are often able to supply a variety of glazing types in their windows.

WINDOW PERFORMANCE RATINGS

Before expanding on the idea of orientation tuning with examples, it will be useful to understand how window performance is rated. Most windows are rated for thermal performance by the National Fenestration Rating Council (NFRC; see Resources). The NFRC sticker on a window states all of the performance criteria you need to compare window efficiencies and make an informed selection. The following are the performance criteria to look for on the sticker.

U-factor is a measure of the insulating value of a window (including the framing material). Chapter

◀ Sample NFRC window performance label

LOW-E LOWDOWN

You've probably heard something about "low-E" windows or coatings and wondered what all the fuss and confusion are about. The "E" is for emissivity. This describes a material's ability to emit, or radiate, energy. Reflective materials have a lower emissivity, while materials that absorb heat have a higher emissivity.

Low-E coatings are used on double- or triple-pane windows to improve their energy performance. These selective coatings allow visible light to pass through the window, but they reflect heat energy (infra-red radiation) away. The Low-E coating in a window is applied to a single layer of glass that faces the space between two panes or (in the case of a triple-pane window) to a thin film in between panes.

Typically, the window is oriented so that the glass with the low-E coating faces the warmer zone. For example, in a heating climate the low-E coating is on the outward-facing surface of the inside pane, so that heat radiating from the house is reflected back into the living space. For cooling climates, the coating is on the inward-facing surface of the outside pane, so that solar heat is reflected to the outside.

There are two types of low-E coatings, and windows from different manufacturers differ in the type, location, and number of surfaces with coatings. *Sputtered*, or soft, coat is used in windows with low or moderate levels of solar heat gain, while *pyrolitic*, or hard, coat is used in windows with higher solar heat gain. One is not better or worse than the other; it all depends on how you want your window to perform.

---

3 explains that U-factor is essentially the inverse of R-value (see page 62), so a lower U-factor number indicates a better-insulated window. A single pane of glass has a U-factor of about 0.91, translating to an insulating value of about R-1.1. Most new double-pane windows have a U-factor between 0.50 and 0.30 (or an R-value between 2 and 3.3). New high-performance windows with three or four glazing layers can offer U-factors near 0.1 (R-10).

Solar heat gain coefficient (SHGC) is a measure of how much solar radiation (heat energy) is admitted through the window, via direct transmission and absorption. An SHGC of 0.32 means that 32 percent of the solar energy falling on the window (including the framing material) is transferred through it. Lower SHGC windows (<0.30) can reduce the air conditioning requirements of a home in a warm climate.

Visible transmittance (VT) measures how much visible light comes through a window (including the framing material). VT is expressed as a number between 0 and 1. A higher VT means more light is transmitted. A VT of 0.51 means that 51 percent of the sun's visible light passes through the window. Lower VT values are found on tinted windows that can be beneficial in hot climates to reduce solar heat gain.

Air leakage (AL) is expressed in cubic feet per minute of air passing through a square foot of window area. Heat loss and gain also occur by air infiltration through small air leaks in the window assembly. A lower AL means less air will pass through these cracks. A 15-square-foot window with an AL of 0.2 means that 3 cubic feet of air will move through the entire assembly at the tested pressure differential (75 pascals) in one minute. AL is an optional rating and is not always included.

Condensation resistance (CR) measures the ability of a window to resist the formation of condensation on its interior surface. The higher the CR rating, the better the window is at resisting condensation formation. CR is expressed as a number between 1 and 100. CR is an optional rating and is not always included.

HOW ORIENTATION TUNING WORKS

An orientation-tuned window strategy puts the right glazing properties in the right place. An efficiency consultant can help with specifying the right products and approach for your home and climate. Generally speaking, in northern heating-dominated climates, the sun's angle is low in the

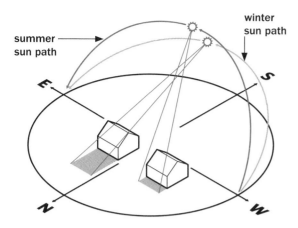

Example of window tuning by orientation for a heating-dominated climate: As the sun rises, high-SHGC glazing in the east helps to warm the house after a cool night. As the sun moves to shine into the south-facing windows, high-SHGC glazing allows winter sun to warm the house, while the higher summer sun does not shine as deeply into the house. Low-SHGC glazing in the west helps to reduce long summer afternoon heat gain. All windows should have a low U-factor for maximum insulation value.

winter, flooding the south side of the house with sunlight. Choosing a window with a *high* SHGC and a *low* U-factor allows you to gain heat from the sun without losing too much heat back out the window. If the sun strikes floor and wall materials, such as tile or stone that have thermal mass (meaning they can absorb and hold heat), the stored heat will be released later as the sun sets and the house cools.

The west-facing side of that same house will be exposed to long afternoons of direct sunlight in summer, potentially overheating the house. Choosing a *low* SHGC window (again with a low U-factor) can help reduce this overheating. In cooling-dominated climates, windows with a low SHGC can help control air conditioning costs.

The Roof

DER STRATEGIES FOR roofs vary significantly depending on what's directly underneath the roof: If it's unfinished attic space, the improvements focus on the ceiling between the attic and living areas below. If the roof directly covers living space, such as with a finished attic or a cathedral ceiling, the improvements are made to the roof itself.

Ceiling improvements follow the same approach of thorough air-sealing and insulating described in chapter 3. With a cathedral ceiling over finished attics or living spaces, often the approach is to create a deeply insulated, unventilated roof system, sometimes called a "hot roof."

UNVENTED ROOF ASSEMBLIES

An unvented sloped, or cathedral, roof retrofit can be very similar to the wall retrofits described on pages 82–84. Think of a sloped roof as a slanted wall. There will be insulation in the rafter bays, then roof sheathing, followed by continuous (taped and sealed) insulation over the sheathing, and finally a moisture barrier (drainage plane) that drains away any water that gets through the roofing. On top of all that go the shingles or other roofing materials. The exception is that roofs may need an additional layer of sheathing under the shingles.

As discussed in chapter 3, conventional roof assemblies over unfinished attics must incorporate ventilation to keep the roof deck and attic space cooler and to help prevent ice dams, excess moisture, and other problems. With an unventilated roof, aggressive insulating and air-sealing creates a sufficient thermal barrier between the roof surface and the living space. How much insulation is enough? Lots. But it depends. An energy consultant must be involved to ensure a durable, efficient, condensation-free design.

Air Leakage

AIR LEAKS IN a building are often the largest component of heat loss, and reducing air leakage is generally a low-cost improvement with substantial energy savings potential.

In addition to heat loss from outdoor air infiltrating a building, air movement from inside to the outside of a building (exfiltration) can sometimes lead to moisture and mold inside the walls or in the attic. This happens during the winter when warm, moist air from inside your home travels into a colder wall cavity, where the moisture in the air condenses. The reverse is true in the summer when hot, humid, outside air tries to make its way into a cool, dry, air-conditioned interior (remember that heat moves from warm areas to cold areas).

To achieve significant energy savings, adding a continuous, durable air barrier is an extremely important element in reducing air leakage. It also serves to keep the spaces hidden inside walls dry and mold-free. Air leakage control can be especially challenging in certain areas:

- Stone foundations
- Junctures between foundation walls and above-grade walls
- Around windows
- Connections between roofs and walls
- Complex roof structures

Be Thorough

Air-leakage control and thermal boundary integrity are essential to any energy efficiency home improvements, but with a deep energy retrofit, you want air leakage to be at an absolute minimum. This requires thorough knowledge of the correct products and how to install them, as well as extraordinary diligence with the installation details. Every last crack, hole, or penetration between conditioned space and unconditioned space must be sealed — from the hose spigot in the basement to the chimney chase through the attic. Air leakage control products include building wrap, caulk, expanding foam, rigid foam board, high-quality tapes, peel-and-stick membranes, and weatherstripping.

Heat loss calculations used to select the right size heating and cooling sytems usually overestimate the air leakage of a home because they are doing just that — estimating. Since air leakage is such a significant factor when it comes to heat loss and gain in a home, it's important to quantify the leakage rate accurately so that space conditioning systems can be properly sized. With a DER, heating and cooling loads are so much lower than a typical home that proper mechanical system sizing becomes extremely important in achieving maximum performance and comfort.

Quantifying air leakage can be done using a blower door test (see page 53), typically conducted by a heating contractor or energy auditor. This test not only helps in sizing a heating or cooling system, it also pinpoints air leakage areas that need to be sealed.

Remember, the goal with a DER can be considered "end-game" planning. Go as deep as possible so that you don't need to revisit or redo.

Ventilation

ALTHOUGH A HERMETICALLY sealed house is ideal in terms of energy efficiency, living in such a home would not be pleasant. If the house is too tight, both occupant and building will suffer ill effects. So, how do you ensure a sufficient supply of fresh air to create a healthy environment and prevent physical damage to the building materials? How much air is enough?

People need good air quality inside their homes. This requires maintaining acceptably low levels of allergens, humidity, and pollutants, such as carbon dioxide (CO_2) exhaled by people and the off-gassing of nearly every manufactured product brought into the home. In a high-performance building, uncontrolled air leakage is reduced to a minimum through air-sealing; therefore, fresh air needs to be provided by a mechanical ventilation system that is controlled to optimize both indoor air quality and energy efficiency.

As mentioned earlier, the builder's phrase for this is "build tight, ventilate right." And for the record: Ventilation systems do use energy, and some conditioned air is lost to the outdoors. But a tight, efficient building with good ventilation will use far less energy than a leaky building with no ventilation, and the energy cost is a good trade-off for a healthy home.

The generally accepted standard for ventilation in homes is 15 cubic feet per minute (cfm) of fresh air per person. This can vary according to the activity level in the house. In addition, combustion appliances (cooking, heating, water-heating), and

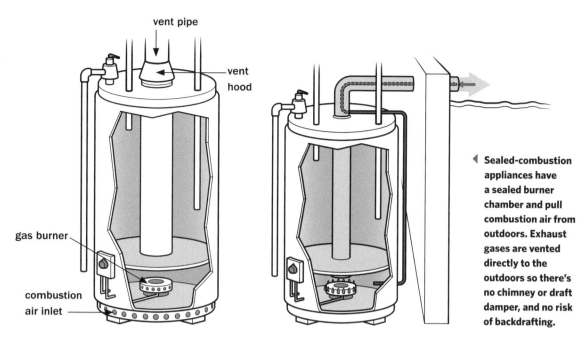

With a conventional water heater (or other combustion appliance), combustion air is drawn from inside the home, and the burner's exhaust naturally flows up and out of the flue, taking conditioned air from inside the home with it. Backdrafting occurs when exhaust is directed into the home by a pressure imbalance, flue blockage, or outdoor wind currents. So-called atmospheric draft equipment needs indoor air for both combustion and chimney draft. ▶

vent pipe

vent hood

gas burner

combustion air inlet

◀ Sealed-combustion appliances have a sealed burner chamber and pull combustion air from outdoors. Exhaust gases are vented directly to the outdoors so there's no chimney or draft damper, and no risk of backdrafting.

chimney drafts all require air. If all of the home's heating appliances are sealed-combustion (with make-up air provided from the outdoors; see below), then it's safe to tighten up the house to reduce heating and/or cooling energy use without creating flue backdrafting.

FLUE DRAFT

Fossil fuel heat and hot water equipment needs combustion air (air to burn fuel). In a typical installation, this air comes from inside the house, but there are better ways to address air requirements. In all homes — especially airtight, energy-efficient homes — it is very important to consider where the combustion air is coming from and how the flue gases are getting out of the building. Failure to pay attention to this simple detail can result in poorly performing equipment or potentially deadly carbon monoxide poisoning.

Fossil fuel–burning equipment needs a chimney or other venting system to remove poisonous combustion byproducts from the house to the outdoors. It is important to make sure the chimney or venting system is in good condition and properly sized for the equipment. There are very strict codes for venting combustion appliances. To ensure both proper operation of combustion equipment and occupant safety, two considerations of venting exhaust gases must be understood: draft and backdraft.

Draft is the flow of air into the fuel burner and out the flue. As the fuel burns, air is drawn from the surrounding environment into the combustion chamber. The flame creates heat, and the hot, buoyant combustion gases rise up the flue leading to the outdoors. Draft can be created naturally by the combustion process, or it can be induced by a fan in the venting system to ensure removal of combustion byproducts.

Backdraft occurs when exhaust moves backward through the flue and into the home due to inadequate draft. Inadequate draft can occur when the interior of a house is under negative pressure in relation to the outdoors, and the flue gases reverse flow within the chimney: Instead of flowing up and out, the gases flow down and in.

Backdrafting in New Homes

Building airtight new homes and upgrading older leaky homes have exacerbated the problem of backdrafting. In a tightly constructed home, heat and hot-water appliances must get all the air they need for combustion and draft from the outdoors, not from inside.

This can result in continuous spillage of poisonous combustion gases into the house.

Backdrafting can be caused by several conditions: if the chimney is blocked or is the wrong size for the equipment, if high winds force a backdraft down the chimney, or if the air pressure drops in the area where the heating equipment is. This pressure drop can be caused by exhaust fans (including those in clothes dryers and high-powered kitchen vent hoods), or even from other combustion equipment with a stronger draft.

Sealed Combustion

When selecting a space or water heating system for your home, the best choice is to install *sealed combustion* equipment. This means the entire combustion process is sealed off from the surrounding indoor environment. All combustion air is supplied by a duct from the outdoors, and the exhaust is vented outside with a separate duct or duct passage.

Cooking with Gas

The exception to venting the exhaust from fuel-burning equipment has always been gas cooking equipment. Unless you cook for more than a half hour a day, gas ranges, ovens, and cooktops typically do not cause problems with excessive moisture or combustion pollutants, but it's always a good idea to use a vent hood for exhaust while cooking with gas. For very tight homes, recirculating range hoods with filtration are recommended, but again, if you cook more than a half hour a day, you may need to investigate a custom approach to balanced ventilation that allows fresh air into the home to make up for the exhaust air that leaves through the fan. Recirculating fans do not create pressure imbalances in the home, and they do not require a hole to be cut in the wall for the fan's exhaust duct (any hole in the wall will always increase air leakage). In conjunction with a well-designed, balanced ventilation system (and average cooking skills), you should not experience any issues with cooking-related pollution.

TYPES OF MECHANICAL VENTILATION

Mechanical ventilation systems fall under two general categories: exhaust-only ventilation and balanced ventilation.

Exhaust ventilation includes kitchen vent hoods and bathroom exhaust fans. They remove stale air at the source and bring fresh air inside by creating a slightly lower pressure inside the house with respect to outside. This low pressure causes outside air to be pulled into the house by way of the path of least resistance, such as an open window, chimney, or leaks in the house. A high-CFM fan (such as a downdraft range fan or commercial range hood), used in a very tight house with a fossil fuel heating system or hot water system that is *not* a sealed combustion type, can present a very dangerous backdraft situation.

Balanced ventilation offers separate exhaust and supply ducting, and continuously removes stale air from inside while introducing fresh air from outdoors. This is the best option for effectively ventilating a home because it removes stale air at the source (bathrooms, kitchens, and bedrooms) and replaces it with fresh air where it's needed (living rooms and bedrooms). It is called "balanced" because an amount of air is supplied to the house equal to that exhausted from it.

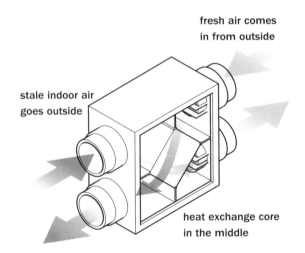

fresh air comes in from outside

stale indoor air goes outside

heat exchange core in the middle

▲ An HRV allows air streams to exchange heat without mixing the streams.

A **heat recovery ventilator (HRV)** is the most effective type of balanced ventilation system. An HRV allows heat energy to be transferred from stale (conditioned) outgoing air to the fresh (unconditioned) incoming air as both air streams pass through a heat exchanger — without mixing together. This heat recovery process minimizes the energy penalty of introducing unconditioned outside air into the home. HRVs have the following advantages over other types of ventilation systems:

- The heat transfer process works in both heating and cooling seasons.

- Effective ventilation occurs evenly throughout the entire home, greatly improving indoor air quality.

- Pressure imbalances are eliminated, so there is a lower potential for ventilation-induced backdrafting of combustion equipment.

- The home can be significantly air-sealed, offering maximum energy savings while still providing adequate fresh air to occupants.

- Enhanced indoor air quality leads to increased occupant health and comfort.

Another type of HRV is called an **enthalpy recovery ventilator (ERV)**. An ERV does what an HRV does but also allows for the transfer of water vapor, thereby recovering both latent and sensible heat (see Sensible Heat and Latent Heat on page 93). It accomplishes this with a permeable heat exchange core or a desiccant wheel revolving between incoming and outgoing air streams. An ERV may be the right choice if the home tends to be too dry, or if the home is air conditioned and you want to control the humidity.

VENTILATION CONTROL

Ventilation systems must be controlled because they use energy and they remove conditioned air from the house, so you don't want to run them more than you need to. Underventilating a home compromises indoor air quality and can lead to high levels of pollutants such as carbon dioxide, which is exhaled by humans.

Exhaust Fans

For exhaust-only ventilation, choose a quiet, low-wattage fan, coupled with a programmable 24-hour timer. The fan's run time depends on how tight the house is. Start with a timer setting that allows the fan to run 20 minutes every hour while the house is occupied (no need to provide fresh air if nobody is home). Adjust the operating time up or down from that setting depending on the results. An efficient, quiet bathroom exhaust fan can provide economical, reasonably effective ventilation (50 to 100 cfm is required in most average-sized homes). Install the fan on an upper floor to take advantage of the building's "chimney effect" that makes air want to move upward.

If you have condensation problems on your windows, it indicates that the window has reached the dew point temperature. Poorly performing windows will have colder indoor glass surfaces, exacerbating condensation. If condensation is a consistent problem on windows, ventilation will help by eliminating moisture at its sources, such as showers.

Relative humidity can be an indication of ventilation or air leakage rates in your home. Humidity levels below 25 or 30 percent in a northern climate heating season probably indicate too much air movement between indoors and out. For most of us, comfortable relative humidity in a home is in the 35- to 55-percent range, but we can certainly tolerate a much wider range. Keeping an eye on humidity levels with a hygrometer can help you determine how long to run your exhaust fan.

HRV/ERV

Heat recovery ventilators include their own dedicated controls. Fan speed and run time can be set with a timer, or they can be automatically adjusted in response to indoor humidity or carbon dioxide levels. This is called "demand-controlled" ventilation and is common in commercial buildings to ensure a balance of good air quality and energy efficiency. CO_2 detection and demand control is now making its way into the residential market.

Atmospheric (outdoor) CO_2 levels are approaching 400 parts per million (ppm). For indoor spaces with ventilation systems, a good balance between economy and efficiency is between 800 and 1,000 ppm. Poorly ventilated buildings may have CO_2 levels that exceed 2,000 ppm, a level that will cause humans to feel drowsy.

Heating and Air Conditioning

TODAY THERE ARE many types of heating systems available that use various fuels, as well as very efficient central and room air conditioners. Typically, very large heating and cooling systems are installed in homes, and often these are oversized in terms of their space conditioning capacity.

Heating and cooling a low-energy-use home requires an extra degree of care in the sizing of mechanical systems to ensure comfort and efficiency. In other words, you will not be shopping for conventional space conditioning systems. Rather, you want systems that efficiently deliver low quantities of space conditioning and that allow you to modulate those levels according to the weather and indoor comfort needs.

CHOOSING A SYSTEM

When researching and selecting the best heating and/or cooling system for your home, be sure to assess the heat requirements of the house, installation costs, and local fuel costs (and carbon content). Get help from a professional heating contractor or energy consultant to find an appliance that matches the new heating and cooling requirements of your improved home, or one that can meet your needs now and still be appropriate after your next stages of energy improvements. Following are some popular equipment options for low-load homes.

Gas-Fired Heating

Many highly efficient homes can be effectively heated with a wall- or floor-mounted gas space heater or with a high-efficiency, modulating gas furnace or boiler. These heating plants adjust their flame to deliver varying amounts of heat to the air or hot water circulating in baseboard or radiant floor distribution systems. Look for efficiency ratings in the mid-90 percent range.

Heat Pumps

There are two main types of heat pumps: ground source and air source. Ground source heat pumps remove heat from the ground by way of a working fluid flowing through tubes that are in contact with the ground. The fluid goes through a compression and evaporation cycle to extract heat from the ground and deliver it to the house.

One advantage of a heat pump is that it can also work in reverse and act as a space-cooling system, removing heat from the house and transferring it to the ground. Disadvantages of ground source heat pumps are that they are quite complex, can be costly to install, and can use quite a lot of electricity. They are generally most effective in cooling-dominated climates, and some studies show that in heating-dominated climates

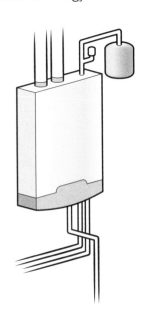

Modern, high-efficiency furnaces and boilers are compact enough to hang on a wall. Some new systems incorporate space heating, water heating, and heat recovery ventilation into a single package. ▶

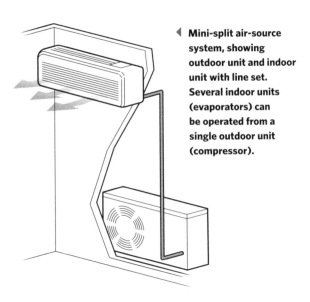

◀ Mini-split air-source system, showing outdoor unit and indoor unit with line set. Several indoor units (evaporators) can be operated from a single outdoor unit (compressor).

SENSIBLE HEAT AND LATENT HEAT

Air conditioning obviously helps to manage summer heat, sometimes referred to as "sensible" heat. Another important function of air conditioning is to control humidity, or "latent" heat (heat held in water vapor). Proper equipment sizing (in sensible Btus per hour), along with a **sensible heat fraction** (SHF) that matches your climate's humidity, will help to ensure efficiency and comfort.

SHF is a measure of the percentage of an air conditioner's total capacity that is available to remove sensible heat from the air. This value ranges from 0 to 1 (typical between 0.5 and 1.0). The remaining capacity is available for the removal from the air of moisture (in which the latent heat is contained). When selecting cooling equipment in humid regions, knowing the SHF will help you choose a system that will properly cool and dehumidify your home. Work with a knowledgeable contractor to understand the regional recommendations for air conditioner ratings. Generally, the higher the efficiency, the higher the SHF. Unfortunately, this means that the highest-efficiency air conditioners may not do a good job of dehumidification in a very humid region.

---

the actual efficiencies are somewhat less than rated levels.

Air-source heat pumps can be efficient and versatile for both heating and cooling a home. They operate by transferring, or pumping, heat between indoor and outdoor air. Sometimes called "mini-splits" or "ductless mini-splits," these units are similar in operation to central air conditioners, but they do not use ductwork to deliver the conditioned air. Instead, one or more indoor units are connected (via a refrigerant line set) to a single outdoor unit. This setup allows for different temperature zones within the house. The absence of ductwork eliminates the problem of leaky ducts, a potentially large loss of energy.

Other Heating and Cooling Options

Additional options to consider for heating include cordwood wood stoves, pellet boilers or stoves, and masonry heaters. These options require more frequent attention by the owner than automatic, fossil fuel systems.

Evaporative coolers work well in dry climates by taking advantage of the same principle that our bodies use to remove excess heat by sweating. Evaporating water absorbs and dissipates heat energy. When it's humid, moisture can't evaporate from our skin, and we feel hot and sticky. In hot, dry climates, moisture added to the air evaporates, absorbing heat energy and cooling the surrounding air in the process.

Evaporative coolers are sometimes called "swamp coolers" because they use a steady supply of water to cool the air while also making it more humid. Evaporative cooling equipment requires no compressor and therefore consumes much less electricity than a conventional air conditioner. Though not as effective as conventional air conditioners, swamp coolers can provide sufficient comfort under the right conditions. Because they rely on the evaporation of moisture, they work only in dry climates where relative humidity levels are below 40 percent.

More novel approaches to cooling very efficient homes include a ground coupling system, where fluid is circulated in underground pipes and then through a "fan coil" (something like a car radiator) that lives inside the ductwork of a space conditioning or ventilation system. Air blows over the cool coil and delivers cool air to the house. Another innovative system, called the NightBreeze (see Resources), allows you to use cool nighttime air to ventilate your home at night through a balanced ventilation distribution system.

Such systems can be found by exploring the cutting-edge work being done to promote the Passive House Institute U.S. efficiency standard, which primarily addresses new homes (see Resources). This extremely efficient standard has taken a strong foothold in Europe and is just beginning to gain traction in the United States.

Hot Water

PART OF YOUR DER may include getting your house renewable-ready for a solar hot water system. The basic efficiency improvements to start with are getting the most efficient water heater, using less hot water, and modifying your behavior around waiting for hot water (see pages 44 and 45).

Here are some other equipment options for highly efficient hot water heating:

Heat pump water heaters extract heat from the surrounding air and "pump" it into the water, much as a refrigerator removes heat from the air inside its compartment and pumps it out into the room. These systems have a **coefficient of performance** (COP) of over 2, meaning that they deliver twice as much heating energy to the water when compared to the electrical energy required to operate the heat pump. Because they remove heat from the surrounding area, they also help cool the room they are in.

Condensing water heaters are very efficient at getting every last bit of energy out of natural or propane gas and converting it to hot water. They have efficiencies in the mid- to upper 90-percent range.

On-demand water heaters, especially condensing models, can be extremely efficient when used conscientiously. For example, if you need only a small amount of hot water, you're better off using cold instead. This is because it takes a minute or two for the heater to heat the water and another minute to get it to the tap. If you shut off the faucet right away, any unused heated water ends up sitting and cooling in the pipes. An on-demand heater can also make it tempting to take longer showers because the hot water will not run out as with a conventional tank heater.

Drain water heat recovery system (DWHR). When you shower, the water temperature is somewhere around 104°F. The heat contained in the water is used only briefly before going down the drain. A DWHR device (see Resources) allows you to capture that waste heat and reuse it. The system consists of a length of copper drainpipe wrapped with a long coil of ½" copper plumbing supply pipe. Cold water is fed into one end of the coiled pipe. As warm water travels down the drainpipe, it warms the water in the supply coil, and this preheated water is routed to the cold-water inlet of the hot water heater.

SMART PLUMBING

A significant challenge in hot water efficiency lies in transporting the heated water from the heat source to the point of use (the fixture). Long or indirect runs of pipe can sap a lot of heat from the water, especially when the piping passes through unheated areas, like basements and crawlspaces, or within exterior walls. Here are a few ways to minimize this energy penalty:

- Insulate all pipes with appropriately sized foam pipe insulation.

- Install direct plumbing runs from the water heater to the fixtures (sometimes called "home-run" plumbing).

- Plan the layout of plumbing runs so that all (or most) rooms with hot water are back-to-back or stacked. This keeps plumbing fixtures close together for an efficient pipe layout, with the water heater central to the fixture locations.

- Use smaller pipe diameters, as appropriate, to get water to fixtures more quickly.

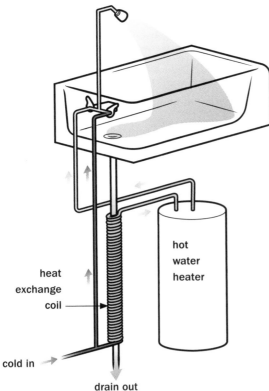

Drain water heat recovery system ▶

heat exchange coil

hot water heater

cold in

drain out

Home Energy Monitoring

WHILE YOU'RE PROBABLY already in the habit of tracking household energy costs, prices change over time. A more meaningful and precise indicator of energy use is consumption: how many units of energy you use over a given period of time. You want to understand your usage in terms of kilowatt-hours, therms, cubic feet, gallons, pounds, or cords; you want to know how much of each fuel you use and when. With this knowledge you can begin to see why you use what you do and then make a plan of action to save.

The specifics of how you monitor your home's energy use depends on the fuel you want to measure, how much detail you want, how you will use the information you gather, and your budget. The information you get from monitoring your energy use can be invaluable in discovering energy wasters in your home and identifying savings solutions.

This chapter introduces you to some of the popular products and approaches to real-time energy monitoring and data logging. Keep in mind that this is a dynamic, emerging market, and there are many additional options out there, as well as smart-grid products and applications used by electric utility companies.

Electric Energy Monitoring

THROUGH THE HELP of technology (and perhaps your electric company, if the "smart grid" has come to your neighborhood), you can monitor energy use by the minute, day, week, month, or year. You can look at whole-house information or just the use of a single appliance. By far the most advanced and user-friendly home-based energy monitoring technology has been applied to electrical usage. There are easy-to-use products for monitoring electricity at the appliance (point of use or plug-load) level and at the whole-house level.

POINT-OF-USE MONITORING

As discussed briefly on page 40, plug-in electric meters, such as the Kill A Watt and Watts Up? (see Resources for chapter 2), have become affordable and widely available. With these devices you can record how much power (in watts) an individual appliance is using in real time, as well as how much energy (in kilowatt-hours) it uses over a period of time.

There are even some power strips available that let you monitor the electric use of everything plugged into them, such as your home office equipment or entertainment center. With this information you can discover the most efficient operating modes of various plug-in appliances and electronic equipment as well as determine whether or not a more efficient appliance would be cost-effective.

These meters accumulate the data and report the electrical consumption over time in kilowatt-hours so that (with a little programming of the meter) you can see how much that appliance is costing you. The power draw of some appliances varies depending upon what that appliance is doing. For example, a refrigerator has a compressor and fans that cycle on and off, and periodically enters into defrost mode. Each of these modes has a different power requirement. If you want to know the energy consumption of a refrigerator, or anything else that cycles on and off, a plug-in meter that measures kilowatt-hours offers the most accurate reading.

You can sort out nearly all the electrical use in your house with great specificity by moving a point-of-use meter from one appliance to the next, making notes, and adding things up. However, one limitation of plug-in meters is that they are limited to 120-volt appliances; they won't work with 240-volt appliances, such as electric clothes dryers or water heaters.

WHOLE-HOUSE ELECTRICAL METERS

Whole-house electrical meters measure and monitor real-time electric usage and report the information to a remote display that can be brought to any room in the house. This allows you to see how much power your entire home is using in real time, as well as the immediate effects of your conservation efforts.

One meter, the Power Cost Monitor (see Resources), uses an optical sensor that attaches to the outside of the utility's electrical meter and transmits data wirelessly to the display monitor indoors. Another, The Energy Detective (see Resources), uses a current sensor attached to the home's electrical panel to sense the real-time power consumption and uses the home's electrical wiring to transmit data to a wireless transmitter. From there, the energy use information is sent to a handheld display, computer, or smartphone.

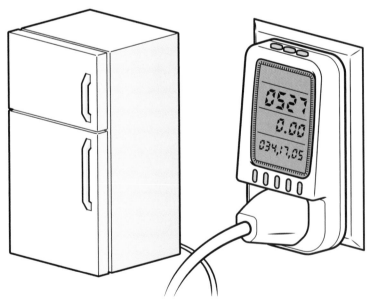

▲ **A plug-in wattmeter records electrical energy used over time, allowing you to evaluate consumption and cost.**

monitoring and identifying savings opportunities in the home. A current sensor is attached to each wire in the main circuit breaker box, and each sensor is connected to the monitor. The monitor is connected to a router and sends data to the company's server, allowing you to view real-time data on a Web browser.

If you have a renewable power system, the eMonitor will display power production as well as consumption. The monitor device has a very basic display on the unit itself, but the Web-based information center, or "dashboard," offers many layers of information. In addition, the system allows you to view power use remotely and control wireless electronic thermostats via a smartphone app. A similar product called Agilewaves is made by Serious Energy (see Resources).

▲ Whole-house electrical monitors. Power Cost Monitor (above) with outdoor sensor attached to meter, and indoor monitor. The Energy Detective (below) with transmitting unit and current sensors. ▼

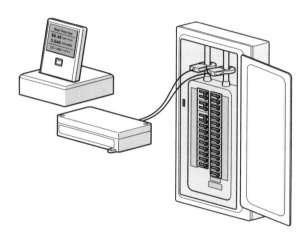

Circuit-Level Monitoring

The eMonitor (see Resources) senses the electrical use of each individual circuit in your home so that you can get a better idea of exactly how much power is used and where. Circuit-level metering takes a lot of the guesswork out of energy-use

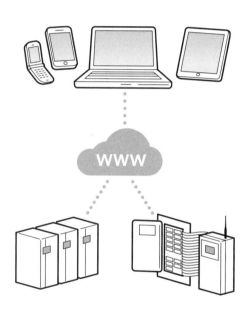

▲ Web-based circuit-level energy monitor

Remote Monitoring and Control

Some whole-house monitors have the ability to extend monitoring and control to individual appliances by interfacing with home automation and control software, often using the ZigBee Alliance wireless communication standard (see Resources). A review of the ZigBee website will give you an idea of the growing number and variety of products available for home automation, energy management, and control. In addition, many information products have matching smartphone applications that allow you to monitor and manage energy use from afar. See Resources for more products and companies leading the way in the industry. For example, the folks at Plot Watt and Bidgely seek to take the perceived nerdiness and complexity out of home energy monitoring.

Gas Monitoring

MONITORING NATURAL OR propane gas use is somewhat less user-friendly than electrical monitoring. This is primarily because metering gas usage for each appliance involves cutting the gas supply line for that appliance and installing a dedicated meter with a dial or pulse output.

For example, Itron (see Resources) manufactures various sizes of gas meters with integrated pulse output modules. Every turn of the meter dial produces an electrical pulse that can be captured, counted, and electronically manipulated, and the results can be displayed graphically on a dashboard or on a spreadsheet. There are ultrasonic meter products that simply clamp onto gas supply lines, but currently these are quite costly.

For accurate metering of a specific gas appliance in your home, you can use a pulse-enabled gas meter in conjunction with a pulse data collection logger. Many different kinds of sensors and data loggers are available from data logging product manufacturers, such as Onset, Omega Engineering, and Campbell Scientific (see Resources).

Reading Your Gas and Electric Meters

Electric and natural gas utilities provide you with a service meter that you can learn to use for monitoring the various appliances in your home. Contact your utility if you need help deciphering dials and numbers, and the quantity of energy each spin of the dial represents. Using gas as an example, the concept is simple: Inventory all the gas appliances in your home, and understand what makes them turn on and off. Then turn off all of the appliances. Read the gas meter's dials to get a starting point, then turn on the one gas appliance you wish to monitor.

For example, if you want to know how much gas your clothes dryer uses, make sure that the heat (furnace, boiler, etc.) and hot water heater are turned off, and don't cook during the test. Don't worry about the pilot lights still being on; they use a negligible amount of gas. Now, operate the clothes dryer, and when it's finished, read the gas meter again. Subtract the initial reading from the final, and the result is the gas consumption of drying a load of clothes. If the meter reads in cubic feet, and the dryer used 30 cubic feet, you've used 30,000 Btus of gas (there are approximately 1,000 Btus in a cubic foot of natural gas). Multiply this by the number of dryer loads you do each month, and that's your total Btu energy consumption for that appliance. Unfortunately, many gas meters read in increments of 100 cubic feet, so you might need to lengthen the duration of the test.

Finally, you need to know how your gas company bills you. It charges for gas by volume (in cubic feet, hundreds of cubic feet, or thousands of cubic feet) or energy content (therms or Btus). Once you know how many Btus are in one billable unit of gas, a little more math will show how much each dryer load costs. You can follow this routine for any gas appliance in your home, or use the disaggregation methodology described in chapter 2. This same approach works equally well with electric meters.

Bottled Gas

If you use bottled gas, typically there is no meter involved. Your only clue may be the percent-full gauge on the tank, and that's not accurate enough for the metering method described above. If you're determined to monitor bottled gas consumption, you can have a whole-house gas meter installed for a few hundred dollars.

Environmental Monitoring

YOU CAN MONITOR both indoor and outdoor environments for any number of parameters. Sensors are available to measure temperature, humidity, water flow, rainfall, wind speed, solar radiation, carbon dioxide, and volatile organic compound levels, to name a few. You can assemble a weather-monitoring station and integrate it with your home energy data through products and services from companies like Rainwise, APRS World, and PowerWise Systems (see Resources).

CALCULATING HEATING ENERGY

If you want to meter heating energy use, it may be useful to evaluate heating energy consumption in the context of weather. Your heating energy consumption will likely change from year to year, and this might be due to changes in occupancy, behavior, equipment, or weather. Assuming everything else is equal and the only thing that changes is the intensity of the heating season, you can compare your energy use based on seasonal **heating degree days**.

A heating degree day (HDD) is the difference between a balance point (usually 65°F), against which calculations are based, and the outdoor temperature. As an example, if the average temperature over the course of a day is 50°F, then that day has accumulated 15 HDD. If it took 1 therm of

DIY HOME ENERGY MONITORING

For those of you who are inclined toward hands-on gadgetry building, there are at least two options for you to explore. For intrepid do-it-yourselfers, the OpenEnergyMonitor explores interfacing various input and output modules using the Arduino micro-controller platform (see Resources).

A less intimidating option (for most of us) is the Web Energy Logger (WEL; see Resources). This is a versatile data logger that reads temperature sensors, wattmeters, pulse inputs, contact closures, and analog voltage. The WEL system connects to the Internet (via your router) where data is uploaded, stored, and managed on the company website. There you can view and label data, build charts and graphs, and manipulate information to calculate things like Btus produced by your solar hot water system or the efficiency of your geothermal heat pump. You can even load your own graphics to illustrate your system for others to view.

Third-party software developers are able to work with product manufacturers and even DIY system sensor outputs to develop customized and sophisticated dashboard displays so that you can measure, record, view, and manage any number of connected devices and systems in your home.

▲ Example of a home energy and environment dashboard developed by PowerWise Systems (www.powerwisesystems.com)

natural gas (1 therm is equal to 100,000 Btus) to keep your house warm that day, the house has a **heating energy intensity** of 6,667 Btus per HDD (100,000 Btus ÷ 15 HDD). If you live in a climate with a total of 5,000 HDD per heating season, you can expect to use about 33 million Btus of heating fuel each season (6,667 Btus x 5,000 HDD). If your gas company bills you in therms, this translates to about 330 therms of natural gas (33,000,000 ÷ 100,000) over the course of the heating season.

In reality, there are many variables to all of this fussing with numbers. Ultimately, if you want to manage your energy use, you need to measure it; otherwise, you're just guessing. For example, the heating energy intensity as described above will increase as the outdoor temperature decreases because your house will lose heat faster when the temperature difference is greater.

Another potentially large variable is "internal gains," or how many heat-generating people and appliances are in your home. Add this to the solar heat gain you get through your windows, and you may not even turn on your heat until it's 50°F or lower outside. You can find local weather HDD data online at the Weather Underground and Degree Days websites (see Resources).

My Hot Water Heating Story

OUR 4-KW SOLAR electric system keeps 50 kWh of batteries charged quite well when the sun shines, with wind energy often picking up the slack when it doesn't. Solar power generation is monitored through the Outback Power MX80 charge controller, and wind data is collected with an anemometer feeding an NRG Systems data logger. Happily, when we compare annual power production graphs of both solar and wind, they are nearly mirror images, with wind providing more power in the winter and sun taking over in the summer.

MY EXPERIENCE

I live off the grid with my family, using solar and wind for electricity, and wood for heat. We have a backup diesel generator (often it's biodiesel) for those occasions when the sun doesn't shine and the wind doesn't blow. My family uses about 7 kWh a day, with each electrical circuit monitored through a Powerhouse Dynamics eMonitor (see Resources for this and other products mentioned).

Last summer, shortly after feeling quite smug about having an electricity surplus, I received a propane bill for over $1,000. Most of that propane is used to heat water, while some is used for cooking and some as a source of backup heat if we're away from the house in the winter and can't load the wood stove. This presented a challenge that I could not resist: how to get off the propane "grid." I knew that part of the answer was efficiency and conservation, and part lay in harnessing excess summertime solar electricity production — at least for now. Conventional solar hot water is not a practical option for us due to the distance

between the water heater and a shade-free place for the solar collectors.

To reduce hot water energy use, I took out my 10-year-old, 40-gallon, sealed-combustion propane gas water heater (which has an efficiency factor, or EF, of 0.59) and replaced it with a Navien NR180 on-demand, condensing propane water heater with an EF of 0.98. Since most of our hot water is used for showers, I also installed a GFX (gravity film heat exchanger) drain water heat recovery system (DWHR). I was thrilled to put my hand on the coil and feel the 20°F temperature rise in the water circulating through it. Now the cold water entering the

Integrated water heating system incorporating DWHR, on-demand water heater, and solar electric pre-heat option when enough sun is available ▶

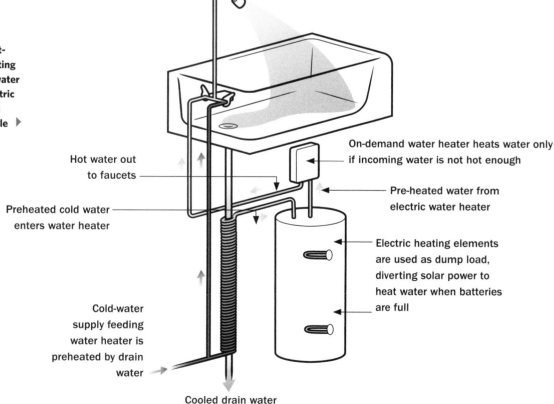

Hot water out to faucets

Preheated cold water enters water heater

Cold-water supply feeding water heater is preheated by drain water

Cooled drain water

On-demand water heater heats water only if incoming water is not hot enough

Pre-heated water from electric water heater

Electric heating elements are used as dump load, diverting solar power to heat water when batteries are full

My Hot Water Heating Story

water heater is about 70°F instead of 50°F, saving on water-heating fuel use.

But I didn't stop there. I did something I would not necessarily recommend for anyone else, simply because it is not terribly cost-effective. However, in my case (off-grid with too much power production) the numbers worked a little bit differently, and it was in fact quite cost-effective. Admittedly, I will do things that get me off the fossil fuel mainline even if they aren't cost-effective — I can't help myself, I'm just wired that way, and I won't try to talk anybody into trying this at home. But for those of you who are curious, here's how it went:

I needed a "dump load" for excess electricity generation. I considered a ductless, mini-split air source heat pump for space conditioning, but decided against it because most of my excess power comes in the summer. It doesn't really get too hot here in New England, so I didn't need the cooling benefit — I needed heat.

I also considered a heat pump water heater (HPWH), but my off-grid power system limits me to 120 volts, and the HPWH requires 240 volts. Also, HPWHs are a bit loud, and since I don't have a basement I wanted something quiet. So I bought the most efficient electric hot water heater I could find, a 40-gallon Marathon, and swapped out the 240-volt heating elements with elements from HotWatt (see Resources for this and other products) that were suitable for use with the 48-volt battery bank that would now provide electricity to heat water in the Marathon. Marathon water heaters are not exactly conventional, and I needed to have a couple of bushings custom-machined to accept the replacement heating elements. (Of course, the lifetime warranty on the unit is now void.)

Finally, I put together a control system that uses a signal from the solar charge controller. This activates a solid-state relay that connects the battery bank's DC voltage to the water heating elements once the batteries are fully charged. Excess electricity is automatically diverted to heat the stored water in the Marathon heater.

Did I say finally? Since I know how much propane gas I've used for the past 15 years, and I will of course want to see the effects of the changes I've made, the only choice (really, it couldn't be helped) was to install a data monitoring system. This is not such a difficult or expensive thing anymore with high-speed Internet, nearly free online data storage, and some neat innovative tech products.

An Itron gas meter with a pulse output allows me to monitor how much gas is being used in the home. In addition to the dials and numbers indicating rate and quantity of use, the meter delivers one electronic pulse for every cubic foot of propane moving through it. Cooking energy is almost negligible, so most of the gas used in summertime is for water heating. Therefore, I can establish a baseline to which I add winter gas consumption to determine the propane room heater's consumption.

To monitor how much hot water the family is using, I installed an Omega in-line water meter with a pulse output. Half a dozen temperature sensors complete the water monitor sensor array. All these sensors are plugged into a WEL (Web Energy Logger) data logger that uses a Web interface, which allows the user to see real-time and cumulative data on a computer or smartphone. This combination of monitoring devices allows me to see how much, and when, gas and hot water are being used. I can also determine the energy contribution (in Btus) of the DWHR and solar backup system.

The result is that we use about one-half of the amount of gas we did before this retrofit,

Last summer, I received a propane bill for over $1,000. Most of that propane is used to heat water. This presented a challenge that I could not resist: how to get off the propane "grid."

and even less during long sunny stretches when the sun provides most of our hot water. The financial payback is under 7 years, but in terms of satisfaction, the payback was immediate. On a sunny day, the batteries are charged before noon and the water overheats before dinner. It's nice to see the electrons going to good use. I may soon be looking for a dump load for excess hot water. Hot tub, anyone?

The graph at right, generated by the Web Energy Logger, shows the hot water system temperatures during a shower, with a DWHR unit placed in the shower drain. Here's how to interpret the data:

- The cold water inlet temperature is 47°F, represented by the gray line.

- Cold water feeds into the DWHR coil, and you can see that the water temperature drops as the water starts to flow through the coil (it had warmed to the ambient room temperature).

- The water is warmed in the coil by shower water going down the drain. The preheated water leaving the DWHR coil, represented by the dark blue line, is 67°F.

- The outlet of the coil feeds the cold water inlet of the storage tank, which also serves as a dump load for excess solar electric power. There was no solar-preheated hot water at this time.

- The light blue line shows the temperature of the water leaving the storage tank and entering the on-demand water heater: 66°F. Without any electricity supplying the heating elements in the storage tank, the temperature remains relatively constant in the tank as the cold water inlet is fed with water preheated by the DWHR unit.

- The black line indicates the temperature of the water coming out of the on-

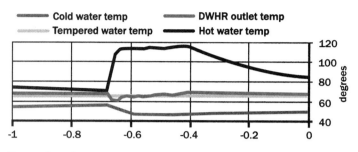

WEL TEMPERATURE GRAPH: DHW SYSTEM TEMPS

History in Hours. 1 Min. samples

▲ Hot water system temperature graph generated by the Web Energy Logger

demand water heater and feeding the shower as 114°F.

The WEL captures data at one-minute intervals and can be downloaded to a spreadsheet for further analysis. In this case, I was able to determine that an 11-minute shower used 15.4 gallons of water and 0.11 gallons of liquid propane gas. Propane gas contains about 91,500 Btus per gallon, so my water heating system requires 659 Btus to heat one gallon of water. At $3.50 per gallon for propane, each shower is costing me $0.385 in propane, plus the cost of a small amount of power to run the water well pump. Since I know my showerhead's flow rate is 1.75 gallons per minute, I've used just over 19 gallons for the shower. To calculate the percentage of hot and cold water is fairly straightforward:

$$15.4 \text{ gallons} \div 11 \text{ minutes} = 1.40 \text{ gallons per minute of hot water}$$

So, hot water makes up 80 percent of the water in the shower. (See another formula for this on page 45.) This discussion could go on for a very long time in many interesting ways, but this should be enough to give you the idea of the power of data collection and analysis.

The result is that we use about one-half of the amount of gas we did before this retrofit, and even less during long sunny stretches when the sun provides most of our hot water. The financial payback is under 7 years, but in terms of satisfaction, the payback was immediate.

PART TWO

RENEWABLE ENERGY

ONCE YOU'VE TURNED your home into a comfortable, energy-sipping dwelling, it's time to invest in renewables to offset a portion of your now smaller energy bill. This section can help you to understand the opportunities you have to generate energy for your home. When you've passed the decision-making, technical, and cost hurdles, living with renewable energy becomes part of a lifestyle that involves a keen awareness of the availability of those resources. Before you know it, you'll be predicting the weather and looking at your garbage (and coveting your neighbor's garbage) as valuable energy sources. You'll also find yourself taking advantage of times of abundance while conserving when resource availability is low.

The value of producing your own energy goes way beyond dollar savings and leads you to a place of empowerment, to a place where you can actually take matters into your own hands and meet a good part of your household needs without assistance from the energy industry. Start slow, start small, gather parts and information. Think about your energy future. Then take action.

6

Solar Hot Water

HEAT FROM THE sun can be used to heat water for your showers or swimming pool, or to provide heat for your house. To be clear, solar thermal energy is an altogether different technology than solar electric power (discussed in the next chapter), in which light from the sun is converted into electricity.

Using the sun is the easiest way to gain and use free heat and can be very cost-effective. Simply leaving a garden hose out in the sun can provide useful hot water; from there, it's not such a big step to moving and storing that water with a simple controller and a small pump. This chapter provides an overview of solar thermal systems for water heating.

Types of Systems

SOLAR THERMAL HOT water systems can have many different configurations, depending primarily on the climate, how much hot water is needed, and when it's needed. All systems have one or more **collectors** (the "panels" where the water is heated) and a storage vessel that holds the heated water. The stored water can be used directly, or it can be delivered to a secondary water heater to boost the temperature before use.

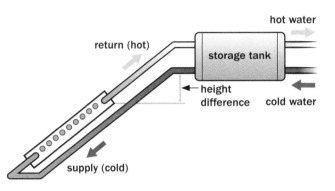

▲ Passive solar water-heater layout taking advantage of a thermosiphon's pumping action to move hot water from collectors up to the storage tank. Hot water is less dense than cold and naturally moves upward, while cooler, denser water sinks to the lowest level. The liquid in the collectors and storage tank can be water or a working fluid.

PASSIVE SOLAR

Passive solar thermal can be described as simply allowing the sun to heat something. It could be heating the air and floor in your living room, or water held in a black barrel outside. A passive hot water heating system often relies on a naturally induced **thermosiphon** to transport heated fluid between a solar collector and a water storage tank.

A thermosiphon requires a temperature difference to work: Hotter fluid (being less dense and therefore more buoyant) will rise to the top of the system, while cooler fluid is moved toward the bottom of the system. This means that the storage tank must be at a higher elevation than the collectors, since the collectors will be hotter than the storage tank when the sun is out and you want hotter fluid to move upward into the tank.

In practice, thermosiphon systems are somewhat uncommon, because typically the water storage tank is in the basement and the collectors are on the roof. In this situation, hot water stored in a tank in the basement will naturally want to thermosiphon up to the collectors on the roof when the collectors are cool. Unwanted thermosiphoning must be stopped with a check valve in the plumbing, ensuring that water flows in only one direction.

ACTIVE SYSTEMS

Active systems use the sun's heat to warm a fluid in the collector, and the fluid is moved by a pump so that the heat in the fluid can be used or stored. Pumps are turned on and off by a controller that responds to the temperatures at the collectors and the storage tank. Active systems can be directly or indirectly heated.

In a **direct** system, the water that's heated in the collectors is the same water you use at the faucet, and it's pumped between the storage tank and collectors. Also called "open-loop" systems, these are often used in climates where freezing temperatures are not a concern. Hard water can be problematic for directly heated systems because minerals in hard water build up on hot surfaces, restricting or even stopping water flow. In such cases, a water softener is used to reduce mineral content in the water supply.

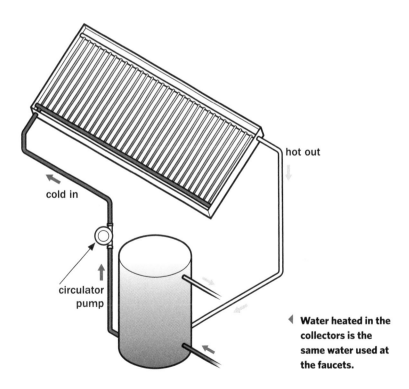

◀ Water heated in the collectors is the same water used at the faucets.

Examples of directly heated systems include **integrated collector storage** (see page 110) and **drain-down** systems, where water is drained out of the collectors at night and pumped back into the collectors when solar energy is available.

With an **indirect system**, a heat transfer fluid, or "working fluid," is heated in the collectors and circulated passively or actively in a closed loop between the collectors and a heat exchanger. In a separate loop, water that will be used at the faucets is circulated between the heat exchanger and the hot water storage tank, taking heat away from the working fluid in the heat exchange process. Indirect systems are common in cold climates, where the working fluid is an antifreeze solution.

Examples of indirectly heated systems include active systems with antifreeze, and **drainback** systems, where the working fluid (either distilled water or antifreeze) is drained out of the collectors once the desired temperature is reached in the storage tank. This prevents overheating and deterioration of the working fluid caused by stagnation of the fluid. Stagnation occurs when the storage tank is hot, the controller turns the circulating pump off, and the fluid stops circulating; however, the sun is still shining on the collectors, causing the working fluid to get hotter and hotter, with no place for the heat to go.

TYPES OF HEAT EXCHANGERS

The heat exchanger is where heat is transferred between the working fluid and the water in the storage tank. The fluids exchange heat without coming into contact with each other. Heat exchangers can be inside of the tank or external

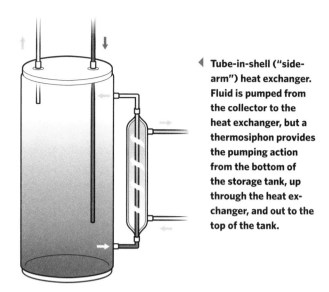

Tube-in-shell ("side-arm") heat exchanger. Fluid is pumped from the collector to the heat exchanger, but a thermosiphon provides the pumping action from the bottom of the storage tank, up through the heat exchanger, and out to the top of the tank.

to it. Internal heat exchangers are in contact with the water to be heated and offer the benefit of a simpler installation that requires only a single pump. External heat exchangers can be either **flat plate**, where fluids flow through small holes in flat plates, with the working fluid and water in separate internal channels; or they can be **tube-in-shell**, or sidearm, units in which a pipe (tube) carrying the water to be heated lives inside a larger pipe (shell) carrying the hot working fluid.

Flat plate external heat exchangers require two pumps — one to pump working fluid between the collectors and the heat exchanger, and another to pump water between the heat exchanger and the tank. Tube-in-shell heat exchangers need only one pump, because the circulation of cold water from the bottom of the tank, through the shell, and out to the top of the tank relies on a thermosiphon. Using an external heat exchanger offers the advantages of more tank choices and of being replaceable separately from the tank, potentially reducing maintenance costs.

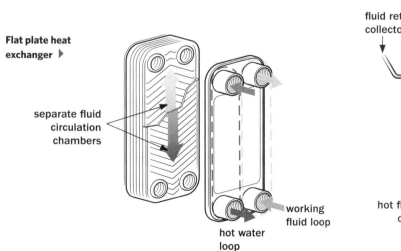

Flat plate heat exchanger ▶

separate fluid circulation chambers

working fluid loop

hot water loop

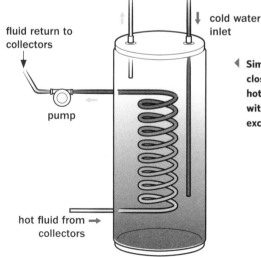

fluid return to collectors

pump

hot fluid from collectors

cold water inlet

◀ Simplified active, closed-loop solar hot water system with internal heat exchanger

Solar Hot Water Collectors

EFFICIENTLY CAPTURING AS much of the available solar resource as possible requires the right technology, proper installation, and intelligent design. Choose the right equipment for your specific needs and location, and install for maximum efficiency.

Solar thermal collectors are designed to absorb as much of the sun's heat energy as possible and transfer the heat to a liquid, heating it to 180°F or more, depending on the available solar energy and type of collectors. Collectors are ideally installed on a shade-free roof or ground rack, facing within 30 degrees of south. More easterly or westerly orientations may be acceptable but will require more collector area to meet the hot water demand.

The **tilt angle** of the collectors (the slope at which they're positioned) is set according to the season that provides the best performance. For example, if it's always cloudy in your region during the winter, you won't have much production then, so it might be best to set the tilt angle to maximize summer production. This may sound counterintuitive: Why not tilt the panels so that you can gain more in the winter and maximize what little sun there is? But the general rule is this: Increasing a small amount of energy by a small percentage will not amount to much, so focus your efforts on where the energy *is* rather than where it is not.

There are three common types of collectors — flat plate, evacuated tube, and integrated collector storage (ICS) — any of which can be part of either an active or passive system. Choosing the best collector depends primarily on your climate and how you will use the hot water.

FLAT PLATE COLLECTORS

Flat plate collectors are insulated boxes covered with low-iron, tempered, textured glass that provides a durable, highly transparent and absorptive solar window. Inside the box, to absorb solar heat, are flat, dark-colored (usually nickel-plated) copper plates to which copper tubes are fused. The heat transfer fluid flows through the tubes,

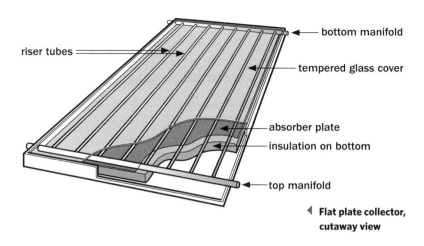

Flat plate collector, cutaway view

(labels: bottom manifold, riser tubes, tempered glass cover, absorber plate, insulation on bottom, top manifold)

removing heat from the absorber plates along the way. Flat plate collectors are simple and compact (typically measuring 4 x 8 feet or 4 x 10 feet and several inches thick), and multiple units can be plumbed together to increase the collector surface area. As they heat up, flat plate collectors lose efficiency because they lose heat to the surrounding environment.

EVACUATED TUBE COLLECTORS

Evacuated tube collectors do not hold fluid in the same way as other collectors. They consist of a series of annealed glass tubes, each covered with a selective coating that absorbs solar energy while inhibiting reradiation heat loss. During manufacture, air is evacuated from the tubes to create a vacuum; like a Thermos, this eliminates conductive and convective heat loss. In the center of each tube is a copper absorber plate.

From here, there are several variations on the specifics of getting heat out of the tube and into the working fluid. Often there is a heat pipe attached to the center of the absorber plate. This pipe contains a liquid that vaporizes when heated, rising to a metal bulb at the top of the tube. The bulb is immersed in the working fluid in a header that is common to multiple tubes. Working fluid flows through the header, absorbing heat from the hot bulbs along the way. This cools the bulbs and condenses the fluid in them, allowing it to flow down to the bottom of the heat pipe to be heated and vaporized again.

The size of evacuated tube collectors is easy to customize because each tube is separate, and

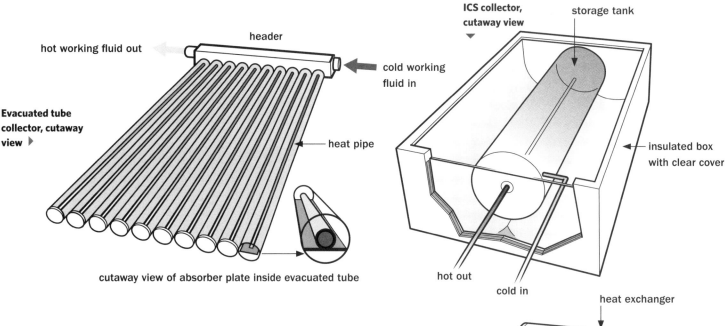

hot working fluid out

header

cold working fluid in

Evacuated tube collector, cutaway view ▶

heat pipe

cutaway view of absorber plate inside evacuated tube

ICS collector, cutaway view ▼

storage tank

insulated box with clear cover

hot out

cold in

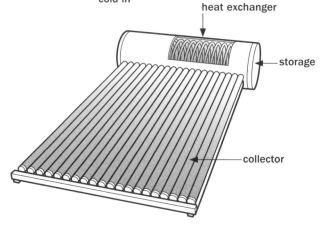

heat exchanger

storage

collector

▲ **Indirect ICS with separate collection and storage**

you can add tubes to the header to increase the collection area, up to the maximum number of tubes the header can hold. Evacuated tube collectors offer several potential advantages over flat plate collectors; they are:

- Modular and lightweight
- Less sensitive to orientation
- Provide potentially hotter water (making them a good choice for space heating)
- Tend to be more efficient in very cold temperatures
- Don't reradiate heat

However, evacuated tubes are more fragile than flat plate collectors, and they don't shed snow as well because they don't reradiate their heat. The vacuum in the tubes is a great virtue but also a great liability because it's difficult to maintain a vacuum forever in the real world — so look for a long warranty.

INTEGRATED COLLECTOR STORAGE (ICS)

ICS systems, sometimes called "batch" heaters, heat a batch of water in a storage tank inside of an insulated box with a clear cover. These simple systems heat the water to be used directly, as cold water flows in and hot flows out, feeding the faucets or another water heater used as a backup. They are used in situations where there

is no danger of freezing and are most effective if hot water is needed primarily in the evening, after the sun has had time to warm a large volume of water. ICS systems are heavy, due to the weight of the water they contain, and sturdy mounting structures are mandatory. One efficiency disadvantage is that heat is lost from the water at night.

A variation on ICS integrates evacuated tube collectors into an insulated water (or other working fluid) storage tank. A heat exchange coil is immersed in the working fluid, through which the water to be heated flows, making this an indirectly heated ICS.

SOLAR COLLECTOR RATINGS

Solar hot water collectors are rated for thermal output in terms of Btus per square foot per day.

You can get this information from the manufacturer or from the Solar Rating and Certification Corporation (SRCC). This nonprofit, independent, third-party organization (formed by the solar industry, state energy officials, and consumer advocates) exists to certify and rate solar hot water equipment. Performance is listed in a directory published on the SRCC website (see Resources).

Note that a certified *collector* carries the SRCC OG-100 label, while certified water-heating *systems* carry the SRCC OG-300 label. These certifications offer a level playing field for comparing the relative performance of collectors and systems but may not predict actual performance on your rooftop. Think of the ratings like the fuel economy ratings for cars.

While you may be attracted to the most efficient collector, a more important factor is how well the system will perform to meet your needs in your specific situation and application. This requires looking closely at the performance characteristics of various collectors.

Hot Water Storage

A SOLAR WATER heater is typically set up as a way to preheat water and deliver it to a secondary (backup) electric or fossil-fueled heater to boost the water to the desired temperature. The storage tank for the solar-heated water can be the same tank as the secondary heater, or a separate tank can be used to provide greater storage volume.

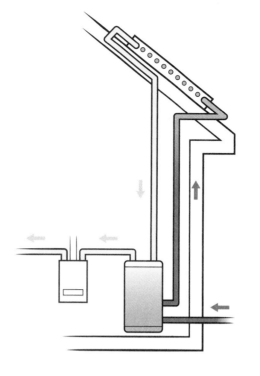

◀ **Simplified view of an active solar water heater with collector on roof, storage tank in basement, and an on-demand gas water heater to boost the temperature when needed**

Typical storage capacity can range from 40 to 120 gallons for domestic hot water use (water used at faucets, showers, and other fixtures or hot water–consuming appliances) and 500 gallons or more for space-heating systems. The appropriate storage capacity depends on collector output and hot water demand. Storage tanks can be made of steel or fiberglass. Be sure the tank comes with ample insulation, or add more to the outside to keep the heat in.

SINGLE- VS. TWO-TANK STORAGE

Whether or not you use a single water heater — one that both stores solar heat and has an additional heat source — depends in part on your

SOLAR POOL HEATERS

Pool heating is the most popular application of solar water heating. Pool heating collectors are made of chemical-resistant, copolymer black plastic with a header on each end, and an absorber plate with integrated riser tubes in between. Pool water is circulated directly through the collectors. These simple and inexpensive collectors perform well for their intended purpose but are not designed for freezing conditions.

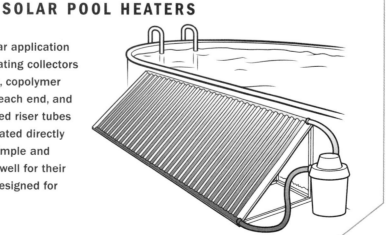

hot water usage patterns. A single tank may save space and money, but you can use up all your solar-heated water after the sun goes down, and then the backup heat source will engage to heat the water in the tank. In the morning, when the sun comes out, the hot water produced in the collector will have nowhere to go.

With a two-tank system, the collectors heat the water in a solar storage tank, which in turn feeds a backup (fossil fuel or electric) water heater with preheated solar hot water. The backup water heater turns on when water is not hot enough. In a strange twist of the right actions yielding the wrong results, some research has shown that where a two-tank system is used in a very efficient household, not much water is drawn from the solar storage tank into the backup tank, causing most of the water to be heated by the backup heater. If this is your situation, a pump can be used to circulate solar-heated water to the backup tank depending on the water temperature in each tank.

Hot Water on Demand

An on-demand (tankless) water heater can be a good option to use in conjunction with solar hot water. It's important that the water heater has an adjustable, or "modulating," flame that can change its heat output in response to the incoming water temperature. With the right tankless model, there will be no additional heating energy used if the incoming solar-heated water is hot enough. When the solar water is not hot enough, the on-demand unit will supply just enough heat to bring it up to the set temperature.

Additional System Components

IN ADDITION TO the collectors and storage tank(s), a solar hot water system requires "balance of system" components to make everything work correctly and effectively. It's important to note that a quality installation is critical to ensuring the system performs as intended with minimal maintenance over the long run.

WORKING FLUIDS

Working fluids include water (in direct systems) and nontoxic antifreeze, such as propylene glycol solution rated for high-temperature (for indirect systems where freezing is a concern). Fluids leak, and propylene glycol will deteriorate over time, becoming more acidic with age and, if subjected to high temperatures, it stagnates. If the fluid doesn't stagnate, it should last for 10 to 15 years. Stagnation and deterioration are minimized by allowing the antifreeze to circulate continually when the sun is out; this requires having enough storage capacity to accept the heat and thus not overheat the working fluid in the collectors.

Collectors overheat when the water heater is hot enough and the circulator pump stops operating while the sun is still shining on the collectors. It is extremely important to remove all air from the system when it's being filled, or "charged," with working fluid, and to prevent air from getting into the system. Air in the system can cause fluid circulation problems that prevent the system from working.

PLUMBING

Plumbing pipes should be copper for the hot water leaving the collectors but can be PEX for the cool water return. Avoid connecting PEX directly to the collector; use copper pipe instead. Some installers avoid PEX altogether because it softens when heated. Don't forget to insulate the pipes with high-temperature–rated insulation, such as fiberglass or Armaflex (see Resources), and cover all exterior insulation with a PVC or aluminum jacket to protect it from UV radiation and other environmental hazards.

EXPANSION TANK

An expansion tank contains an expandable air bladder that responds to, and compensates for, pressure changes in the collector loop as water (or working fluid) alternately heats and expands, then cools and contracts. The air inside the expansion tank compresses as pressure in the system increases with temperature.

Without an expansion tank, pressure inside the system would get too high, causing something to burst. The size of the expansion tank depends on the quantity and thermal expansion properties of the working fluid, as well as the temperatures the fluid will reach.

PUMPS

Pumps can be AC (standard household current) or DC (battery or solar-powered current), and should be made of brass, bronze, or stainless steel suitable for high temperatures.

The pump size and required flow rate depends on the collector area and the plumbing layout, and must be sized according to system design. As an example, flow rate through flat plate collectors typically is around one gallon per minute for each 4 x 8-foot collector.

Some DC pumps (such as the Laing D5 series and those made by El Cid, which are available from plumbing suppliers or solar dealers) can be powered directly from a small solar electric panel mounted next to the solar thermal collectors. When the sun shines, the pump turns on and moves fluid through the collectors. This is a very simple and elegant solution to pump control because if it's sunny enough to make electricity to operate the pump, it's sunny enough to make hot water. This approach also addresses the problem of power grid failure, when an AC pump will stop working. No circulation on a sunny day means no hot water collection, leading to collector fluid stagnation, overheating, and deterioration.

CONTROLLER

A controller is used with active systems to turn the circulating pump on and off depending upon the relative temperatures of the collectors and storage tank. Responding to sensors that monitor collector and storage tank temperatures, a **differential temperature controller** turns on the pump when the temperature of the collectors is higher than that in the storage tank, and then turns off the pump when the collectors cool. A high-limit setting prevents the tank from overheating.

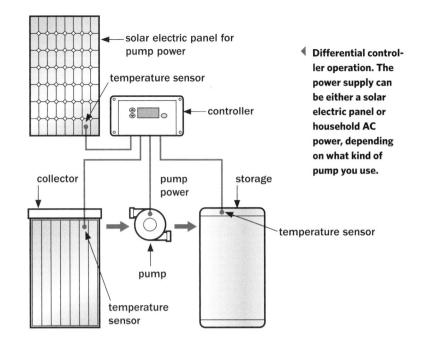

◀ **Differential controller operation. The power supply can be either a solar electric panel or household AC power, depending on what kind of pump you use.**

GAUGES AND OTHER DEVICES

Additional parts and materials are required to complete your system, and good plumbing knowledge and skills are essential for success. Gauges for temperature and pressure are useful to monitor the system's performance. Other parts you may need include check valves, isolation valves, air vents, an air eliminator, an aquastat (water immersion thermostat), thermocouple temperature sensors, drains, temperature and pressure relief valves, a tempering (mixing) valve, and a vacuum breaker.

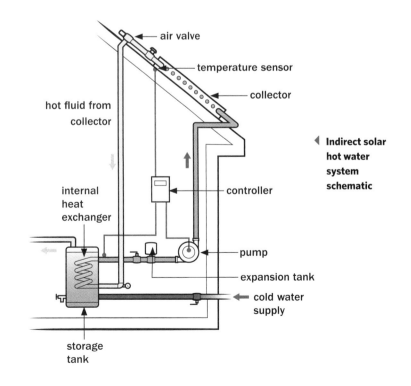

◀ **Indirect solar hot water system schematic**

Sizing the System

COLLECTION AND STORAGE must be carefully matched so that the heat produced by a solar hot water system can be used and stored effectively. Too much collector area relative to storage means that the water will overheat, the pump will stop circulating, and the fluid in the collectors will stagnate, causing it to overheat and break down. Too little collector area means that the stored water will seldom, if ever, get hot enough.

Whatever you do, don't undersize the system. Doubling the size of a system doubles the output, but it doesn't double the cost. The best plan is to design the system to meet your average daily needs under sunny conditions.

DETERMINING YOUR HOT WATER DEMAND

The average American uses about 15 to 20 gallons of hot water every day. Most system designers use 20 gallons per person as the design requirement, while your actual use will vary depending upon your needs and habits. From this starting point, you can calculate the collector square footage necessary to provide all the hot water you need on a sunny day. This means you'll get what you need in summer but probably less than you need in winter. Of course, the sunnier your location, the less collector area you need to satisfy your hot water load.

Calculating Btus

Here's an example using a three-person household consuming 60 gallons of hot water each day, in a location with lots of summer sun but cloudy winters. A backup water heater provides hot water during cloudy periods.

You can calculate water heating energy requirements in Btus. Remember that a Btu is the amount of energy it takes to raise the temperature of 1 pound of water by 1°F. A gallon of water weighs 8.3 pounds. To calculate the Btus required to heat our daily demand of 60 gallons, we start with the difference between the cold water ground temperature and the desired hot water temperature. We'll assume the ground water temperature is 55°F, and we want 130°F water, yielding a temperature difference of 75°F. The formula looks like this:

Temperature difference x weight of water in pounds = Btus needed

75 (temperature difference) x 8.3 (pounds per gallon) x 60 (gallons per day) = 37,350 Btus

Being realistic, we'll add 10 percent to account for system inefficiency, bringing our demand up to about 41,000 Btus per day. This is what we need our collectors to deliver to us each day.

SIZING THE COLLECTORS

Once you have a target Btu output for your system, you can research available collectors using manufacturer and SRCC data. This will help you to determine the total square footage of collector area to meet your needs. To get there, we need to dig a little deeper.

Knowing the solar resource available in your location helps in sizing a system accurately. Solar resource maps and data are available from the National Renewable Energy Lab (NREL). Explore the links listed in Resources to find the information you need.

Solar radiation data is often presented in terms of kilowatt-hours per square meter per day (kWh/m²/day). For solar thermal applications, it

---

Solar Sweet Spot

Depending on your climate, a solar hot water system typically provides about 50 to 75 percent of a household's annual hot water needs. This sizing scheme offers a reasonable compromise between system size, cost, and energy savings.

is useful to convert this to Btus per square foot per day (Btus/ft²/day). To do this, multiply the square-meter value by 317:

$$kWh/m^2/day \times 317 = Btus/ft^2/day$$

For Instance . . .

Here's an example using NREL's "redbook" data for Boston, Massachusetts, where we want to install a solar water heater for a family of three. We know that we need to heat 60 gallons of hot water and that this requires 41,000 Btus every day. The data indicates that for a flat plate collector mounted at an angle equal to Boston's latitude of 42 degrees, we can expect a summertime solar radiation average of 5.5 kWh/m²/day. That converts to 1,743 Btus/ft²/day. Solar radiation, and therefore solar hot water production, will be somewhat less in the wintertime.

We've found a 32-square-foot, flat plate collector rated by SRCC to deliver 22,200 Btus per day, given a solar radiation level of 1,500 Btus per square foot. Two of these collectors (64 square feet) will produce over 44,400 Btus, which would more than cover our needs.

SIZING STORAGE SPACE

Regarding storage tank size, a very general rule of thumb to prevent overheating is to provide 1.5 gallons of storage volume for every square foot of flat plate collector area in an average seasonal climate. For sunny regions, raise that to 2 gallons for every square foot, and for cloudier regions, lower it to 1 gallon. Applying the standard storage tank sizing rule to our Boston example, we would need:

$$64 \times 1.5 = 96 \text{ gallons of water storage}$$

MAINTENANCE

A properly installed solar hot water system using high-quality components can last up to 40 years and require only minimal maintenance. Spending a little more up front for a better pump or hardware will be well worth it in the long run. Here are some regular maintenance items to keep things running smoothly:

- Inspect the system components and plumbing once a month. Check for leaks, corrosion, and missing pipe insulation. Listen for unusual noises from the pumps.

- Keep the collectors free from debris and shade.

- Check the temperature gauge to confirm the collector fluid is hot on a sunny day.

- Check the pressure gauge to confirm the system is maintaining pressure.

- Check the controller to confirm the system is operating in accordance with the indicator lights. If the controller tells you the pump should be running, but the temperature is low and you don't hear the pump, you may have a damaged pump or controller, poor wiring connection, or faulty temperature sensor.

- Plan on replacing the circulator pump about every 10 years, and expect to replace the expansion tank every 15 to 20 years.

- Test the antifreeze solution after 10 years of use or after any significant stagnation events. Check for proper pH (no lower than 7.5) with litmus paper, and check for freeze protection using a tester that is suitable for propylene glycol (*not* an automotive antifreeze tester).

Homemade Hot Water

HAVING LIVED OFF-GRID for a while, Lori was already in tune with her daily energy use and was motivated to improve the 100-plus-year-old farmhouse she's lived in for 14 years. She heats her house with wood and has made many efficiency improvements, including adding insulation, air-sealing, and drying out the dirt-floor basement.

LORI BARG

is a small-scale hydro-electric power developer and consultant. She is the founder of Community Hydro (see Resources) and works with communities to identify hydropower opportunities and develop appropriate solutions. Lori is personally driven to reduce her home's energy use and produce her own energy using simple techniques and accessible resources.

While working in her berry patches one summer, Lori became intrigued with the idea of capturing the solar heat that was just lying around inside the hoses she used to water her gardens. The water inside those hoses always ran hot for the first few minutes. This inspired her to develop a simple water-heating scheme using her existing water heater, an additional water heater, the sun in the summer, and the wood stove in the winter. She succeeded in her goal of putting something together herself for under $1,000, having less than a five-year return on the investment.

The solar hot water collector Lori built consists of an old 3 x 10-foot tin roof panel painted black, with 75 feet of ¾" flexible black hose laid out in a serpentine pattern on the panel and fastened to it with UV-resistant black plastic cable ties (zip ties). The collector is secured to two pressure-treated 2x4s mounted on the roof. To avoid penetrating the roof (to eliminate the possibility of leaks), Lori extended the 2x4s beyond the ridge and added two more boards that extend down the opposite side, so the pieces "hook" over the top of the roof. A third 2x4 spans across the joined ends of the two boards on each assembly to strengthen the hook. (Lori chose the roof for better solar access, but a ground-mounted rack would work equally well).

An old 30-gallon electric water heater serves as a storage tank that provides solar-preheated water to feed a 40-gallon propane water heater. If the water coming out of the solar preheat tank is too cold,

the propane fires up the heater. The electric heating elements of the preheating tank were removed and replaced with adapters that connect ¾" copper pipe to the heating element ports on the side of the tank. Both tanks are on the second floor of the house, higher than the solar collector.

To deal with the inevitability of leaks, each tank is set in a drain pan, with drains plumbed down to the basement sump pump. The sun heats the water in the collector, and as long as the collector water is hotter than the water in the preheating tank, a thermosiphon effect passively circulates water between the collector and the preheating tank. Because there is water (rather than antifreeze) in the collector, the system has valves to isolate the collector loop so it can be drained in cold weather.

The top pipe of the solar collector leads to the port of the former top heating element of the preheating tank (which has been removed). Another pipe leads from the bottom port of the tank to the bottom of the solar collector. As the water in the collector heats up, it rises to the top of the collector and continues to move upward into the water heater, while cooler water at the bottom of the tank feeds the collector. As long as there is a temperature difference, thermosiphoning provides continuous circulation between the collector and the heater. The greater the temperature difference, the higher the flow rate.

Next, Lori built an air-to-water heat exchanger for the wood stove out of tube-and-fin

Lori became intrigued with the idea of capturing the solar heat that was just lying around inside the hoses she used to water her gardens. The water inside those hoses always ran hot for the first few minutes.

baseboard radiators, the kind used for boiler-based (hydronic) home heating. She cut several lengths of fin tube to the width of the stove, then plumbed them together in a serpentine pattern and backed them with a shroud to retain the heat and to support the heat exchanger. This "reverse-radiator" heat collector is fastened to the chimney (which won't move), not the wood stove (which will). The fins fit snugly against the stove to absorb its heat and transfer it to the water inside the pipe. A thermosiphon is created between the fin-tube collector and preheating hot water tank upstairs.

In operation, cold water from the main house water supply feeds the preheating tank, and that water is heated with solar or wood heat when available. Water from the preheating tank feeds the propane water heater, where it is further heated if needed. If the preheated water feeding the gas water heater is hot enough, then no gas is needed to heat the water. Hot water from the gas heater feeds the home's hot water fixtures.

After the first year of operating the system, Lori provides this report:

"The thermosiphon works. I put some inexpensive temperature sensors on the inlets and outlets, and water is preheated [by the solar collector or wood stove] to around 105 or 110°F by each system. While I still use some gas to heat the water, I like that I don't have to worry about the water getting too hot or building up high pressure. I added a check valve to the solar loop to keep the hot water from recirculating at night

and cooling off. When I drained the system for winter, I propped open the check valve to let that pipe drain. Otherwise, the pipe could possibly freeze and burst with the first hard frost.

"Overall, I am quite pleased. I had thought of solar hot water for a long time but could not afford it. I had looked at ways of preheating hot water off my wood stove and discarded most of them. I did not want to take heat off my chimney, and possibly increase creosote and the risk of chimney fires; I did not want to take heat off the top

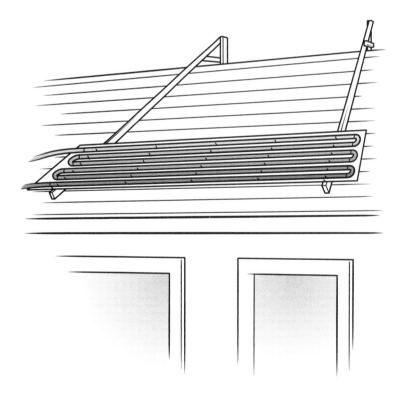

▼ **Lori's solar collector mounted to her roof with the 2x4 hook system**

Homemade Hot Water

of my wood stove, because I like to cook on it in the winter, keep a teakettle on, and allow water to evaporate to keep the house from being so dry. The chimney-mounted, serpentine, high-Btu fin tubing used in typical baseboard hot water systems works great.

"I am thinking of adding additional supports to my rack and mounting some photovoltaics above the hot water panel."

Water circulation diagram. Shutoff valves allow each system to be isolated, drained, and turned on or off according to the season and which heat source is active. ▶

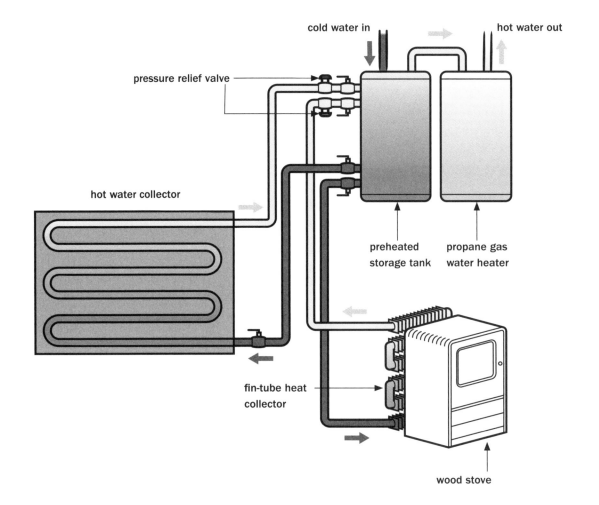

cold water in

hot water out

pressure relief valve

preheated storage tank

propane gas water heater

hot water collector

fin-tube heat collector

wood stove

Safety Considerations

When the water in both the preheating tank and collector loops are very close in temperature, the thermosiphon action slows or stops. If heat continues to be collected in this situation, it's possible for pressure in the system to reach dangerous levels. For this reason, it's critical to have a temperature and pressure relief valve (TPR valve) in both the solar and wood stove preheat loops to eliminate the potential for a burst pipe, which could release scalding hot water or steam. A small expansion tank would be required in any isolated loops, such as for an indirectly heated system with a heat exchanger, or if there is a chance that a fluid-filled preheat loop would be isolated from the rest of the system by valves.

Build a Solar Hot Water Batch Heater

This is a very simple design for a batch-type solar water heater that uses a thermosiphon loop to move water between a solar thermal collector and a storage barrel. While it's not the most efficient solar water heater and works rather slowly, it effectively demonstrates the operation of a thermosiphon and its connection to hot water storage. Better yet, the system can provide useful amounts of hot water for an outdoor shower, washing garden produce, keeping a biogas generator warm (see chapter 13), or any number of other uses you can think of. Because it uses water, the system must be drained when the weather turns cold.

Buying all new parts for this system might cost around $300 or $400, but there are many ways to modify the design to incorporate salvaged parts or for enhancing the system; see Ideas for Upgrades on page 122. For longevity and durability, the parts should be rated for UV exposure and high temperatures. Most plastic (PVC and polyethylene) will start to deform above 140°F, and typical garden hose will not hold its shape much over 100°F.

The parts list includes recommended materials and temperature ratings. Use materials that can withstand temperatures up to 180°F, such as metal, CPVC, or polypropylene fittings, as well as high-temperature water hose material, such as EPDM rubber. Black-colored hose will assist in absorbing solar heat. If you can't find what you need at your local hardware store, plumbing supplier, or home center, shop online through industrial supply companies (such as Grainger or McMaster-Carr; see Resources).

MATERIALS

Four 8-foot 2x4s
Five 8-foot 1x4s
2½" deck screws
¾" roofing screws or galvanized sheet metal screws
Two 2 x 8-foot corrugated metal roofing panels
High-temperature, flat black spray paint
One 100-foot, black EPDM rubber garden hose, ¾" I.D. (inner diameter), rated for 200°F
One hundred 8" black plastic cable ties, UV-resistant, heat-stabilized, rated for over 200°F
One 55-gallon barrel, black plastic, preferably with wide-mouth top
Two ¾" NPT (National Pipe Thread)/GHT (garden hose thread) brass faucets, rated for 180°F
Teflon tape
Two ¾" polypropylene bulkhead fittings, rated for 180°F
Two ¾" brass garden hose-to-tubing adapters
Two hose clamps
One 36" length foam pipe insulation

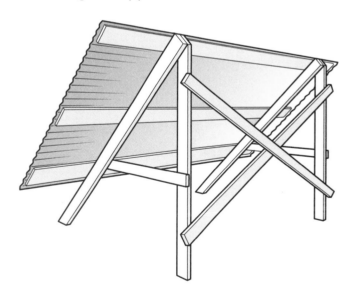

▲ Completed collector frame with panels installed (rear view)

Solar Education

A smaller version of this system makes an excellent science project. Use clear plastic tubing for the collector loop and add food coloring to the water to demonstrate circulation.

1. Build the collector structure.

Construct two A-frame supports using 2x4s and 2½" deck screws: Position the front leg of each support at the desired tilt angle for the collector panel (your latitude angle is a good starting point), and attach the rear leg at an opposing angle for stability. Join the two legs (front and back) of each side with a horizontal cross piece.

Space the A-frame supports about 6 feet apart, and join the two back legs with two diagonal 1x4s to keep the frame from racking.

Install three 8-foot pieces of 1x4 horizontally across the supports. Locate one piece at the tops of the supports, one 24" down from the tops, and one with its bottom edge 48" from the tops. Fasten the 1x4s to the supports with the deck screws.

Mount the roofing panels to the wood frame, using ¾" roofing screws or sheet metal screws. Install the lower panel first, fastening it to the bottom 1x4. Install the upper panel so it overlaps the lower panel, then fasten through both panels into the center 1x4. Fasten the upper panel to the top 1x4.

Paint the top surface of the roofing panels with high-temperature, flat black spray paint (the kind used for painting wood stoves). Let the paint dry completely.

2. Install the collector hose.

Lay out and mark the hose path on the collector panel, using full-length horizontal runs back and

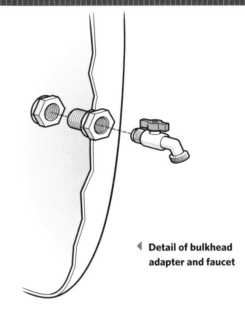

◀ **Detail of bulkhead adapter and faucet**

forth, working from the bottom of the panel to the top. Do not exceed the bending radius of the hose at the ends, as kinks in the hose will stop the flow and can lead to trapped air bubbles. A ¾" I.D. hose should allow for eight horizontal runs across the 4-foot-tall panel. Be sure to leave extra hose at the beginning and end for connecting both ends to the storage barrel.

Note: As the water in the hose heats up, air will be released from it. Keeping the hose runs reasonably level, with no kinks or sags, allows the thermosiphon to work effectively, eliminates trapped air bubbles, and facilitates draining the collector.

Secure the hose by drilling pairs of holes through the collector panel, one above and one

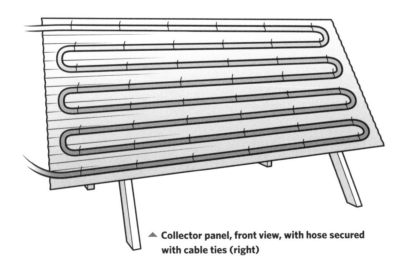

▲ **Collector panel, front view, with hose secured with cable ties (right)**

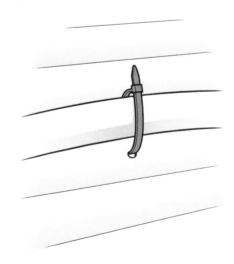

below the hose, then loosely fastening the hose with a plastic cable tie (zip tie). Add a tie about every 12" along the entire path of the hose. Make sure the hose is in full contact with the collector panel for best heat transfer. You will tighten the ties later, after the hose is connected to the barrel and all fits well.

3. Prepare the water storage barrel.

The barrel gets two threaded bulkhead fittings to provide a watertight connection through its side. The fittings receive the tapered threaded ends (not the garden-hose ends) of the faucets that will connect with the collector hose. Make sure the faucets are compatible with the bulkhead fittings and the adapters for connecting the collector hose.

Cut a hole through the barrel for each fitting, using the appropriate size of hole saw. Position the lower hole as close to the bottom of the barrel as possible, and locate the upper hole about one-third of the way down from the top of the barrel. Choose areas with no raised markings and little curve in the barrel surface to ensure a watertight seal.

Wrap the threaded end of each faucet with Teflon tape, and thread it into the exterior half of a fitting. Fit each fitting into its hole and secure it inside the barrel with its nut.

4. Set the barrel and connect the hose.

Position the barrel on a sturdy stand that is tall enough so that the faucets on the barrel are higher than their corresponding hoses mounted on the collector. Position the collector close to the barrel. Extend the lower end of the hose to the lower faucet, and cut the hose to length so it makes a smooth upward arch toward the faucet. Make sure the hose doesn't sag, which can trap air or create a "heat trap," stopping the thermosiphon action. Cut the upper hose end to connect to the upper faucet.

Install a tubing-to-faucet adapter on each end of the hose, securing it with a hose clamp.

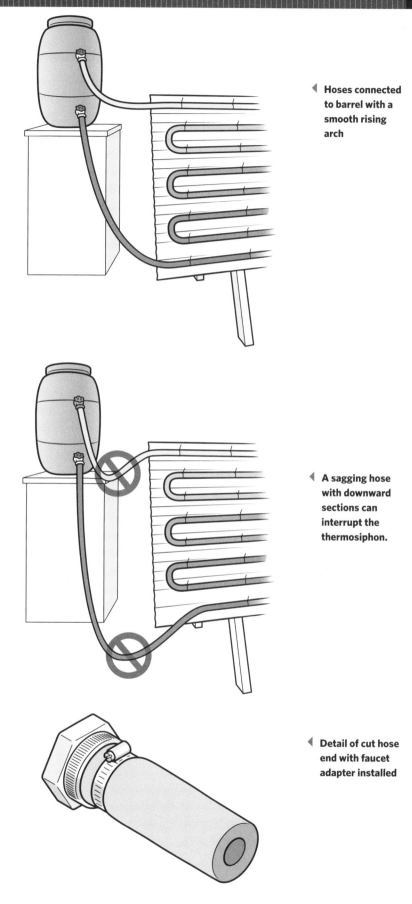

◄ **Hoses connected to barrel with a smooth rising arch**

◄ **A sagging hose with downward sections can interrupt the thermosiphon.**

◄ **Detail of cut hose end with faucet adapter installed**

Thread the hoses onto the faucets. Tighten the cable ties on the collector panel so they hold the hose securely in place. Put a piece of foam pipe insulation on the hot water (top) hose to help reduce heat loss and increase the effectiveness of the thermosiphon.

5. Get started making hot water.

Open the faucet valves, and fill the barrel to the top, leaving 1" or 2" of air space for expansion. You want to be sure that there is no air in the collector loop. Tilt the collector back and forth (one tilt for each hose bend) after the barrel is filled to be sure that all the air is out of the hose.

You'll start making hot water as soon as the sun comes out. Cold water sinks to the bottom of the barrel and continues down the hose to the bottom of the collector. As the water is heated it rises up through the hose, through the top faucet, and into the barrel, where it rises to the top. There will be a noticeable temperature stratification within the barrel until the water is completely heated.

As an example, on a sunny 45°F day, I achieved a 40°F temperature rise through the collector loop using about 75 feet of ¾" hose laid out on the collector, plus another 8 feet leading to and from the storage barrel. This was at a fairly low flow rate, and resulted in a 10°F per hour temperature rise within the stored water during the hours just before and just after noon.

IDEAS FOR UPGRADES

With some additional effort, you can increase the efficiency of the system and even bring the hot water you make into your home, where it can preheat the water in your existing water heater. Check local plumbing codes before modifying your water heater. Using a heat exchanger in between your solar collector and water heater means you won't be putting potentially nonpotable water into the water heater.

- Insulate the water storage barrel to keep the water hotter for longer periods.

- Enclose the collector in a box for hotter water. Insulate the box on the bottom, and cover it with a piece of UV resistant clear Plexiglas or flexible fiber-reinforced plastic (FRP).
  NOTE: Enclosing the collector may make temperatures exceed the materials' ratings, requiring the use of all metal components.

- Collect and store more solar energy by increasing the collector area and the length of the circulation hose.

- Switch to an active system with the addition of a small solar-powered 12-volt DC pump that moves less than 2 gallons per minute; this frees you from the constraints of a thermosiphon system.

Optional equipment for water circulation is available through local and online renewable energy dealers. This includes a low-wattage, 12-volt DC circulator pump that can operate from a 5- or 10-watt solar panel. If you make this investment, you will also want to spend a bit more to insulate the storage barrel and enclose the collector.

Solar Electric Generation

P **HOTOVOLTAIC (PV) DESCRIBES** the electrochemical process of using the energy delivered in photons of light to create an electric current. *Photo* means light, and *voltaic* means related to producing electricity. (A photon is a particle of electromagnetic energy.) When a photon of light strikes one side of a solar **cell**, the photon's energy causes an electron to jump to the other side of the cell. From there, the electrons travel through a circuit to where the electricity performs the desired work, such as lighting a lamp or charging a battery.

A PV **panel**, or **module**, is made up of many small PV cells wired together to produce the desired voltage and power. The cells are assembled into a sturdy frame and covered by a strong, clear, waterproof elastomer or thermoplastic layer that is resistant to breakdown by UV rays from the sun. Each module can range in power output from 10 to over 200 watts.

Two or more PV modules wired together are called an **array**. The size of the array needed for any given task depends upon the power requirements of the particular site. Arrays can be installed on a roof or a ground-mounted rack.

Solar Power Potential

DEPENDING ON THE technology and materials used, commercially available PV cells convert light energy to electrical energy with an efficiency rate of between 8 and 18 percent. Current work in laboratories is producing cells with efficiencies of over 30 percent.

The material in PV cells that converts the energy of photons to electricity is called a **semiconductor**. The most common semiconductor used in cells today is the element silicon (as in Silicon Valley; not *silicone*, as in tub and tile caulk). However, other materials, such as germanium, gallium, and cadmium can also be used. Organic solar cells using polymers are also being developed.

Semiconductor materials can be formulated in a solid crystal for use in rigid cells, or in sintered form, where powdered semiconductor material can be sprayed onto flexible products, such as roof shingles (see Space Requirements, page 134) and clothing. Each technology has advantages and disadvantages in terms of cost, efficiency, and flexibility. Multicrystalline silicon PV cells are the most common type used for stationary residential and commercial installations and have an efficiency of about 16 percent.

Assuming an average efficiency of 15 percent, a PV panel can deliver about 14 watts per square foot (150 watts per square meter). Keeping your panels aimed directly at the sun throughout the year and throughout the day will increase overall output. You can adjust panel position manually, but an easier and more reliable way is to incorporate a tracking system that automatically moves the array to the optimum position (see pages 131 to 132).

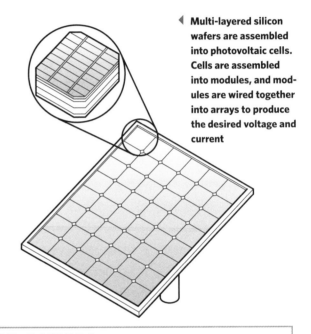

◀ **Multi-layered silicon wafers are assembled into photovoltaic cells. Cells are assembled into modules, and modules are wired together into arrays to produce the desired voltage and current**

HOW MUCH POWER?

On average throughout the world, solar energy striking Earth from directly overhead delivers about 1,000 watts of energy per square meter, or about 93 watts per square foot. In reality, though, how much energy your solar collector "sees" and absorbs depends on several factors:

- Your location on the planet
- Season
- Temperature
- Level of cloudiness
- Light reflection from the ground to the solar panel
- Angle of incidence between the collector and the sun (see page 130)
- How much light the panel reflects away from itself

On a clear day, you can expect the sun to deliver between 700 and 1,400 watts of raw solar energy to each square meter of solar collector area that is aimed directly at the sun.

AC vs. DC

The power produced by PV systems is **direct current** (DC), not the **alternating current** (AC) you need to power your home. The conversion from DC to AC is handled through an electronic device called an **inverter**. See chapter 10 for more information about inverters and other components common to renewable electricity generation.

Planning for a PV System

YOU CAN DETERMINE how much electricity you use simply by looking at your electric bill, which tells you how many kilowatt-hours (kWh) you used during the past month. Some bills present daily usage, and some utilities let you look at how much energy you're using at any given time by way of a "smart meter" that may even have a Web interface. You can also learn to read your own electric meter on page 98, and see chapter 5 to learn about focused monitoring of specific power users in your house.

As you examine your electrical use, you may find patterns where consumption during certain times of the year is greater than at other times. But the general rule is to size your PV system for average daily use, knowing that you will make more than you need on some days and less on others. Solar electric systems can be expanded over time, so you can start small and add more as your budget allows. Making your home ready for renewables requires long-term planning; if you prepare for future expansion, there will be less work and expense when the time comes to upgrade.

ASSESSING YOUR SITE

Let's say that your home uses an average of 20 kWh of electricity each day, and you want to supply 50 percent of that electricity using solar power. Your solar-generating capacity needs to provide 10 kWh per day. Depending upon where you live, there are some months when you can count on the sun more than other months. It would probably not be practical to size your system to generate 10 kWh on cloudy days, because on sunny days you would have far more than you need. So let's keep things simple and use a sunny day as an example to design a solar power system.

How Much Sun?

Once you've determined how many kilowatt-hours you want to generate, the next step is to understand how many hours of sunlight you can expect each day at your location. The greatest power output occurs during **peak sun** hours, when the

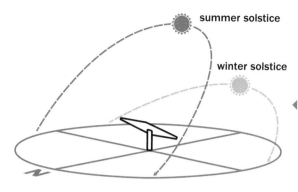

Seasonal sun trajectories show the sun's path across the sky at winter and summer solstices. Any obstacles between the sun and the PV array will cast shadows and reduce power.

sun is high in the sky, typically spanning the three hours on either side of noon (that is, 9:00 a.m. to 3:00 p.m.). Outside of that ideal "solar window," power production starts to fall off unless you have a tracking system that allows the array to follow the daily movement of the sun across the sky. The intensity of sunlight will vary throughout the year (unless you're on or near the equator), so the actual power output of your PV panels will vary further with the seasons.

You can find average daily sun hours from a local weather station, or you can research weather data online through the National Climate Data Center or the National Renewable Energy Lab's Renewable Resource Data Center (see Resources). The latter website includes an analysis tool called *PV Watts* that helps you predict the output of a PV system in your area. These tools are a good place to start, but you still need to perform a solar site survey to assess the daily and seasonal solar resource available at your specific site.

Solar Site Survey

A site survey takes a close look at the path the sun takes across the sky at different times throughout the year. There are two tools commonly used by professional solar installers for site surveys and power production analysis (see Resources for websites).

Solar Pathfinder. This tool lets you see where the sun will be at any given time in any season. It also provides a visual indication of when and where the panels will be shaded. Shaded PV panels produce power about as fast as you get a suntan sitting under a shade tree, so this is an important part of any solar site assessment. The user (typically a professional solar installer) sets up the

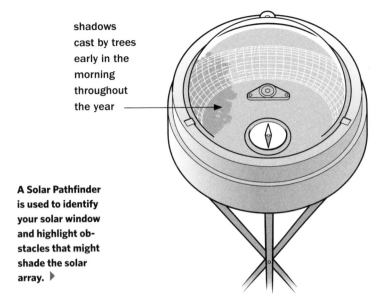

shadows cast by trees early in the morning throughout the year

A Solar Pathfinder is used to identify your solar window and highlight obstacles that might shade the solar array. ▶

Pathfinder at the prospective PV site and observes the sun's path (along with shadows) against a grid that shows latitude and longitude. The company's software helps to predict annual power output and modify the results, should the homeowner decide to eliminate some of the shading.

SunEye. Made by Solmetric, the SunEye is an electronic analysis tool that may be used by professional solar installers. It does everything the Pathfinder does but with the accuracy and additional bells, whistles, reporting, charting, and computer interface that you would expect from this professional electronic analysis tool.

SIZING A SYSTEM

There are two ways to determine how many kilowatt-hours a PV array of a given size will produce over the course of a year.

Math Estimating

This sizing method starts with learning how many hours of sunlight you can expect each day, on average, throughout the year. Then, factoring in how much power you need, you can work backward to estimate the capacity of your PV array in watts.

Back to our example, to deliver 10 kWh (10,000 watt-hours) per day. Your site survey indicates you can expect an average of 5 hours of unshaded sunlight each day of the year. This average includes cloudy days and seasonal variations; your actual daily power production will vary quite a bit.

The math required is simple:

Watt-hours needed ÷ daily sun hours =
PV system size (in watts)
10,000 watt-hours ÷ 5 hours = 2,000 watts

Map Estimating

The second way to estimate the size of a PV system is to find the average annual power that can be produced for each watt of PV installed at your location. Most solar estimating maps (such as those available from the National Renewable Energy Lab) present solar insolation (exposure to sunlight) in terms of raw solar power available in kilowatt hours (kWh) per day. This represents how much solar energy can be harvested by a solar collector and does not account for the efficiency of the PV cells. In terms of system design, it's more useful to convert available solar insolation into kWh (energy produced over time) that can be produced for each kilowatt (kW) of installed PV power (rated electrical power production). This is expressed as the kWh/kW factor.

Refer to the solar energy estimation map (see facing page) to see the approximate daily kWh of solar energy available at your site for conversion to solar electricity. The value shown is for each square meter (10.7 square feet) of PVs installed on a fixed-mount (non-tracking) rack with a tilt angle equal to the latitude. The sun shines more frequently and intensely in some places of the world, and less so in other places. Solar electric panels are tested and rated at a light level of 1,000 watts per square meter (93 watts per square foot). At 14 percent efficiency, one square meter of PVs will produce 140 watts (13 watts per square foot) under equivalent sunlight.

A more meaningful interpretation of the chart for our purpose is to understand that the kWh per square meter per day figure is the same as the number of available daily sun hours, adjusted for insolation intensity. Using this value as our guide allows us to more easily estimate how many kWh can be produced for every kW installed. Here's an example:

Let's say you live in a place where the sun delivers 5 kWh each day for every square meter of PV array or, in other words, the sun shines

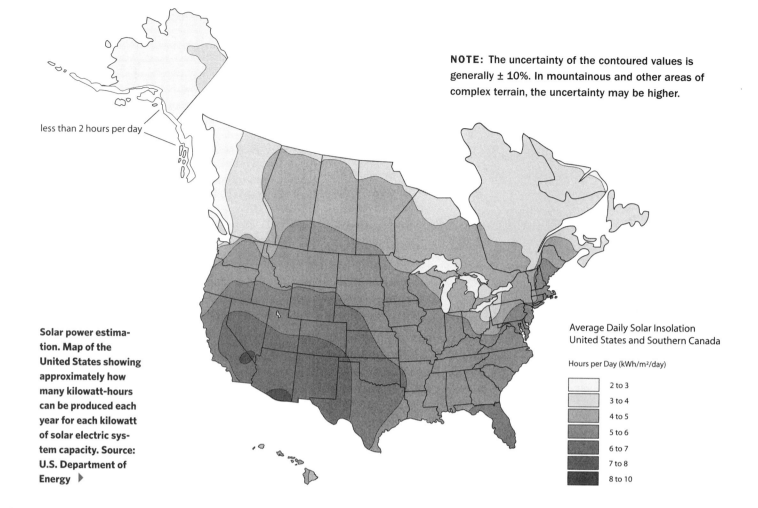

less than 2 hours per day

Solar power estimation. Map of the United States showing approximately how many kilowatt-hours can be produced each year for each kilowatt of solar electric system capacity. Source: U.S. Department of Energy ▶

Average Daily Solar Insolation
United States and Southern Canada

Hours per Day (kWh/m²/day)

2 to 3
3 to 4
4 to 5
5 to 6
6 to 7
7 to 8
8 to 10

5 hours on average each day. Every 1 kW of PV capacity can deliver:

$$1 \text{ kW} \times 5 \text{ hours per day} \times 365 \text{ days} =$$
$$1{,}825 \text{ kWh per year}$$
The kWh/kW factor in this case is 1.825

Keep in mind that this map is highly generalized and does not account for local conditions, system inefficiency, or tracking rack adjustment factors, but it gives you a general idea as to what you can expect in terms of annual power production. As a comparison, the 2,000-watt PV array calculated in the math example above will deliver approximately 3,650 kWh per year (2,000 x 1.825).

Accounting for Inefficiencies

The above examples do not account for any inefficiencies in the system. You may experience up to 25 percent power loss through wiring, connections, power handling equipment, panel derating (they don't always operate in factory-test conditions), temperature (cooler panels deliver more power), and dirt on the panels (even a little bit of shading hurts a lot). A 15 percent reduction from the published PV panel rating is a good value to use when estimating net power delivered from your PV system. Dividing the power needed by the total efficiency brings our 2,000-watt PV contribution requirement up to an array size of 2,353 watts (2,000 ÷ 0.85 = 2,353). This lowers the effective kWh/kW factor from 1.825 to 1.55.

Bigger Is Usually Better

Now you can choose your PV panels. PV panels are rated in watts of power output, as well as voltage (volts) and current (expressed in *amperage*, or amps). Larger panels typically are a bit less costly (in terms of price per watt) than smaller panels. They also require less wiring because fewer panels are needed in the array. Years ago, PV arrays were assembled from 50-watt panels, but today 200-watt panels are common.

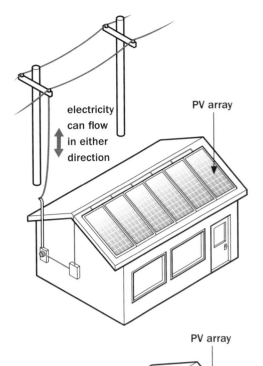

On-grid solar electric system with utility intertie. Electricity can flow in either direction, depending on how much power the house needs relative to how much is being produced by the solar panels. ▶

electricity can flow in either direction

PV array

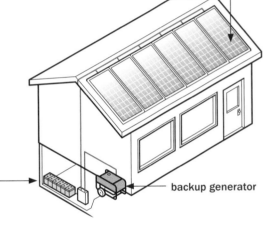

Off-grid system where electricity is stored in batteries and no utility power is available. ▶

PV array

battery bank →

backup generator

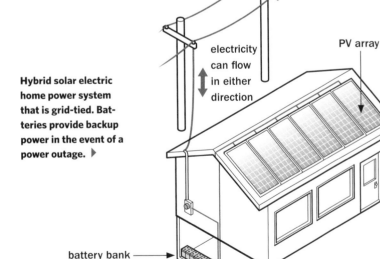

Hybrid solar electric home power system that is grid-tied. Batteries provide backup power in the event of a power outage. ▶

electricity can flow in either direction

PV array

battery bank →

PV System Wiring

PV PANELS TYPICALLY produce a nominal 12 volts of DC electricity, while current (amps) depends upon how many cells are in the panel and how bright the sun is. Power output is expressed in wattage, which is a product of volts and amps (volts x amps = watts). In reality, a 12-volt PV panel will produce up to 20 volts in full sunlight with no load, meaning that the wires are not connected to anything. This is known as **open-circuit voltage**.

Nominal voltage is used in system design to match voltage output with other components. Open circuit voltage is higher than nominal voltage because the electrons need to overcome wiring losses, along with other voltage-derating factors, such as heat. As an example, a 12-volt battery may require 15 volts to charge fully, and this voltage can be expected from a nominal 12-volt panel. To prevent overcharging of batteries, a charge controller is used.

A PV system can be designed to charge batteries or feed power to the utility grid, or to do both. Each approach requires different voltages to maximize the efficiency for each scenario and to match the needs of the power handling equipment. Higher-voltage equipment generally is more efficient and allows for the use of smaller diameter (and less expensive) wires without significant **voltage drop** — loss of power resulting from low voltages over long wire runs.

WIRING CONFIGURATIONS

It's acceptable to combine different panels in a system, but for best efficiency and performance the panels in each array should be closely matched for voltage, while individual current output can vary. In order to deliver the desired voltage and power, the panels can be wired in three configurations: **series**, **parallel**, or **series-parallel combination**.

In a series circuit, the positive wire of one panel is connected to the negative wire of another, and the remaining positive and negative wires are connected to the load. A series-wired array produces a voltage that is the sum of

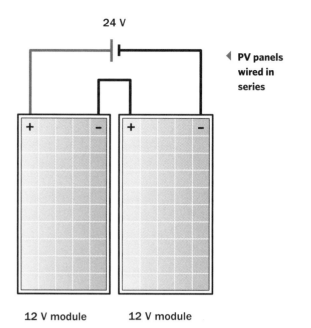

24 V

◄ **PV panels wired in series**

12 V module   12 V module

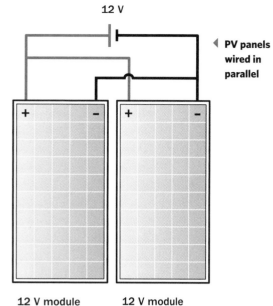

12 V

◄ **PV panels wired in parallel**

12 V module   12 V module

each individual panel's voltage, while its current (amperage) is the average of all of the panels' currents. The diagram above shows two PV panels wired in series. Each panel produces 12 volts, 6 amps, and 72 watts (12 x 6 = 72). The total output is 24 volts, 6 amps, and 144 watts (24 x 6 = 144).

In a parallel connection, all positive wires are connected together, and all negative wires are connected together. With this configuration the voltage stays the same, and the total amperage is the sum amperage of all of the panels. Our two 12-volt, 6-amp panels are now producing 12 amps at 12 volts, with the same total output of 144 watts.

Series-Parallel Combination

Let's look at an example of a battery-charging PV system that's wired for 48 volts. Higher voltage helps to increase equipment efficiency and to reduce the cable size between the PV array, charge controller, and batteries. There are several brands of 200-watt PV panels that produce 24 volts each (and thanks to *Home Power* magazine's *Solar Electric Module Guide,* we know there are several choices for power inverters that meet our needs).

We will need to wire these in a series-parallel combination in which every two panels are wired in series to produce 48 volts. Each set of two panels is then wired in parallel with other sets of two. Using our example system design, we've

determined that our desired power output (wattage requirement) is 2,353 watts. To find the number of panels required, we divide the total wattage requirement by the wattage per panel:

2,353 watts ÷ 200 watts per panel =
11.7 panels

We must have an even number of panels for our 48-volt configuration, so we'll round up to 12 panels, yielding a total output of 2,400 watts. For the practical purposes of mounting location, future maintenance, and troubleshooting, we'll split the system into two separate arrays of 1,200 watts each.

System wired with series-parallel combination, including combiner and main PV disconnect ▼

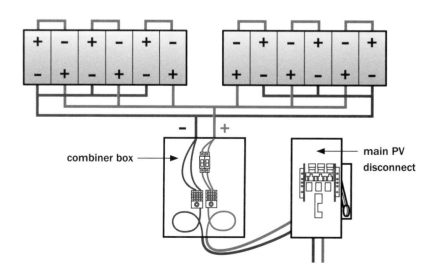

combiner box   main PV disconnect

Because we have more than one PV array, the arrays must be electrically connected, or combined, in a **combiner box** (see chapter 10). Here, each of the array wires is connected to a common electrical distribution block for positive, negative, and ground, so that only a single wire runs to the charge controller (for an off-grid system) or to the inverter (for a grid-tied system). Each array has its own circuit breaker inside the combiner box so that individual arrays can be disconnected as needed without interrupting the entire PV power supply.

The drawing on page 129, bottom, shows our two separate PV arrays feeding into a single combiner box. The positive PV feed wires from each array are connected to a dedicated circuit breaker. The output side of each circuit breaker is connected to a junction block, as are each of the negative PV feed cables. The junction blocks have one large supply cable that carries all of the PV current to the power management center that includes the charger controller and inverter.

The PV supply feed coming from the combiner box is next run to the main PV disconnect switch. The disconnect is required both for electrical protection and servicing and can include a master fuse or circuit breaker. In the case of a grid-tied system, a disconnect is mandatory to protect the line workers, should they need to work on power lines fed by your PV array.

Orientation

FOR MAXIMUM POWER output, the PV array should live in an unshaded location at a perpendicular angle to the sun's rays for 4 or more hours each day. Even the shade of a single leaf can eliminate most of the power output of the affected panel and the series-connected part of the array it's wired into. Maximizing the time of full sun exposure maximizes the power output and increases the rate of return on your investment.

If you're in the northern hemisphere, the sun will be in the southern half of the sky (unless you're within 23 degrees of the equator, in which case the sun will cross over to the north for the summer), and PV panels should therefore face south. When orienting your PV array, be sure to account for magnetic deviation, an effect that causes a compass needle to indicate other than true north. The amount of deviation depends upon where you are in the world, and it changes over time. Consult a geomagnetic map (see Resources) to find the adjustment you need to apply to your compass.

However, a fair amount of offset from true south can be tolerated without substantial loss of power. For example, a fixed PV array facing southeast or southwest will produce about 85 to 90 percent of the power of an array that faces due south. In fact, some utility-sponsored programs support PV installations oriented specifically to help offset peak load conditions (periods of maximum power demand) on the utility. In such cases, you may find PV arrays facing due east or due west. The array orientation is based on where the sun is during the time of the utility's peak demand.

As the earth travels around the sun throughout the year, the sun's altitude changes with the seasons (see How Much Sun? on page 125). At the time of summer solstice, the sun is at its highest point in the sky, which is your latitude

PV and Peak Demand

Peak demand is the period of time when the utility needs to deliver the greatest amount of power to its customers. For example, if peak demand is at 5:00 p.m. in the summertime — when everyone comes home to air conditioning, electric cooking, and water heating — a west-facing PV array will help offset the demand on other power plants that may be close to full production capacity. The overall power output of the PV system may be substantially less than with a south-facing array, but the utility's objectives (which are likely different from yours) are achieved. This approach is often a better solution than building a new power plant which would be needed for only a few hours a day.

plus 23.5 degrees (the tilt of the earth on its axis). So if you live at a latitude of 42 degrees, the sun will reach a maximum latitude of 65.5 degrees above the horizon.

At winter solstice the situation is exactly the opposite, and the sun will be at a latitude of 18.5 degrees (42 *minus* 23.5) above the horizon. When mounting solar panels on a permanently fixed rack, the best tilt angle is the latitude of your location. You may decide to increase or decrease the tilt angle to gain a bit more power in the winter or the summer, depending on either your seasonal needs or on the season that delivers the most sun. If your system is grid-tied, you'll want to orient the array for maximum annual average power production.

Racks and Tracking

PV PANELS ARE mounted on a rack to assemble an array, and the racked array is then mounted on supports either on the ground or on a roof. A ground-mounted PV rack requires a sturdy foundation, typically either a concrete pier supporting a steel pole or multiple pressure-treated posts anchored firmly in the ground. A roof-mounted rack can lie flat against the roof over the shingles, or it can be tilted to a specific angle.

Before mounting panels on your roof, be sure that the roof construction is capable of holding the additional weight. Also consider future shingle replacement: You don't want to have to take down your PV system in 5 years just to replace the shingles.

You can buy commercial racks designed for simple assembly and installation, but you can also make your own with 2" slotted steel or aluminum angle stock. Regardless of the rack you use, don't skimp on the foundation, which must support the weight of the PV array as well as keep it from sailing away in high winds.

Consult with an engineer and/or mounting system manufacturers (such as Direct Power and Water; see Resources) for specifications on pole and foundation requirements based on the size and weight of your array and the type of soil you have.

TRACKING RACKS

The previous examples of power production assume a **fixed** PV array, where the array is permanently mounted in a fixed orientation. If the array is mounted on a pole, the tilt can be manually adjusted a few times each year to match the seasonal angle of the sun, but essentially this is a fixed installation. A **tracking** rack automatically follows the sun throughout the day and year.

Tracking extends the number of hours during which you can capture peak sunlight by adjusting the horizontal and vertical orientation of the array throughout the day as the sun moves across the sky. For a tracking rack to make economic sense, you must have clear access to the sky from shortly after dawn to nearly dusk; no more than a couple of hours of light should be lost in either direction. Tracking racks require space for the array to pivot around the pole upon which they are mounted and are not suitable for roof-mounted arrays.

Single-Axis vs. Dual-Axis

Tracking racks are available with single- or dual-axis tracking capability. Longitude, or azimuth (east and west), is the most important axis to track for maximum power collection. A single-axis, longitudinal tracking rack can increase your average annual power production by 24 to 33 percent over a fixed rack. A dual-axis tracker (adding altitude to azimuth tracking) will increase output by 7 to 9 percentage points over a single-axis tracker. The lower ranges of these power increases occur in locations with relatively less sun.

Power gains are always greater in summer, when the sun is in the sky for longer periods. This can be particularly beneficial for grid-tied systems, since excess summer power generation can help offset winter use, should your utility offer net metering (which allows solar power to, in effect, turn your electric meter backward; see page 175).

Is Tracking Worth It?

You may wonder whether it's worthwhile to spend money on a tracking rack to increase power, or

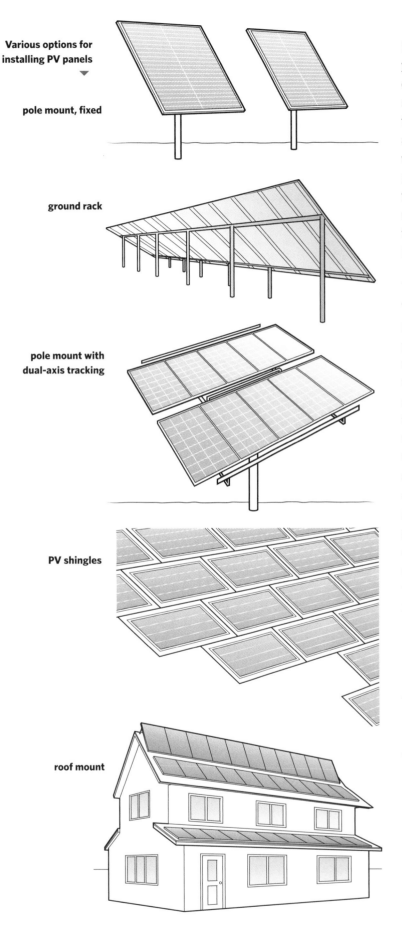

**Various options for
installing PV panels**
▼

pole mount, fixed

ground rack

**pole mount with
dual-axis tracking**

PV shingles

roof mount

if it's more cost-effective to increase the size of your fixed PV array. The chart on page 133 presents a comparison of three options using the same baseline cost to buy PV panels and install them on a rack. The costs shown do not include any other system components because these would be common to each system regardless of rack type. A tracking rack uses the same foundation and supporting pole as a fixed rack, so the only variables are the cost of the rack (based on the size needed to hold the panels) and whether it is fixed or tracking.

Tracking Rack Mechanics

Tracking racks can use electric or hydraulic motors, controlled by an integrated clock drive or photo sensors that detect variations in light level and provide feedback to the motor drive. Zomeworks (see Resources) manufactures a thermally operated, passive solar tracker using a refrigerant that changes phase when heated by sunlight. As the sun's heat vaporizes the liquid refrigerant, the vapor moves to the high side of the rack, where it cools and becomes denser, increasing the weight on one side of the rack and causing it to tilt. Thermally operated racks tend to be a bit more sluggish to respond than motorized trackers (especially in colder weather), and therefore not quite as accurate, but they are simple, maintenance-free, and a bit less expensive, and they require no energy.

Economics

YOU MAY WANT to use solar power for many reasons: to avoid connecting to distant power lines, to be independent of power companies, to "go green," to reduce your exposure to risky energy markets. As an investment, solar power will yield predictable returns for many years.

If you are considering the long-term economics of renewable energy as an investment, you'll probably want to dig a little deeper. And remember, using less electricity means that you can meet more of your needs with fewer solar panels, so be sure to invest in efficiency first. Efforts to reduce energy use are almost always less expensive than buying more solar power capacity to support inefficient habits and old energy-hog appliances.

PLUGGING IN THE NUMBERS

Our sample 2,400-watt PV array will produce about 3,840 kWh per year. If electricity costs $0.10 per kWh, you're saving $384 each year. As electricity prices rise, your solar dividend pays even more. Over a useful life of 25 years, the system will have produced 96,000 kWh (or $9,600 worth) of electricity at today's rates. If electricity costs rise just 2 percent annually above the rate of inflation (a 2 percent escalation rate), you'll save about $12,600 over the life of the system. To figure your return on investment, you need to subtract the installation and lifetime maintenance costs from this lifetime savings amount.

At the time of writing, a complete, professionally installed solar power system costs anywhere from $4 to $8 per watt of DC-rated capacity, depending on the size and complexity of the system and whether or not there are batteries to buy. Federal, state, local, and utility incentives can help to bring installation costs down, and sometimes utilities offer a payment for power produced. Maintenance costs with PV are minimal, but you may want to consider potential inverter replacement approximately every 10 to 15 years, and, if it's an off-grid system, batteries will need replacing every 5 to 10 years.

If the PV system costs $5 per watt installed, the 2,400-watt system totals $12,000. Without incentives, lower installed costs, or increasing energy prices, this system just about breaks even over its lifetime with the 2% escalation rate. If you are building a new home far from existing power lines, compare the costs of PV to those of bringing in utility power. Additionally, as I mentioned in the introduction, there are many noneconomic reasons to buy energy efficiency or generation. How much is energy autonomy worth to you? Only you can put a "payback" value on that.

The Shelter Analytics website (see Resources) includes an energy-improvement analysis tool that allows you to enter details about PV systems and costs, incentives, utility costs, and escalation rates, as an aid to understanding lifetime economic and carbon impacts of a renewable energy system.

COMPARISON OF FIXED AND TRACKING RACKS

In the chart, the *Fixed* column assumes a 2,400-watt PV array on a fixed rack. The *Tracking* column assumes the same size of array on a dual-axis tracking rack. The *Fixed Plus* column shows how many watts a fixed array would need to be if it were sized to provide the same power as the tracking rack. Keep in mind that, in reality, costs and power output will vary widely based on your situation and location, so be sure to get accurate costs and solar insolation (solar radiation) data for your particular site and project. Be sure to read Living with Solar (and Wind) Power (page 135) for another perspective on tracking.

In the example here, a dual-axis tracking mount delivers 38 percent more power at a slightly lower cost when considered in the context of lifetime power production. Both higher PV prices and larger system sizes improve the economics of tracking racks. With larger, multi-array systems, there's potential for significant savings if tracking increases power production enough to eliminate the need to install an additional rack with foundation and supporting structure.

Type of Rack	Fixed	Tracking	Fixed Plus
Array size, watts	2,400	2,400	3,300
Rack cost	$2,184	$5,376	$3,003
Annual power output, kWh	3,840	5,280	5,280
Adjusted kWh/kW	1.6	2.2	1.6
Installation price per watt (array and rack)	$5.91	$7.24	$5.91
Lifetime kWh (25 years)	96,000	132,000	132,000
Lifetime cost per kWh generated	$0.148	$0.132	$0.148

If you're using PV panels made with multicrystalline silicon cells, you'll need about 1 square foot of rack space for every 10 watts of PV capacity. The panels don't need to be contiguous, but the electrical wiring layout needs to work with the physical layout. You can electrically combine several arrays mounted in various locations to feed a common load. The 2,400-watt array from our sample system will require about 240 square feet, along with a roof and rack that are sturdy enough to carry the weight.

Alternatively, if you use **thin-film** PV technology, you'll need about 50 percent more area, since these panels are generally less efficient than those made from crystalline cells. Thin-film PV is used in some PV roofing shingles, such as those available from Dow Solar (see Resources). A solar-shingled roof costs a bit more per watt than framed PV modules, but the shingles have a much lower profile, blend in well with a dark roof, take the place of standard shingles, and provide power for your home. This makes your roof a real asset, rather than just another maintenance item for your home.

Safety

MODERN EQUIPMENT MAKES it fairly easy for a skilled do-it-yourselfer to install a solar power system safely and successfully. This does require working around potentially lethal voltage and current levels, so it's critical to take all possible precautions for the installer's safety and the safety of others affected by the work, such as utility line workers. In all cases, national and local codes must be adhered to. Plan on hiring a licensed electrician to advise you along the way.

Here are just a few of the essential safety considerations for any solar power installation:

- Improperly selected or installed equipment can be a shock and fire hazard.

- Cables that are not in conduit present serious hazards for anyone digging holes in the ground or driving nails or screws into walls.

- Ungrounded or improperly grounded equipment can be troublesome at best, deadly at worst.

PV panels produce high voltages when exposed to sunlight, with enough energy to kill a human who touches the bare conductors. During installation and wiring, cover the panels with a tarp to decrease or prevent electrical generation, and cap the conductors so that no bare wire is exposed. PV power is a "soft" power, in that the panel's electrical output connections can be shorted (positive and negative terminals connected together) without hurting anything. The electrons simply continue on their way around the electrochemical process. Just don't get in their way!

It's important to remember that PV produces DC power, and all electrical components must be rated for use with DC power. Using electrical equipment that is not DC-rated fails to provide suitable electrical protection, resulting in early (if not immediate) failure of the components and possibly creating extreme hazards. This is especially true for circuit breakers and fuses. Solar electric power system installations must follow the National Electrical Code (NEC) Article 690 safety standards. The NEC covers all requirements for wiring, grounding, fuses, batteries, and grid-tie systems.

Proper system grounding requires connecting the frame of every PV panel in each array to a grounding rod driven into the ground. Grounding hardware should be bronze or stainless steel, rather than aluminum, for best weather resistance. All electrical equipment must be grounded as well. This is required for system safety, and it also helps to protect components from getting fried by lightning.

Maintenance

Solar electric systems rely primarily on electronic components to do most of the work and therefore require minimal attention. In the case of solar electric modules, the only maintenance you may ever need to do is check the electrical and mechanical connections every few years to be sure they are clean, free of corrosion, and secure.

Living with Solar (and Wind) Power

A **DARK AND OVERCAST** day delivers about 10 percent of peak output, but a sunny day with snow on the ground may increase peak output by about 10 percent. When the sun shines for 5 hours, I gain 4,000 watts x 5 hours = 20,000 watt-hours, or 20 kWh. However, since my panels are on a fixed (nontracking) rack, peak output happens only for a couple of hours either side of noon. Since I live in the North, wintertime output is much less than summertime output because the sun doesn't deliver as much energy to the northern hemisphere in the winter, there are more cloudy days, and, of course, the days are shorter.

MY EXPERIENCE

The second question people ask me about living with solar electricity (right after "Does it really work?") is, "How many panels do you have?" It's not how many panels that matters (they come in all sizes), but rather how much power those panels produce. I have 4,000 watts of peak solar electric generating power; in other words, when the sun shines, the modules produce about 4,000 watts of power.

We also have a wind generator, and the question I get most often is, "How big is it?" I never know exactly how to answer this because "big" is a subjective term, and it depends on who's asking. So I provide multiple answers, and that's probably why most people's first question is often their last.

- The tower is 115 feet tall (tall is one kind of big); tall enough to get the turbine 30 feet above the treetops, which is necessary to avoid wind turbulence.

- The blade diameter (another kind of big) is just under 10 feet.

- What blade diameter really tells you is the bigness of the "swept area," which essentially is the wind collection area — in my case about 76 square feet. This is the kind of big that really counts when it comes to wind energy.

- The final "bigness" quotient is how much power it can generate, and of course that depends on the wind speed. My Kestrel e300 will start to spin in a 7 mph wind, producing only a few tens of watts, and produces a maximum of 1,000 watts in a 25 mph wind. (See chapter 8 for more about wind power.)

Our household uses about 7 kWh per day, with storage capacity of about 50 kWh in batteries. That gives us about one week of power storage if there is no sun or wind. "No sun" conditions occur for us about two months out of the year. During that time, our wind generator helps to cover some of the loss, but we usually need to rely on a backup generator to keep the batteries fully charged. It would not be cost-effective to add more PV panels to meet our needs during those two months because "no sun" is just that; it wouldn't matter if we had one watt or one megawatt.

You may wonder what we do with 20 kWh of daily power generation when the household only uses 7 kWh per day. The answer is "dump load." Also known as a "diversion load," this is a place to use excess power that keeps coming in even after the batteries are charged. Normally, this power is simply not used, and not to use available solar power feels like wasting it. For us, an electric water heater is the recipient of excess electrons. On a bright summer day our batteries are charged by lunchtime, and by dinner we have 40 gallons of 120°F water. Of course, this only works during sunny periods, so during the winter months most of our hot water is heated by gas.

I decided not to install a tracking rack because being at 45 degrees north latitude, the main benefit would be reaped in the summertime, when we don't need extra power. During the winter, the sun angle is so low that tracking would not provide appreciably more power. After crunching the numbers, I found that it would be more cost-effective to buy more PV panels than

Living with Solar (and Wind) Power

> The best thing about PVs is that they simply sit in the sun and quietly do their job with no moving parts, requiring little or no attention for years at a time. Sometimes I go outside just to look at them — it's quite amazing to see something that does so much work with so little fuss.

to invest in a tracking rack. This argument works only because we are off-grid. If we were on-grid and able to sell electricity back to the utility, the extra summer power generated would be beneficial. A careful examination of local conditions for solar power generation, utility costs, and the incremental costs and benefits of a tracking rack will tell you which racking system is more cost-effective in your situation.

The best thing about PVs is that they simply sit in the sun and quietly do their job with no moving parts, requiring little or no attention for years at a time. Sometimes I go outside just to look at them — it's quite amazing to see something that does so much work with so little fuss.

The real maintenance for us is in the batteries. They're heavy, corrosive, and smelly, and they need regular care and attention. For best operation, they need to be kept between 60 and 90°F, they must be filled every couple of months with distilled water, and the terminals need to be checked for corrosion (and cleaned, if needed). If they're not charged just right, batteries don't last.

Our batteries live in a shed next to the house, in a well-insulated box that has a vent fan to remove hydrogen gas that is released during charging. Make-up air for the fan comes from inside the house, through a wiring conduit, allowing for warm air to be moved around the batteries during colder weather. I keep a remote thermometer in the box so I know if they're too hot or too cold.

Before the box was well insulated, I used a couple of 100-watt electric battery heating pads during the coldest part of winter. These were only marginally effective on subzero nights, and of course they came with an energy penalty when we could least afford the power. Once the box was insulated and air-sealed against drafts, I found the heat generated by the batteries during sunny-day charging to be far more substantial than what the heating pads could provide. With the insulated and ventilated box, the battery temperature is acceptable without the heating pads.

In general, I'm really happy with our current system. The wind and sun complement each other quite well, in that the windiest months are also the least sunny, and sunny summer months are not very windy at all. The chart here shows the relative monthly energy resource available from both sun and wind monitored at our site. Solar power is expressed in how many kilowatt-hours of energy we collect over the course of a month for each kilowatt of charging capacity. Wind energy is expressed in terms of power density, or how many watts are generated for each square foot of swept area (wind collector area represented by the area covered by the blades as they spin in a circle), based on the average monthly wind speed. The chart shows at a glance how PV and wind power can be seasonally complementary.

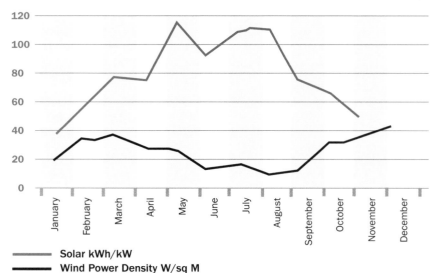

MONTHLY ENERGY RESOURCE AVAILABLE FROM SUN AND WIND

—— Solar kWh/kW
━━ Wind Power Density W/sq M

8

Wind Electric Generation

WINDMILLS HAVE BEEN around for a very long time. Relatively short towers once hosted very large rotors with many blades to produce the power and torque needed for things like grinding grains and running machinery. During the 1930s, wind electric generators made their way into rural areas where there were no electric power lines. These low-voltage machines were primarily battery chargers and were used to power low-voltage DC home appliances. Some were dedicated to pumping water.

Today there are a handful of manufacturers producing electricity-generating wind turbines for both grid-connected systems and off-grid battery-charging applications. The term *turbine* generally refers to the combination of blade set and generator assembly, while the term *generator* refers specifically to the electricity-producing unit.

Modern wind machines use high-efficiency generators or alternators and highly refined blade designs and materials for maximum efficiency. There are also some interesting and novel devices on the market that can capture energy in the wind. These range from small rooftop wind machines to **vertical axis wind turbine** (VAWT) designs. While there are niche markets and encouraging research in some new areas of design, current best practices for harvesting wind energy center on the **horizontal axis wind turbine** (HAWT). This chapter will focus on wind fundamentals and how they apply to the tried-and-true HAWT.

Using Wind Energy at Home

WIND TURBINES GENERATE electricity by capturing the wind's energy as it moves around two or three propeller-like blades. The blades are attached to a generator that produces electricity when it spins. The turbines sit high atop towers, taking advantage of the faster, stronger, and less turbulent wind at 100 feet or more above ground.

Many locations have some potential for capturing wind energy, but the resource varies widely with location, season, and time of day. Your neighbor down the road may have more wind available than you do, due to local conditions such as elevation, exposure, terrain, and trees or other obstructions.

Most small wind turbines employ an "upwind" configuration, meaning the rotor (the blades) points into the wind and spins in front of the tower, and the assembly is oriented by a tail vane that is downwind of the rotor. A notable exception are the Kingspan (formerly Proven) wind products (see Resources), which are downwind and do not have a tail. Downwind turbines may have an unconventional look, but they can perform just as well as their upwind counterparts.

Large, utility-scale generators and residential grid-tied systems produce alternating current (AC; the same current used by household electrical systems), but there are a number of direct current (DC) wind generators that can be used as battery chargers for off-grid applications. In either case the power must be managed and manipulated before it can be used. Residential wind generators range in peak power generation from 50 watts to 10 kilowatts or more and may cost between $3 and $5 per peak "rated" watt to buy.

As you'll see, however, while the peak power rating of a wind generator may help you get your head around the relative size and capacity of the unit, it's not the best measure for comparing different machines because it has little bearing on how much energy will be delivered over time.

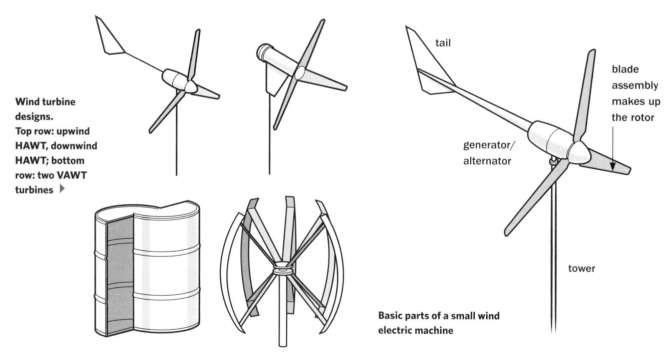

Wind turbine designs. Top row: upwind HAWT, downwind HAWT; bottom row: two VAWT turbines ▶

tail

blade assembly makes up the rotor

generator/ alternator

tower

Basic parts of a small wind electric machine

Wind Test

My friend Hilton Dier III, a renewable-energy teacher and consultant, uses his own *"Sound of Music test"* as a crude indicator of wind potential: "If the breathtaking, panoramic view of the valley below makes you want to sing like Julie Andrews in *The Sound of Music,* the site may have possibilities."

Units of Measurement

Keeping measurement units clear and consistent is extremely important when assessing wind equipment. Mixing meters and miles or pounds and kilograms quickly leads to trouble. Wind power has been long established in the European market, so many wind equipment manufacturers use metric units when describing their machines, while some U.S. manufacturers also give values in imperial units. Just be careful not to mix them up.

ASSESS YOUR SITE

Before buying and installing a wind machine you must assess your site for wind power potential, determine how much energy you hope the wind will produce, and research which models will deliver what you need based on your site and the generator specifications. You will likely find a few options, and you'll need to compare differences in cost, quality, durability, sound level, and ability to produce the most energy at your site.

Also be sure to know what's covered by each manufacturer's warranty, as this is a good indicator of a company's trust in its own products. *Home Power* magazine publishes a wind turbine buyer's guide every year or two that surveys units worth buying in North America (see Resources).

Estimating Energy in the Wind

THE AMOUNT OF energy that can be captured from moving air is a function of wind speed, wind "collector" area (called the **swept area**), and (to a lesser extent) air density. Swept area is the circular area covered by the blades as the rotor spins. It can be expressed in square feet (sq. ft.) or square meters (sq. m).

SWEPT AREA TRUMPS ALL

Longer blades mean a greater swept area, allowing for more wind energy collection. Doubling the length of the blades quadruples the area from which the wind's energy can be captured. Swept area is reported on manufacturer specifications, but you can calculate it yourself using the rotor diameter and applying the formula for the area of a circle:

Swept area = π x radius²

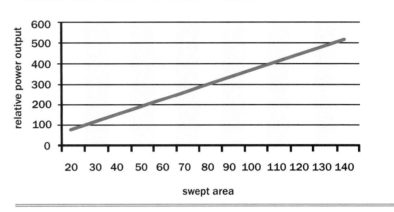

POWER RELATIVE TO SWEPT AREA

For example, if your wind turbine has a rotor diameter of 10 units (the units may be any measure of length), the radius is 5 units. Therefore, the swept area is:

3.14 x 5² = 78.5 square units

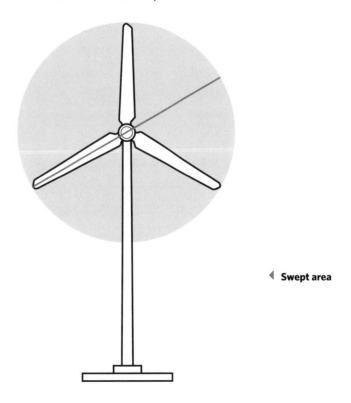

◄ **Swept area**

Swept area increases exponentially with rotor diameter. Compare the above value for a 10-unit rotor with the swept area of a 12-unit-diameter rotor:

$$3.14 \times 6^2 = 113 \text{ square units}$$

As these examples show, a 20-percent increase in rotor diameter results in a swept area increase of 44 percent. In terms of evaluating potential performance of a wind turbine, swept area is the most important factor. Simply put, the greater the swept area, the more energy a wind generator will produce, given the same wind resource. The chart below shows that doubling the swept area doubles the potential power output.

AIR DENSITY

Designated by the Greek letter *rho* (ρ), air density decreases with increasing altitude, temperature, and humidity. It is expressed in pounds per cubic foot (lb/cu. ft.) or kilograms per cubic meter (kg/cu. m). Manufacturers rate the output of their machines at a standard temperature of 59°F and air density at sea level.

Air density has a relatively small effect on available energy when compared to wind speed, but for a basic understanding, there's more energy available in a flow of cold, dry air at low altitude than from warm, humid air high in the mountains. All other things being equal, change in air density is roughly 3 percent for every 1,000 feet in elevation change.

WIND SPEED AND POWER

Wind speed is the air velocity and is expressed in either miles per hour (mph) or meters per second (mps). When working with formulas, the values must be in the same units (metric or imperial) as the density and swept area. 1 mph is equivalent to 0.447 mps; conversely, 1 mps is equivalent to 2.24 mph.

Power increases as the cube of velocity, so doubling the wind speed increases the available energy eightfold. Small changes in wind speed yield dramatic changes in energy produced. The Power Relative to Wind Speed graph (at left) shows how power output increases cubically compared to wind speed. Keep in mind that the actual power output of a turbine varies with the swept area and generator capacity, but the relative comparison between wind velocity and power always follows the same relationship.

Useful Range of Wind Speeds

Many wind turbines do not start turning until the wind speed reaches the point of overcoming the inertia of the system, often between 7 and 10 mph. This is called the **cut-in speed**. There is very little energy in wind speeds below 6 to 8 mph, so it's not worthwhile to try to capture them. On the other end of the spectrum, most turbines will not produce additional power when wind speeds increase beyond 25 or 30 mph, having mechanisms to limit speed and protect themselves in high winds. Be wary of advertising that shows energy performance values in winds below 6 mph — it just isn't going to happen! Likewise, performance claims above about 30 mph indicate that

AIR DENSITY VARIATIONS

This table shows a few examples of air density variations. The difference in air density between 0°F at sea level and 70°F at 2,000 feet is 24 percent, translating into about a 15-percent change in the power of a moving air mass.

Temperature °F	Elevation (feet)	Air Density (pounds per cubic foot)	Air Density (kilograms per cubic meter)
70	0	0.074	1.191
70	2000	0.069	1.107
0	0	0.086	1.379
0	2000	0.080	1.282

POWER RELATIVE TO WIND SPEED

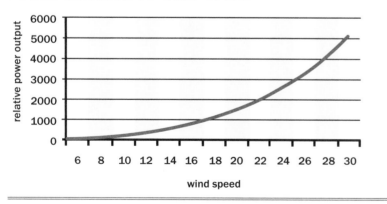

relative power output vs. wind speed

someone wants to sell you a machine that may end up tearing itself apart in high winds.

WIND ENERGY

Energy, expressed in watt-hours for our purposes, is a quantity of power (wattage) produced over time. The energy produced by a wind generator is a function of:

* Average wind speed at the tower location

* Tower height (taller towers provide access to higher and more consistent wind speeds than those available closer to the ground)

* Wind speed frequency distribution, based on data showing how many hours during the year the wind blows within a certain speed range (this range of combined data points is called **bin data**)

* Wind turbine power curve, indicating the power produced by the generator at various wind speeds

Kinetic energy, or power, available in the wind can be can be expressed by the following relationship:

$$\text{Power} = (\text{air density} \div 2) \times \text{swept area} \times (\text{wind speed}^3)$$

POWER CURVE AND ENERGY CURVE

Manufacturers publish **power curves** showing the power output, in watts, at various wind speeds. However, this information will not tell you how much energy (watt-hours) the machine will produce at your site given your wind resource. Some important information a power curve does provide is whether, when, and to what extent the machine will protect itself in high winds. Look at the wind speeds over 25 mph on the curve and notice if they drop off or flatten out. A steep drop-off indicates that the wind may have reached a speed where the turbine's over-speed protection mechanism has activated, and the rotor furls, or turns itself out of the wind, and stops producing. A small drop or flattening of the curve may indicate

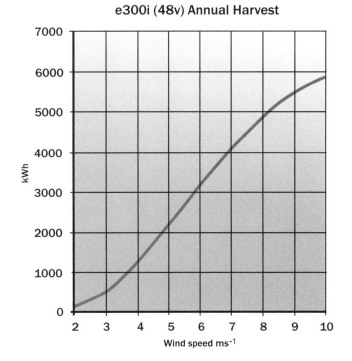

e300i (48v) Annual Harvest

▲ Average energy produced as a function of average annual wind speed for a Kestrel e300i. Image courtesy of Kestrel Renewable Energy, kestrelwind.co.za

at what speed the blades pitch, limiting the rotor to a maximum speed.

More useful than a power curve is the **energy curve**. This is the real nugget of information you want to use when estimating the value of a wind turbine at your site. The energy curve indicates how many kilowatt-hours are produced over a specific period of time given the average wind speed at the turbine's location. The Annual Harvest chart (above) shows how many kilowatt-hours are produced at various average annual wind speeds.

Estimating Wind Speed

WHEN IT COMES to getting power out of the wind, it's all about how fast the air is moving. A wind generator's maximum output power is rated at a specific speed, usually about 25 mph, give or take 5 mph, depending on the manufacturer (until recently, this number was even more arbitrary). This lack of consistency is important to understand when comparing wind machines.

One manufacturer might rate the output of its machines at 24 mph, while another might rate output at 28 mph. The additional power produced at higher wind speeds might make the 28 mph machine look somehow "better" than the other. However, the rated power output is *not* an indication of energy delivered!

Until recently, there was no standard for small wind turbine performance ratings. A new American Wind Energy Association (AWEA; see Resources) performance and safety standard specifies 24.6 mph as the speed at which output power is rated. The Small Wind Certification Council (SWCC; see Resources) is working independently to verify test results and to certify and label wind machines to the AWEA standard so that consumers have a better understanding of performance ratings and comparisons.

RATED ANNUAL ENERGY

Rather than being overly concerned with the maximum power (watts) rating, what you want to know is how much energy (kilowatt-hours) a generator will deliver at *your* site and in *your* wind conditions. For this reason, it's best to compare wind machines on the basis of swept area and the manufacturer's (or SWCC's) test results of energy production at various average wind speeds, rather than the "rated" power output. When using the SWCC ratings to compare machines, the important value is the **rated annual energy**.

Lower average wind speeds produce less energy than higher average speeds. It's worth repeating that the cubic relationship between wind speed and watts means that doubling the wind speed available to your wind turbine increases the available power eightfold. Therefore, cutting the wind speed in half results in one-eighth the performance.

Increasing the turbine's height above ground offers access to faster, more consistent, less turbulent wind streams. In general, a minimum average annual wind speed of 10 to 12 mph is the point at which wind power generation makes economic sense, depending on the cost of electricity in your area. (There are, of course, other values you might want to place on wind power.) This does not mean that the wind blows 10 mph or more all the time; it is an average of the various wind velocities occurring at each hour throughout the year. Sometimes the wind is at 0 mph, while other times it may blow at gale force.

GATHERING WIND SPEED DATA

Before building an expensive addition to the list of things you need to maintain, it's important to understand the wind resource available at your site. This is best done by obtaining long-term (at least one year) wind speed data for the specific site under consideration. There are several ways to get this information, and they vary in ease and accuracy.

Internet tools. The U.S. Department of Energy's *Wind Powering America* website (see Resources) offers a wealth of information about wind power. The National Climatic Data Center (NCDC) maintains long-term records of wind speeds in various locations around the United

Where's the Wind?

When planning a solar electric system, the sunny spot is fairly easy to find, and shadows are obvious. Wind being both invisible and variable, siting is not so easy. It may not be windy where you stand, but 100 feet in the air, above the treetops, the situation changes. Not only is it windier up high, but there is less turbulence and variability created by obstacles that the wind needs to move around.

THE TROUBLE WITH TURBULENCE

Buildings, crops, and trees create drag and turbulence in the wind. Local wind speed depends in part upon the "roughness" of the surface below the wind turbine. For example, smooth water and ice have a very low roughness coefficient, which means little drag is created as the wind blows across these surfaces. Every doubling of turbine height can increase the wind speed up to 7 percent over water, while over woodlands or typical suburban areas, this increase can jump to nearly 20 percent. To see turbulence in action, all you need to do is fly a kite and watch how the tail behaves at different heights. Turbulence is very hard on wind turbines, creating lots of stress on bearings and the tower without much power to show for it.

States In addition to this historical data, there are a few Internet resources that can get you started with a reasonable estimate of wind energy availability in many places around the country. The National Renewable Energy Laboratory (NREL) has produced some excellent wind resource maps that others are using to build estimating tools. One such effort is the Distributed Wind Site Analysis Tool that guides you through selecting your site and a wind machine to predict energy production. See Resources for links to NREL's maps and the wind assessment tool. It's important to note that data for these maps are estimated for a turbine height of 80 meters above the ground. For most residential systems, a more common tower height is about 30 meters, where the wind speed will be somewhat lower.

Subjective assessment observations, such as use of the Beaufort scale, show how the wind affects your surrounding environment, and these observations give you clues about available wind energy. Of course, this works only in the moment, while effective wind power assessment requires long-term observation. Subjective scales are more recreational: they can help you get a feel for what the wind speed might be at any given moment, but they are not at all useful at predicting annual wind generator energy output.

The Griggs-Putnam Index of Deformity illustrates average wind speed based on the wind's flagging effects on trees and shrubs. Conifers are especially susceptible to permanent growth deformations in consistently strong winds. If you have class 3 flagging or better, you probably have a reasonable wind resource. However, if you don't see any flagging, you don't necessarily have a poor wind resource. There is only one way to be sure of your wind resource, and that is to measure and monitor.

Recording anemometer. The most accurate way to measure site-specific wind resources is to use a recording anemometer. This device measures wind speed along with the duration of specific speed ranges (bins) and stores the data electronically, allowing you to quantify the "wind regime" at your site accurately to estimate long-term energy production. To ensure accurate measurements, the anemometer must be located at the same height as the proposed wind generator. Data logging systems, such as the NRG Systems Symphonie data logger, are available from professional wind equipment suppliers. Recreational-level equipment includes products from Horizon Fuel Cells Technologies, Inspeed, and Talco Electronics.

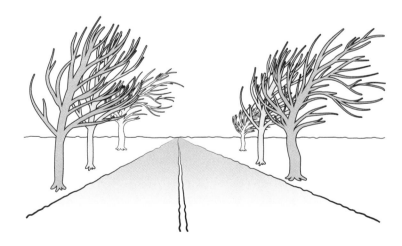

▲ **Windswept trees.** If you're driving down the road and see trees that look like they've had a bad haircut, it could indicate a good site for wind energy.

One drawback to installing wind monitoring equipment is cost. A wind data acquisition system can cost from a few hundred dollars up to $2,000, a price tag approaching that of a small wind machine. Depending on the height and design, the cost of buying and installing a tower can range from a few thousand to tens of thousands of dollars. A few places in the U.S. offer anemometer loan programs. Ask your local Extension service, technical college, or wind equipment installer whether they provide such a service. Another option might be to install a small wind machine on the tower and keep tabs on power produced over the course of a year.

Efficiency and Power

AS WITH ALL things, harnessing energy in the wind is not without its inefficiencies. There are limitations to how much wind the rotor blades can capture, along with losses in wiring and controls.

A fundamental limitation that applies to all wind collection devices, known as the **Betz limit**, states that a wind turbine cannot harness more than 59.3 percent of the energy in the wind as a theoretical maximum. To understand this limitation, imagine the wind blowing at a brick wall. The wind speed abruptly drops to zero, and any energy not absorbed by the wall is diverted as wind in a different direction. A wind turbine's rotor is not a brick wall, but wind velocity on the downwind side of the rotor will be slower than on the upwind side, due to the energy harvested out of the wind stream by the rotor blades.

Capturing 100 percent of the energy in the wind stream would require reducing the wind velocity to zero. Stopping the wind is impossible if you want it to blow through and spin a set of turbine blades. In reality, modern small wind turbines will capture 20 to 40 percent of the maximum energy in the wind. This "capture" efficiency varies with the blade design and wind speed. Of course, the generator, wiring, and electronic controls in the system are not 100 percent efficient, so there are additional efficiency penalties inherent in the rest of the system.

HOW MUCH POWER CAN YOU MAKE?

We have enough information and understanding now to put the power equation together with an example using my own wind turbine, a Kestrel e300i. According to the manufacturer specifications, this turbine will produce 1,000 watts at a wind speed of 10.5 mps. The rotor diameter is 3 meters, resulting in a swept area of 7 square meters. We'll assume a capture efficiency of 25 percent and an electrical efficiency of 85 percent. When metric units are used, this formula result is power expressed in watts. When using imperial units, multiply the answer by 0.134 to obtain watts. Here's the equation written out and then with the metric units (from the example) plugged in:

$$(\text{air density} \div 2) \times \text{swept area} \times (\text{wind speed}^3)$$
$$\times \text{capture efficiency} \times \text{electrical efficiency}$$

$$(1.191 \div 2) \times 7 \text{ sq. m} \times (10.5^3) \times 25\%$$
$$\times 85\% = 1,025 \text{ watts}$$

HOW MUCH ENERGY CAN YOU MAKE?

Instantaneous power generation (watts) is only a small part of the picture. The useful number you need when considering a wind power system is an estimate in kilowatt-hours of **annual energy output** (AEO) that can be produced. Manufacturers have charts that present this data at various wind speeds. There are also several methods for calculating this yourself. All AEO values should be considered estimates. The wind and the operating characteristics of turbines are far too variable for anything better.

One simple AEO calculation comes from Mick Sagrillo, in the American Wind Energy Association's newsletter, and is attributed to Dean Davis of Windward Engineering:

$$A \times V^3 \times 0.085 \times OTE = AEO$$

The formula is the product of the swept area in square feet (A), the average annual wind velocity (V) in mph cubed, the density of air in pounds per cubic foot, and the overall efficiency of the turbine (OTE). OTE accounts for the efficiency of the blades, generator, wiring, and controls — everything between the wind and the electricity that is ultimately used. Most residential turbines operate at somewhere between 15 and 25 percent efficiency in typical wind distribution patterns. Commercial machines used on wind farms typically achieve 35-percent overall efficiency.

Here's an example using my Kestrel turbine, with an *average* wind speed of 6 mph, where OTE is the capture efficiency multiplied by electrical efficiency:

$$75.4 \times 6^3 \times 0.085 \times 21.3\% = 294 \text{ kWh per year}$$

Applying Wind Speed Data

Understanding average wind speed is useful in determining how much energy can be produced at your site and will help you in choosing the most suitable turbine for your wind regime. Higher average wind speeds mean that a higher-power generator would be cost-effective. Lower average speeds call for a smaller (lower power output) generator with a larger swept area to capture more of the available wind. But average

wind speed only tells part of the story. Knowing how wind speed is distributed over time brings you closer to a more realistic estimate of annual energy production. A speed distribution assessment indicates not just the speed of the wind, but how much time the wind spends blowing at that speed over a period of time.

Wind speed distribution assessment. This assessment methodology is often built into more expensive recording anemometer software. Despite the expense, it is the most useful way to quantify the available wind resource and the energy produced by a specific combination of wind regime and wind machine. This assessment is especially important where costs and risks are high and there is a need for detailed technical and economical evaluation of the wind resource.

The occurrence and duration of various wind speeds can be assembled graphically as a **wind speed distribution pattern**. The Weibull distribution curve describes this annual variation in wind resource. It indicates the number of hours each year during which you can expect the wind to blow at a specific speed (or range of speeds) and distributes this information in data "bins." For example, if the wind blows at 10 mph at your site for 600 hours (out of the 8,760 total hours in a year), you can use the manufacturer's output power data to calculate how much energy (in

EXAMPLE OF WEIBULL WIND SPEED DISTRIBUTION

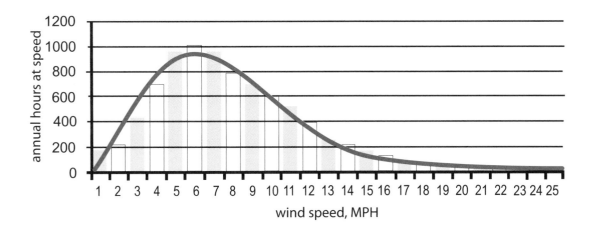

◀ **Weibull Wind Speed Distribution graph. The number of hours in each bin add up to the total number of hours in a year.**

Power Density

The Kestrel's **cut-in speed** (the speed at which the rotor starts to spin and produce power) is about 6 mph. If my average speed is only 6 mph, then how can there be any useful power in the wind at all? The answer lies in the wind speed distribution, which can be used to quantify the wind's **power density**. Power density is a measure of how much wind energy passes through the swept area. It is the average of all wind speeds weighted by the duration of each speed "bin" (as per the Weibull distribution), which is different from the simple average wind speed.

kilowatt-hours) will be produced in that 10 mph wind during those 600 hours.

A variation on the Weibull distribution curve is known as the Rayleigh distribution. This is simply a mathematically derived "shape" of the Weibull curve that is commonly used to estimate wind speeds and thus energy production. The shape changes based on the average wind speed and variability. As average speed increases, the bell on the curve moves to the right, indicating more overall wind energy.

THE PROOF IS IN THE WIND

As you can see, there are many ways to estimate wind power and energy that give you an idea of what you can expect from the wind resource at your site. While these offer some level of comparison, such methodologies are very rough estimates, and manufacturers' engineering data are not the same as real-world results. Be aware that, as with all things, your "actual mileage" will vary. Despite manufacturer's claims, tests, and charts, there is no way to predict exactly how much energy your wind generator will actually deliver given your particular site characteristics. If you have a high-cost project in a questionable location, it's especially important to remember that nothing will take the place of real-world, long-term monitoring.

COST-EFFECTIVENESS AND HYBRID ENERGY SYSTEMS

Your goal is to maximize renewable kilowatt-hours, so any comparing of cost must include the total value of the energy produced over the lifetime of the machine, compared to the money invested over the lifetime, including maintenance costs. If I'm only producing 294 kWh per year, and I'm paying $0.15 per kilowatt-hour for electricity from the power company, I'm earning only $44 per year. That's a 227-year payback on a $10,000 system (not including maintenance costs or any monetary incentives). Worse, it can't begin to cover my annual electrical energy needs. However, if my average wind speeds were doubled, I'd generate eight times more energy, and the payback falls to under 30 years.

In reality, I harvest about 1,000 kWh per year from the wind. This is not a lot of energy, but there are five factors that change the payback equation for me:

1. I am off-grid with no utility power available at all.

2. I have a solar electric power system.

3. At my location, there is more sun in the summer and more wind in the winter.

4. If I don't get the power I need from nature, I need to run my generator, at a cost of about $0.75 per kWh.

5. I have a strong desire to reduce my reliance on fossil fuels, and I understand that comes at a cost, which I am willing to pay.

In terms of economics, small wind does not make sense at my location. But in terms of practicality and personal goals, it's a good option and the perfect seasonal complement to solar electric. The graph on page 136 shows the relative available solar and wind resources and how they complement each other seasonally.

Wind Machines and Controls

WIND POWER SYSTEMS require a good deal of management and control, and most manufacturers provide the required electronic controls as a package with their turbines. All wind generators — whether permanent-magnet generators or high-frequency AC alternators — produce AC power.

Grid-tied generators must be synchronized to match the voltage, frequency, and power quality of utility power. This task is handled by the inverter. For battery charging, AC power must be rectified (at the controller or in the generator) to the DC power required by batteries.

When the batteries are full or the grid is down, any excess power generated needs a place to go. If the wind blows when there is no load on the generator, the fast-spinning rotor can be damaged itself or cause damage to generator bearings. For most wind turbines on the market today, controlling speed in these conditions requires a diversion, or dump load, to absorb the energy. (Read more about controllers and dump loads in chapter 10.)

Wind generators typically have rotors with two or three blades made from fiberglass, wood, carbon fiber, or composite material. Fewer blades yields higher aerodynamic efficiency, but more blades offers better rotor balance and may start spinning at lower wind speeds. A well-balanced rotor reduces mechanical strain on bearings for greater longevity. Modern turbines are commonly available with two or three blades, offering a good compromise between efficiency and balance.

SYSTEM COST

When it comes to wind energy, there's no such thing as cheap. Your goal is to achieve cost-effective energy generation over the lifetime of the wind machine and system components. In terms of overall cost of your wind project, the price of the turbine will probably be small compared to the complete system, which includes the cost of the tower, cabling, labor, and site work (crane, excavator, and so forth) that are required. If you buy an inexpensive wind machine, you will likely spend more on maintenance over the long run, unless it is installed at a site with only light-to-moderate winds. More costly models tend to be heavier and more durable. They also spin more slowly, which reduces noise.

TURBINE NOISE

Small wind machines may be relatively quiet, or they can be quite noisy, depending on their design. Often the sound they produce is just a little bit louder than the surrounding wind noise. However, at certain speeds or in certain kinds of winds, a whirring or whooshing can be heard, even to the point of roaring at times. When the wind speed is high and the turbine is furling out of the wind, the noise can be quite loud, depending upon the machine and the mechanism it uses to protect itself.

When shopping for a wind turbine, it's a good idea to listen to the models in operation, preferably in both low winds (when noise often is more noticeable), and high winds (when speed governing can be loud), to be sure it won't bother you or your neighbors. Like a crying baby, if the wind generator is your own, it doesn't sound so bothersome — but if it's not yours. . . . Taller towers improve performance and move the noise farther away but may create a visual encumbrance for

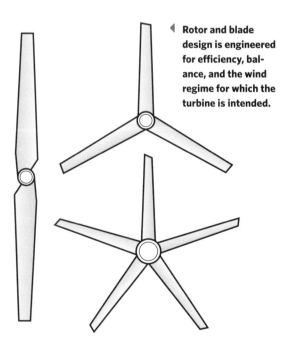

◄ Rotor and blade design is engineered for efficiency, balance, and the wind regime for which the turbine is intended.

the neighbors. Tower height will likely be subject to local zoning regulations, so be sure to look into this issue first.

SPEED CONTROLS

High wind speeds can damage a generator by causing it to spin too fast, so turbines include speed-governing mechanisms to deal with excessive speed. These are automatically engaged, mechanically activated controls that change the pitch of the blades as wind speed increases, or they furl the entire rotor out of the wind (vertically or horizontally) so that the blades do not accept all of the wind's force. Pitch adjustment may allow the turbine to continue to produce maximum power in high winds, while rotor furling may cause the rotor speed to slow, reducing output.

Sometimes it's necessary to stop the rotor from spinning, to facilitate maintenance or simply to turn off the wind generator in the event of an approaching storm. This can be done mechanically, either with a rotor braking disc or a manual furling mechanism that pulls the tail parallel to the blades. Braking may also be accomplished with an electric, or "dynamic," brake that shorts the electrical output wires together, putting the machine under maximum load. Dynamic braking is effective at slowing the rotor, but may not stop it under heavy winds. Such electronic braking may be handled through the charge controller, and is as simple as throwing a switch.

Towers

WIND TURBINES TYPICALLY are mounted atop tall towers so that they can access large quantities of turbulence-free moving air. The towers must be sturdy enough to hold their own weight plus the deadweight of the machine mounted on top, and they must be stable enough

Area of wind turbulence around a building of height H and the relative heights and distances required to avoid turbulent airflow. ▼

to withstand the lateral forces delivered to both the tower and turbine during the highest wind speed that may be experienced at the site.

Turbine manufacturers offer data on the maximum lateral thrust developed by their machines. Commercial towers and turbines often are able to withstand hurricane-force winds of 140 mph or more, but only when proper design and installation techniques are followed.

There are several very good reasons not to mount a wind machine on a rooftop or attach it in any way to a building:

* Location with very low wind resource
* Wind turbulence created by objects near the ground
* Thrust forces acting upon — and developed by — generators that will be transferred to the building
* Vibration created by a spinning turbine

TOWER POWER

You wouldn't put a solar panel in the shade and expect it to generate much power; don't make the equivalent mistake with a poorly sited wind generator. The absolute minimum recommended tower height is 30 feet above any ground feature within 500 feet. If the turbine is located in a wooded area, it must be at least 30 feet above the tops of the trees. In addition to height, there must be no obstructions within a minimum 500-foot radius around the turbine. These are minimums, but winds are influenced by the ground surface at heights up to 300 feet. More tower means more power, because wind movement is faster, more consistent, and less turbulent farther above ground-level influences.

TOWER DESIGN

Towers can be made from a single, heavy-gauge steel pipe, or from steel lattice interlaced between three or four legs. They can be freestanding or guyed (secured by guylines, or guy cables) and are usually assembled on the ground and then stood up using either a crane or a tilt-up kit. In order for the turbine to **yaw** (rotate horizontally to meet the wind), the tower must be reasonably plumb. A guyed tower can be adjusted somewhat for plumb using the guy cables to take minor twists out of the tower. A freestanding tower must have a level foundation and plumb base section, which might require the use of leveling nuts or shims between the foundation and base.

Guyed Towers

Guyed towers can be two types: **fixed** or **tilt-up**. Tilt-up towers and kits are made by the likes of Bergey and NRG Systems (see Resources) and can be used with either pole or lattice towers. Tilt-up towers require four sets of guy cables, while fixed towers generally use three sets, with each set consisting of two or more guys, depending on the tower height and thrust loads. Each set of guy cables will extend out from the tower for a distance of between 50 and 80 percent of the tower height and are attached to a common anchor.

As you can imagine, guyed towers require a fairly large footprint on the ground, with lanes cleared for each guy cable run. Lattice towers can be climbed, and therefore do not necessarily need to be tilt-up. A tubular pole tower is often tilt-up, but it is possible to weld steps onto the pole so that it can be climbed, giving it the flexibility to be a fixed guyed, tilt-up, or even a freestanding tower.

Guyed towers rely upon proper wire tension to stay vertical while maintaining some flexibility to react to the wind. Cable tension is often adjusted by way of a turnbuckle on each cable, located near the anchor. The type of cable and required tension are specified by tower manufacturers. For example, Rohn specifies that stranded **EHS cable** (extra-high-strength; the only type you should consider for guys) be used for their lattice towers and tensioned to 10 percent of its breaking strength (see EHS Cable Breaking Strength at right).

The cable tension will increase under wind load. Too little tension means a wobbly, unstable tower, while too much tension can put excessive compression force on the tower, causing the legs to fail. The combination of guy cables on your tower must be able to withstand the maximum possible forces created, with an additional safety margin.

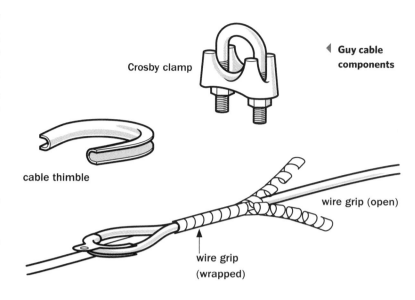

Crosby clamp

cable thimble

◀ **Guy cable components**

wire grip (open)

wire grip (wrapped)

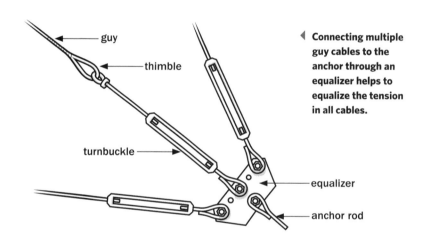

guy

thimble

turnbuckle

equalizer

anchor rod

◀ **Connecting multiple guy cables to the anchor through an equalizer helps to equalize the tension in all cables.**

EHS CABLE BREAKING STRENGTH

Diameter	Pounds
3/16"	3,990
1/4"	6,650
5/16"	11,200
3/8"	15,400
7/16"	20,800
1/2"	26,900

Working with guy cables requires some specialized tools and equipment. Crosby (saddle) clamps or wire grips can be used to secure the cable as it wraps around the tower and anchor attachment points. Thimbles are important to prevent the guy cables from crimping under stress as they bend around connecting points. An equalizer plate can be used to attach multiple guy cables to a single anchor.

TOWER INSTALLATION

Tilt-up towers are assembled on the ground, and a **gin pole** (a pole with a pulley on one end), attached to the tower at a 90-degree angle, is used as a lever to raise the tower with a winch or a griphoist. A griphoist is a hand-operated hoist that is anchored on one end and pulls the gin pole cable through it as you crank the ratcheting handle. The advantage to tilt-up towers is that they can be lowered for maintenance or to protect the turbine from a potentially damaging storm. Tilt-up towers do not require climbing and are quite popular for this reason.

Lattice towers, such as those made by Rohn, are available from wind dealers or equipment suppliers, such as Tessco, Sabre Industries, and Cable and Wire Shop (see Resources for websites). These typically are supplied in triangular, 3-tube (or -rod), 10-foot sections that can be assembled on the ground. They can be lifted by crane, raised with a tilt-up kit, or hoisted up one section at a time using a vertical gin pole, pulleys, and a tag line to steady the section as it is lifted.

Gin Poles

A gin pole for lifting sections of a lattice tower can be made using suitably engineered brackets, pulleys, and a sturdy pipe. The brackets attach to both the tower and the pipe, and the pipe is moved up with the addition of each new section. During this slow process (but less expensive than a crane, if you do it yourself), a climber wrangles each tower section into place while the ground crew pulls the section up with a rope. This rope is attached to the tower section to be moved, runs up to the gin pole pulley, then down to a pulley at

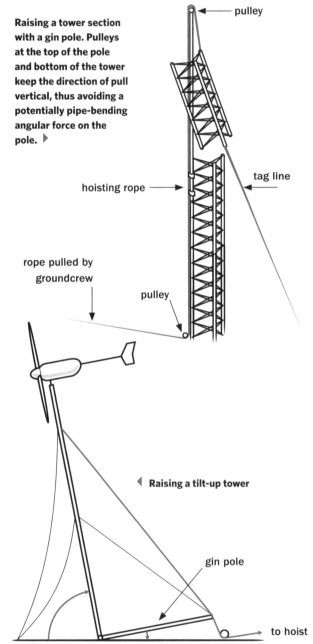

Raising a tower section with a gin pole. Pulleys at the top of the pole and bottom of the tower keep the direction of pull vertical, thus avoiding a potentially pipe-bending angular force on the pole. ▶

pulley

hoisting rope

tag line

rope pulled by groundcrew

pulley

◀ Raising a tilt-up tower

gin pole

to hoist

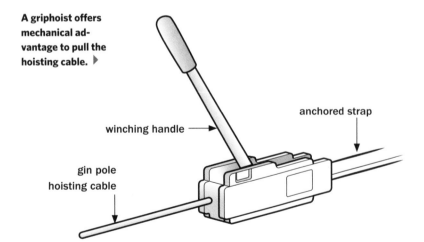

A griphoist offers mechanical advantage to pull the hoisting cable. ▶

winching handle

anchored strap

gin pole hoisting cable

GIN POLE

Used in: All tilt-up and some guyed and freestanding wind tower installations

A.K.A.: Lever arm, lifting pole, falling derrick, davit

What it is: A lever to raise a tilt-up tower or a temporary crane for a non-tilt-up tower

What it ain't: A liquor survey

The phrase "gin pole" is used to identify two very different structures associated with wind generator towers.

In reference to **_tilt-up towers_**, the phrase refers to the lever arm that is used to lift the tower off the ground. Usually it is a steel pipe of the same diameter as the tubular tower pipe, and can be as long as the guy wire radius.

A gin pole makes it easier to raise the tower. Try tilting up a pipe or pole by pulling along the length of it, and you'll find that something may break before anything lifts. Adding a lever at 90 degrees makes it easy to lift the pipe.

In reference to **_non-tilt-up towers_**, a gin pole is a temporary "crane" that sticks up above the tower. It allows you to lift additional tower sections and the wind generator without hiring a $200-per-hour crane.

Generally, two brackets are attached to a tower leg with bolts, providing a sleeve for the gin pole, which is pulled up through the brackets. A davit or block is added on the top, and a lifting line is threaded through before the pole top is raised out of reach.

The gin pole is then bolted securely in place before any lifting is done. After each section is in place, the gin pole and brackets are moved up to the next section. Temporary guy ropes are necessary to keep the tower stable.

— Ian Woofenden
Reprinted with permission.
© 2012 Home Power Inc.,
www.homepower.com

the base of the tower, where the rope changes direction for the crew to pull on.

Freestanding towers can be pole or lattice type and must be anchored in a concrete pier, or foundation, engineered to withstand the design loads of the tower, turbine, and prevailing winds. These are more costly than guyed towers because of the amount of steel in the tower and concrete in the foundation, as well as greater site preparation costs. However, freestanding towers require a much smaller footprint because there are no lanes to clear for guy wires.

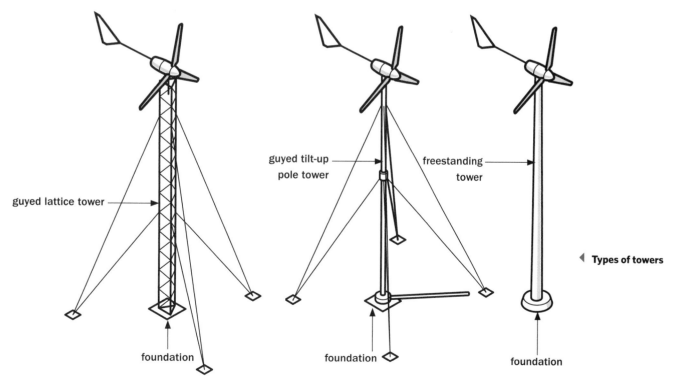

guyed lattice tower

foundation

guyed tilt-up pole tower

freestanding tower

foundation

foundation

◀ **Types of towers**

Foundations and Anchoring

A WIND TOWER must have a solid, stable foundation and must be well anchored to withstand the lateral, compressive, and uplift forces of the wind. Freestanding towers require a single, massive, well-engineered concrete foundation that also acts as an anchor. Guyed towers need a foundation under the tower to resist compressive forces and an anchor for each set of guy wires.

Anchors and foundations must extend below the frost line (the depth to which the ground freezes in winter) to prevent heaving or shifting. A few other critical design factors include:

- Weight and thrust of the turbine
- Wind load on the tower at the design wind speed
- Weight and strength of the tower
- Soil into which the anchor is placed
- Anchor angle (its position in the ground)

Turbine and tower load specifications and anchoring requirements are available from manufacturers and must be consulted for proper anchoring system design.

TYPES OF ANCHORS

Several types of anchors are available to suit various soil types and tower requirements. Chance, a subsidiary of Hubbell Power Systems (see Resources), manufactures a variety of suitable anchors and is a good online resource for information on anchoring. Soil types are classified according to the table below.

Concrete anchors buried underground can provide the weight needed to counteract horizontal and uplift forces, but only if the weight of the anchor is sufficient and the soil is cohesive enough to hold the anchor in place. One cubic yard (volume of a cube with 3-foot sides) of concrete weighs about 4,000 pounds.

Screw-in anchors are rated for use in soil classes 3 through 7. They are available with varying diameters and numbers of helixes, all

SOIL CLASSIFICATIONS

Class	Common Soil-Type Description	Geological Soil Classification
0	Sound hard rock, unweathered (bedrock)	Granite; basalt; massive limestone
1	Very dense and/or cemented sands; coarse gravel and cobbles	Caliche (nitrate-bearing gravel/rock)
2	Dense fine sands; very hard silts and clays (may be preloaded)	Basal till; boulder clay; caliche; weathered laminated rock
3	Dense sands and gravel; hard silts and clays	Glacial till; weathered shales, schist, gneiss, and siltstone
4	Medium-dense sand and gravel; very stiff to hard silts and clays	Glacial till; hardpan; marls
5	Medium-dense coarse sands and sandy gravels; stiff to very stiff silts and clays	Saprolites, residual soils
6	Loose to medium-dense fine to coarse sands to stiff clays and silts	Dense hydraulic fill; compacted fill; residual soils
7(a)	Loose fine sands; alluvium; loess; medium-stiff and varied clays; fill	Flood plain soils; lake clays; adobe; gumbo, fill
8(a)	Peat, organic silts; inundated silts, fly ash, very loose sands, very soft to soft clays	Miscellaneous fill, swamp marsh

(a) Note: It is advisable to install anchors deep enough (using extensions) to penetrate into Class 5 or 6 soil when the soil layer above it is Class 7 or 8

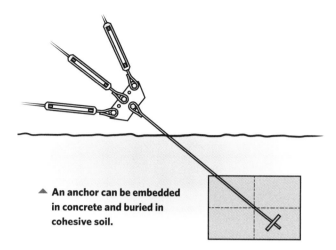

▲ An anchor can be embedded in concrete and buried in cohesive soil.

contributing to holding strength. Some can be screwed into the soil by hand with a lever bar, while harder soils or anchors with more and larger helixes may require the torque of a hand- or machine-operated hydraulic drilling tool. Screw-in anchors are most commonly used for temporary anemometer towers and other light-duty applications.

Expanding anchors allow you to auger a hole into the ground, then place the anchor into the

helix

◀ **Screw-in anchors with various numbers and sizes of helixes are chosen according to anchoring strength requirements and soil type.**

hole. Expanding leaves at the end of the anchor are unfurled, wedging themselves into the surrounding soil, and the anchor hole is then backfilled. These are advantageous in dry, solid soils because the anchor leaves expand into undisturbed (and therefore stronger) soil, but strength of the anchor is also dependent upon the quality of the backfill tamping.

Rock anchors are used when the soil is not deep enough for any other anchor to make sense. A hole is drilled into bedrock using a pneumatic

anchor rod

backfill

expanding anchor closed

expanding anchor open

◀ **With expanding anchors, a hole is drilled into the soil, the closed anchor and rod are dropped into the hole, anchoring helix blades are unfurled, and the hole is backfilled.**

hammer drill. The hole should be at least 3 feet deep so that enough rock is above the anchor. If the anchor is too close to the surface, a piece of rock could crack under pressure or if water enters the hole and freezes. The anchor is inserted into the hole and tightened so that the split wedge is forced against the wall of the hole. As more tension is put on this anchor, the split bolt will want to expand, wedging it more tightly in the hole. The hole should be sealed with brick mortar to keep out water. Epoxy is not recommended because it can prevent the anchor from expanding as it pulls under tension.

▼ **A rock anchor slides into the drilled hole with its split wedge closed, then is tightened to force the split wedge against the rock inside the hole.**

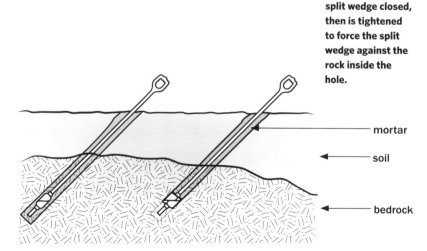

mortar

soil

bedrock

Maintenance

MODERN WIND MACHINES require regular service — they don't sit idly on your roof like solar panels with no moving parts. And while maintaining a wind generator is nothing like changing the oil on your car (which almost anyone can do), maintenance, like oil changes, must be done to get the most out of your investment and to avoid catastrophic failure. For this reason, you need a plan and a tower design that allow you to access the generator for required service.

Annual inspection of tilt-up towers requires lowering the tower, and any maintenance can be performed on the ground. For fixed towers, access requires climbing gear and the skills, strength, and nerve to work 100 feet or more in the air. If you drop a tool from up there, it can be at best a 20-minute round trip to retrieve it and at worst a deadly mistake for someone on the ground. Tether your tools!

MAINTENANCE MATTERS

The turbine manufacturer will have a maintenance schedule that you should follow closely to get the most out of your machine for a long time. The following are some general maintenance concerns to keep in mind.

Environmental Exposure

Consider the elements and their effects on the wind turbine and its tower. Depending on your climate, both of these critical components may need to withstand turbulent thunderstorms, lightning, intense sun, subzero temperatures, pouring rain, sleet, snow, ice, gale-force winds, blowing sand, and dust. Despite this constant abuse, maintenance of modern wind machines is generally minimal (but, as mentioned, mandatory). Turbines that operate in higher average wind speed regimes require more frequent maintenance.

Turbine Sound

Make a note — or even a recording — of the sounds your turbine makes when it's new. As the machine ages, the sounds may change as bearings and blades wear. Notice the vibrations in the tower at various wind speeds. Vibration will worsen if the blades become unbalanced.

Balance

Pay attention to how the turbine behaves in various wind conditions. If it starts to wag its tail (an unbalanced condition), an out-of-plumb tower or excessive turbulence might be taking a toll on yaw or rotor bearings.

Blade Inspection

Get out your binoculars or climb up to the turbine to check the blades for stress cracks or other damage. Also make sure they spin in exactly the same plane. Examine the leading edge of each blade for signs of wear or damage. If water gets into the blades (especially if they are wood), balance will be affected. Be sure that furling mechanisms are still working in high winds; you often can detect a problem by the sound the turbine makes in high winds.

Tower and Hardware

Inspect the tower twice a year to make sure all nuts and bolts are tight and anything attached to it is properly secured. New guy cables will stretch and should be checked every few months for the first year after installation, and once a year after that. If a guy wire becomes loose, the tower can rock back and forth in higher winds, stretching the other cables and eventually bringing the tower down.

A tension gauge, such as one made by Loos & Company (see Resources), makes checking guy wire tension fast and easy. Be aware that a guy cable that continues to loosen may indicate a failing anchor.

Thinking Ahead

Unless you have a tilt-up tower, maintenance will include climbing, so consider whom you might hire to climb the tower when you have arthritic knees in 10 years. Okay, hopefully you won't have arthritic knees in 10 years, but you should have a backup plan that relies on something more concrete than your friends' goodwill (pizza and beer go only so far).

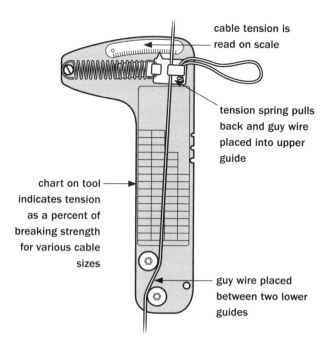

cable tension is read on scale

tension spring pulls back and guy wire placed into upper guide

chart on tool indicates tension as a percent of breaking strength for various cable sizes

guy wire placed between two lower guides

▲ **Loose cable tension gauge**

Safety

WIND GENERATOR TOWERS are tall, and wind machines are heavy. Wind can be unpredictable, trees can fall, and hardware can fail, while gravity *never* fails.

Working on and around wind energy systems requires a hefty "tool kit":

• Well-researched information

• Great attention to detail

• An experienced crew leader

• Mechanical and electrical skills

• Steady nerves

• Quality tools

• Professional climbing gear (if you're not installing a tilt-up tower)

A tower that is properly designed, anchored, and maintained should not fall. And if you've cleared around the tower and guy cable runs, no trees should fall on them. Towers are everywhere in our society today; they should not be feared, but they must be well engineered to withstand the forces of nature and machines.

BASIC SAFETY PRECAUTIONS

Protecting yourself when working on towers is just as important as proper system design and maintenance. Here are the essentials:

Tie off the turbine. Be sure to put the brake on the rotor to prevent it from spinning while you're on the tower. Even with the brake on, assume the turbine can spin or yaw unexpectedly until you have it safely and securely tied off. View the entire turbine as a heavy, two-dimensional spinning weapon ready to knock you off balance when you least expect it. Tie off both the blades and the turbine assembly to stabilize them before doing any work.

Watch the weather. Tower work is safest during calm, pleasant weather. Don't work on the tower when it's too windy, or when it's snowing or raining.

Dress properly. Wear boots, gloves, and a hard hat with a chin strap. Remove all jewelry, tuck long hair inside clothing, and don't wear loose clothing. Wear a full-body harness attached to the tower at at least two different points.

Know your gear. Learn about tower climbing gear and make use of snap hooks, harness shock absorbers, slings, and lanyards. Assume that at some point in your climbing history you will slip and fall; never be at risk of hitting the ground when you do.

Most professional tower workers won't work on a tower unless it has a tower-integrated fall restraint system consisting of a vertical lifeline, running the height of the tower and to which a harness can be hooked (in addition to hooking it to the tower). A locking slider glides up the lifeline, but if you slip the slider catches the lifeline and stops your fall.

Don't work alone. Communicate well with those working around you, using two-way radios for talking with the ground crew. To protect others from dropped tools or parts, make sure no one is working on the ground or tower below you.

Check yourself. Do not climb when you're tired. Never assume you know everything and don't need to double-check.

Plan for everything. Ask yourself how many different ways things can fail and plan for all of them. It is, after all, your life that is at stake.

Wiring and Grounding

WIND ELECTRICAL SYSTEMS now have their own section in the National Electrical Code (NEC). Article 694 covers all aspects related to wiring, fuses, disconnects, grounding, battery charging, utility interconnection, and signage. Refer to the NEC for specific electrical requirements. See chapter 10 for more information on the NEC.

SUPPORTING AND CONNECTING CABLES

Electrical cables leading from the wind generator down the tower must be firmly supported so that they don't pull on the wires coming from the generator. The heavy-gauge copper cables can be quite weighty, requiring a strain relief at the top of the tower to support them. One example of a suitable strain relief is a wire net. Remember that a wind generator tower — and anything attached to it — will vibrate, so be sure to use crimp connectors for electrical connections and cable strain reliefs that will not loosen with vibration over time. Cover all connectors with weathertight, double- or triple-wall heat-shrink tubing with adhesive inside. See chapter 10 for more information about wire sizing.

GROUNDING REQUIREMENTS

Lightning protection is imperative for a wind generator tower. While no grounding system can prevent damage from a direct lightning strike, there are many ways to reduce the static charge surrounding your tower, making it much less attractive to lightning.

Grounding should follow NEC requirements. At a minimum, the tower must be grounded by attaching a wire between the tower and at least one ground rod driven 8 feet into the ground (or other suitable buried grounding array). Guyed towers should have one ground rod at each anchor, and each guy cable must be connected to that rod. Ideally, all ground rods are then tied together with buried copper wire.

Off-grid wind system. A wind generator system also includes many of the same balance-of-system and control elements as a solar-electric setup, such as disconnects, charge controller, inverter, system monitor, and possibly a backup generator. ▶

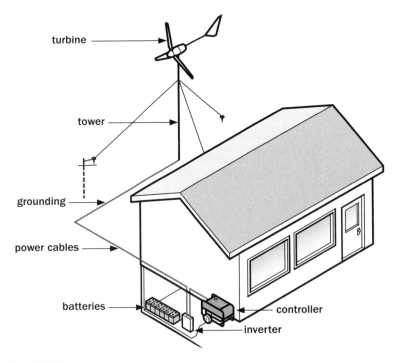

turbine

tower

grounding

power cables

batteries

controller

inverter

Warning!

Wind generators create high-voltage electricity. It does not take very much electrical current to kill a human. Be sure the generator is disconnected from the wiring before performing any work on the turbine or tower.

Wind Wisdom from an Expert

MY LOVE AFFAIR with wind electricity formally began in 1984, when I installed my first machine. An 8-foot-diameter Wincharger on a 112-foot homebuilt tower at my off grid home got me hooked on making kilowatt-hours with wind.

IAN WOOFENDEN

has been working with wind for more than 30 years. An independent consultant and recognized expert in the industry, Ian is also an instructor for Solar Energy International, senior editor for *Home Power* magazine, and the author of the book *Wind Power for Dummies*, along with numerous wind power er articles and equipment reviews (see Resources for Ian's website).

Since then, I have owned and operated multiple machines, learning by successes and failures what works and what doesn't. I had the benefit of informally testing a variety of machines and running up to three machines at once, because of my position in the wind industry, as well as my compulsion to experiment and learn. I've been involved in dozens of installations for others, working with wind contractors and as a workshop coordinator and instructor.

I love the feeling of harvesting free natural forces. When people complain about the weather, I suggest that they harvest it! When it's rainy, my rainwater tanks are filling up. When it's sunny, the solar-electric modules are charging my batteries, and the solar hot water collectors are heating up the tank of water. And when the wind blows, my wind generators charge the batteries, too, staving off fossil-fueled generator use.

Living with wind electricity takes awareness and persistence. It is not for the faint of heart. All wind generators need regular maintenance, which means lowering or climbing a tall tower. Almost all wind generators have "issues" at some point in their lives — it's rare to find a residential wind project that has been trouble-free for decades. If you go into it thinking it will be simple and cheap, you are very likely to be disappointed. If you go into it planning for maintenance and bracing for trouble, you'll

be able to maintain your excitement and not get discouraged, while keeping your machine running for years.

Living off-grid changes the way you think about energy use (unless you have an unlimited bank account). It helps you focus on ultra energy efficiency and on shifting your loads to the times when you have energy coming in. Adding wind electricity to a solar-electric system balances out the low times and also brings you seasonal surpluses that are fun to enjoy.

In my part of the planet, winds come in winter, so I can use the surplus energy for cutting more firewood with an electric chainsaw, moving heat around the house with fans, keeping the fridge cool even though the wood-heated house is warmer than in summer, doing the laundry on windy days, and running plenty of lights during those long winter nights and rainy days. When the calm, dark times come, I have to conserve energy by changing my lifestyle, or turn on "the noise" — the propane-powered generator — to charge the batteries. Changing usage and behavior with the weather is an acquired taste, but one that I enjoy.

How can you figure out if wind energy is right for your site and for you? Those are two different questions with two sets of answers. A good-to-excellent wind energy site will have these characteristics:

I love the feeling of harvesting free natural forces. When people complain about the weather, I suggest that they harvest it!

Wind Wisdom from an Expert

Living with wind electricity takes awareness and persistence. It is not for the faint of heart. All wind generators need regular maintenance, which means lowering or climbing a tall tower.

- Adequate space and permission to put a tall tower well above all obstructions for years

- A reasonable (8- to 14-mph) measured or accurately predicted average wind speed

- Reasonable wire-run distance from the tower to the utility service or battery bank

- Neighbors who are at least happy, if not excited, to have a wind generator in the 'hood

The perfect wind-electric system owner has these characteristics:

- Adequate education before the fact, to understand siting, basic system design, and performance estimating

- Adequate budget to install a robust system on a very tall tower

- Commitment to lower or climb the tower once a year at a minimum (or hire someone to do it) and perform whatever maintenance is needed

- Awareness of the system and a determination to address any changes you notice before they become catastrophic problems

- Patience to deal with the almost inevitable problems that come with being a wind-electric system owner

From my perspective as a 30+-year user of wind-electric systems, student, teacher, climber, and journalist in the residential wind industry, there are some basic lessons that could save you a lot of time, money, and anguish:

- Try to capture the wind only if you can put the wind generator on a very tall tower, well above all obstructions in the area. This is where the good fuel is; trying to capture anything less than the good stuff will only disappoint you.

- Put up a large enough rotor (blades and hub) to generate a significant portion of the energy you need. For most North American homes, this means a rotor diameter of 12 to 50 feet. Small rotors capture only a small amount of energy.

- Ignore "wattage" ratings of wind turbines, and get accurate estimates of how many kilowatt-hours (kWh) the machine you are looking at generates in *your* average wind speed. Get a second opinion.

- Buy turbines from manufacturers that have been around for a while, have a good reputation, offer a solid warranty, and have a track record of good performance and service. Run, don't walk, away from "new," "improved," "exciting," "breakthrough" products — these companies will likely have vanished or be in bankruptcy before too long. If no one else is doing it, ask yourself why. Or, to quote wind geek Dan Bartmann, "Before you 'think outside the box,' find out what's in the box, and why."

- Install your system carefully and don't cut corners. "Cheap" now will turn into expensive later. Build robust infrastructure — wind is a powerful resource that doesn't fool around.

- Maintain your system very regularly, once or twice a year.

Have fun with your wind-electric system! Life is short.

Living off-grid changes the way you think about energy use (unless you have an unlimited bank account). It helps you focus on ultra energy efficiency, and on shifting your loads to the times when you have energy coming in.

Hydro Electric Generation

I **F YOU LIVE** near falling water, a small-scale hydro system may work well for home electricity production. With *micro-hydro* systems, power generation can range from 100 to 2,000 watts of electricity production from streams or small rivers, with minimal damming and water diversion.

A dam serves as a way to create both a reservoir and a flow diversion. The diversion sends water to an inlet pipe at the top of a vertical drop, which is required to develop the necessary pressure to spin a water-powered generator. The reservoir also helps provide a relatively constant flow of clean water with no air pockets. You will need such a reservoir, but substantial dam construction is beyond the scope of a typical home power project and therefore is not covered here.

Most rivers and streams flow all day long and year-round, making hydropower more consistent — and often more cost-effective — than solar or wind energy. Of course, seasonal flow variations, and freezing or drying of the water, may be real issues in your situation. This chapter will help you to evaluate and understand the potential for hydroelectric power at your site.

Home Hydro

IT IS SOMEWHAT rare to find a location that meets all of the practical requirements to make harnessing water power worthwhile. You need access to enough moving and falling water that is reasonably close to a dwelling that can use the power (or to grid power for grid intertie), and you need to secure the required local, state, and federal permits and rights to access the water and harness its power. If such a site is yours, you have struck your own little oil well, and opportunity awaits!

Large-scale, alternating current (AC) hydro systems are generally not practical for homeowners due to cost, complexity, regulations, and the volume of water needed to make the investment worthwhile. Micro-hydro systems don't require much alteration of the stream and so have a minimal impact on waterways and ecosystems.

The power generated by a micro-hydro system may be AC or direct current (DC) but will often be converted to DC so the energy can be stored in batteries. In fact, most micro-hydropower systems can be considered battery chargers. One significant advantage to energy storage is that the generator can be much smaller than your peak power demand, relying on batteries and a power inverter to manage and deliver any power surges required by the loads that exceed the output capability of the generator.

The iconic image of an old-style water wheel is not the modern way of generating electricity that we'll explore in this chapter. Water wheels collected water's energy as it flowed through a river or over a relatively short dam at high flow rates, low pressures, and low speeds. The energy harvested was used for mechanical power, such as grinding grain. Today's modern hydro-power turbines require water to be delivered at high pressure through a nozzle that focuses a jet of water onto a fast spinning wheel, or **runner**, which in turn spins an electric generator.

The most common way to capture the water energy in micro-hydro systems is to divert part of the stream through a pipeline, or **penstock**, downhill to the power-generating turbine, which may be sheltered in a small shed or powerhouse along with the required controls. After the water has done its work, it flows back into the river, often through another pipeline, called a **tailrace**.

HOW MUCH POWER CAN YOU MAKE?

There must be enough energy in the water to justify installing a hydro energy system, so quantifying water resource is the first order of business. To estimate the available power of a stream requires the understanding of a couple of key performance factors:

- How much water is flowing over a given period of time (**flow**)

- How much pressure (**head**) can be delivered to the hydropower generator

Head and flow will determine everything else about the design of your hydropower system, so you must capture these two parameters first — and with reasonable accuracy. Accuracy cannot be overemphasized: If you want to make the most of the time, money, and labor invested in your power system, you must take accurate measurements.

Once you know how much potential energy is available in the water, you can begin to design the

The reservoir upstream supplies a steady water flow through the penstock, which carries water down to the turbine in the powerhouse. Water flows through the tailrace and back into the river. ▶

systems to collect, control, and store this energy. Those details include:

- Determining the type of turbine that best suits your site

- Working with manufacturers to tailor the turbine specifications to your site

- Sizing the generator capacity — usually a compromise between meeting your power needs, what the water can support, and cost

- Determining how much water must be diverted to support the generator

- Sizing and layout of penstock pipe and wiring

- Determining requirements of controls

When you have some understanding of your water power site and requirements for harnessing that power, it's time to contact a water turbine manufacturer to begin fine-tuning the design, based on the specific equipment you choose. Working with a manufacturer is important because there are many turbine variables that can be customized based on your specific site. Manufacturers, along with experienced installers, can offer a wealth of information and can custom-tailor a system to suit your site.

Calculating Hydro Energy

BEFORE MOVING ON to the details, I offer one caveat: There are many different ways to express the values we will be talking about, and it's important to keep the numerical units consistent and clear while performing your measurements

and calculations. Keep this in mind as you work to evaluate the energy in the water by accurately quantifying the head, the flow, and the resulting pressure available for the turbine to generate electricity.

HEAD

Head is the pressure generated by falling water. It is dependent upon the vertical distance that the water falls from the penstock inlet in the stream bed to the turbine. Twice the head means twice the power, so take advantage of all of the vertical drop you can reasonably gain access to. Head is often described simply in terms of feet or meters of vertical drop, or it can be described in terms of pressure. Pressure can be expressed in pounds per square inch (psi), or in kilograms or newtons per square meter.

Each vertical foot of drop creates 0.433 psi of pressure. Therefore:

Total pressure (in psi) = Feet of head x 0.433

Expressed another way: 27.72" (2.31 feet) of elevation drop produces 1 psi of pressure. More head means more power, but *useful* head for a home-based micro-hydro system ranges from 25 to 200 feet. With less than that, you may not have enough power; with more, the costs to manage long pipe runs and high pressure begin to take you out of the micro-hydro realm.

Measuring Head: Method 1

You can evaluate head even if you don't have surveyor's tools, but note that most barometric or GPS-based altimeters often are not accurate enough for head measurement. With two people, a long board or 20-foot section of PVC pipe (much

Water Weight + Elevation Drop = Pressure

Why does a vertical drop create 0.433 psi per foot of head? One gallon of water weighs 8.345 pounds. There are 7.48 gallons in a cubic foot. One cubic foot of water weighs 62.4 pounds. There are 144 square inches in a square foot. Now we can calculate the weight of water in psi:

62.4 pounds ÷ 144 square inches = 0.433 psi

lighter than wood), a level, and a tape measure you can make a reasonably accurate measurement of vertical drop over uneven terrain. Here's how:

1. Set one end of the long board at the high point of your future penstock location and hold the board straight and level.

2. Measure and record the distance between the raised end of the board and the ground. Mark the ground where you took the measurement.

3. Move down the slope and place the board end on the ground at the measurement mark. Hold the board level, measure to the ground as before, and add this dimension to the first measurement.

4. Continue in this manner until you reach the proposed turbine location, adding up all of the heights that you measured; this is the total head.

Tip: You can use a hand-held sight level in place of the pipe and spirit level, allowing you to get a level line of sight over a longer distance.

Measuring Head: Method 2

Another way to measure head is to run a hose from the top of the stream where your penstock inlet will be, downstream to where the generator will be. The diameter of the hose does not matter because there is no flow through the hose

with this test. Allow water to fill the hose, avoiding high spots that could trap air bubbles. Screw a water pressure gauge with an appropriate scale (0 to 30 psi is a good place to start) onto the end of the hose, and read the pressure: each psi indicates 2.31 feet of head. Accuracy is imperative; you will probably be traversing a very long distance, so when you connect hoses, there must be no leaks. Even small leaks can greatly affect accuracy in measuring pressure.

Interpreting the Test Results

What you have measured in these tests is called **static head**, or **gross head** — the pressure of the water in the pipe when the water is not flowing. Friction losses will reduce water pressure as it flows through piping between the inlet and the power generator. Pipe diameter and length, and each twist, turn, and elbow all increase pressure drop (meaning the pressure will be reduced). The pressure you end up with after these friction losses is called **net head**. Net head is the pressure that the turbine has to work with (in terms of useful power) and will be used for system design.

FLOW

Flow is the volume of water that moves during a specific period of time. It is expressed in any combination of a unit of volume over time, such as gallons per minute, liters per second, cubic feet (or meters) per minute, etc. Don't confuse the flow rate of *volume* with the flow rate of *speed*, or velocity. Velocity is a surface-level measure of speed in feet (or meters) per second.

Water flow will likely vary during different times of the year, and power production varies with water flow. Twice the flow means twice the power. It's important to assess your water resource and size your system components for *average* stream

Measuring head with a level and measuring stick requires moving downhill from the intake to the turbine. Total head is found by adding all of the vertical measurements taken along the way: H_1, H_2, H_3, and so on. ▼

measuring stick →

level

H_1

H_2O

H_3

Bypass Flow

A certain amount of water needs to remain in the stream to support aquatic life. Bypass flow is determined by your state agency responsible for issuing the Environmental Protection Agency's *Clean Water Act Section 401 Water Quality Certificate*. Bypass flow must be subtracted from the total stream flow to determine the flow that is available to you.

conditions, not the peak water flow after a heavy rain. The maximum amount of water flow that can be reasonably counted on is called the **design flow**. Design flow, along with net head, will guide your system design and component selection.

Measuring Flow and Estimating Power

YOU MAY BE able to find information about stream flow at your site from the U. S. Geological Survey StreamStats data online (see Resources). If not, there are three ways to measure flow yourself: bucket measure, weir measure, and cross-sectional flow measure.

BUCKET MEASURE

For small streams, start with a bucket or barrel of known volume. Find or create a narrow place in the stream where you can capture most of the water in the container, and use a stopwatch to time how long it takes to fill. For example, if you have a 5-gallon bucket and it takes 4 seconds to fill, your flow rate can be calculated in any number of ways:

5 ÷ 4 = 1.25 gallons per second (gps)

5 ÷ 4 x 60 = 75 gallons per minute (gpm)

5 ÷ 4 ÷ 7.48 = 0.167 cubic feet per second (cfs)

5 ÷ 4 ÷ 7.48 x 60 = 10 cubic feet per minute (cfm)

Metric Conversions for Volume

1 cubic foot = 7.48 gallons

1 gallon = 3.79 liters

35.31 cubic feet = 1 cubic meter

1 cubic meter = 1,000 liters

WEIR MEASURE

In this case, a weir is a temporary dam that allows all the water to flow through a rectangular opening of a known size. A weir needs to be set up carefully but is fairly accurate in small- to medium-sized streams. With a weir, you can easily make flow measurements anytime throughout the year with a simple depth measurement.

The goal is to create an opening, or gate, that allows the stream water to back up into a reservoir behind it, increasing the height of the water in the reservoir while allowing some water to flow through the gate and not over the top of the weir. For accuracy, the bottom of the gate should be level, and the water should flow freely and smoothly through both the reservoir and the gate.

Drive a stake into the reservoir area of the stream, far enough upstream of the weir (4 feet or more) so that the stake is not in the area just behind the weir (where the level may be lower than the rest of the reservoir as the water crests over the gate). The top of the stake should be level with the bottom of the weir gate. With water flowing through the gate, measure the depth of the water from the surface to the top of the stake. The gate width should allow the water to build up some head behind the weir, allowing for a good depth measurement.

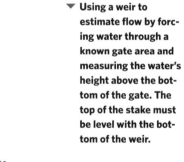

▼ Using a weir to estimate flow by forcing water through a known gate area and measuring the water's height above the bottom of the gate. The top of the stake must be level with the bottom of the weir.

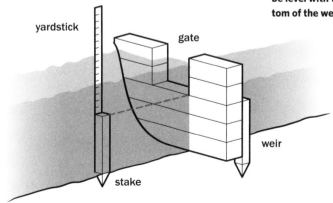

yardstick

gate

weir

stake

Finding the Flow Rate

Once you know the width of the gate and the water depth to the top of the stake, you can use math or a weir table to find the stream's flow rate. Here's the math to measure flow in cubic feet per second (cfs):

D = depth from surface of water to top of stake, in inches

W = width of weir gate, in inches

$3.33 \times D^{1.5} \times W \div 500$ = flow rate in cfs

Let's say we have a 12"-wide weir gate and we've measured 6½" of depth. The formula now looks like this:

$$3.33 \times 6.5^{1.5} \times 12 \div 500 = 1.324 \text{ cfs}$$

Multiply cfs by 60 to get cubic feet per minute (cfm): 1.324 x 60 = 79.4 cfm

Multiply cfm by 7.48 to get gallons per minute (gpm): 79.4 x 7.48 = 594 gpm

To use the Weir Flow Table (right) to determine flow, find the depth by looking down the left column; if you measure an additional fraction of an inch, find the nearest fraction in the top row. The +0 column means that you have measured the depth to an even inch. Move across the inch row to the correct fractional column to find the flow in cfm per inch of gate width. Multiply the value in the table by your gate width in inches to find the flow through the gate.

For example: You measure 6½" of depth, and your gate is 12" wide. Go across row 6 to the +½ column, and you land on the value of 6.62. If your gate were only 1" wide, that's how much water in cfm would be moving through it. Since your gate is 12" wide, multiply by 12 to find your flow rate:

$$6.62 \times 12 = 79.4 \text{ cfm}$$

WEIR FLOW TABLE

CFM flow per inch of gate width

Depth in inches	additional fraction of an inch, depth			
	+0	+¼	+½	+¾
1	0.40	0.56	0.73	0.93
2	1.13	1.35	1.58	1.82
3	2.08	2.34	2.62	2.90
4	3.20	3.50	3.81	4.14
5	4.47	4.81	5.15	5.51
6	5.87	6.24	6.62	7.01
7	7.40	7.80	8.21	8.62
8	9.04	9.47	9.90	10.34
9	10.79	11.24	11.70	12.17
10	12.64	13.11	13.60	14.08
11	14.58	15.08	15.58	16.09
12	16.61	17.13	17.66	18.19
13	18.73	19.27	19.82	20.37
14	20.93	21.50	22.06	22.64
15	23.21	23.80	24.39	24.98
16	25.57	26.18	26.78	27.39
17	28.01	28.63	29.25	29.88
18	30.52	31.15	31.80	32.44
19	33.09	33.75	34.41	35.07
20	35.74	36.41	37.09	37.77

CROSS-SECTIONAL FLOW MEASURE

This flow measurement is useful for assessing larger streams. It involves measuring the average depth of the stream and the speed of an object floating on the surface. To do this, you'll need to find a 10- to 50-foot section of the stream that is reasonably accessible.

To determine the average depth, find a relatively flat section of stream that you can walk through. Mark a board or pipe with 1-foot increments along its length and lay it across the stream from bank to bank. You can also use a rope with a knot tied every foot. The actual distance between measuring points is not critical; what matters is that the distances are the same.

Measure the depth of the water across the stream at equal intervals using the board or rope as your guide. Calculate the average depth by adding up all the measurements and dividing the total by the number of measurements you made. You can increase the accuracy by doing this at several locations across the stream.

Measuring Velocity and Flow Rate

To measure the water's velocity, mark a line across the stream by throwing a rope across to the other side. Do the same thing 10 to 50 feet downstream, and include the section that you used to measure the average depth. Measure the distance between the two ropes.

Next, find something with some weight that will float down the stream (fruit or vegetables work well since they are relatively heavy, they float, and they're biodegradable) and measure how much time it takes to travel from one marker rope to the next.

Repeat the test a few times to get an average. Divide the length of travel (in feet) by the time (in seconds) to compute the velocity of the water in feet per second. However, because the stream bed creates friction, the top of the water will be moving faster than the bottom. An adjustment factor of 0.83 helps compensate for this effect for better accuracy in the velocity calculation. Here's an example:

We measure that it takes 15 seconds for an orange to float 25 feet downstream.

The velocity of the stream is 25 ÷ 15 x 0.83 = 1.38 feet per second (fps), or 83 feet per minute (fpm) (1.38 x 60).

You now have the three things you need to calculate flow volume: average water depth, stream width, and water velocity. Use the following formula:

Flow (cfm) = Area (square feet) x
Velocity (feet per minute)

1. Multiply the width of the stream by the average depth to get the cross-sectional area. Let's say it's 10 feet wide and 2 feet deep, so the area is 20 square feet.

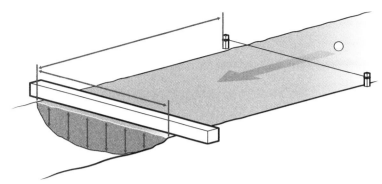

▲ Measure the cross-sectional area of a stream by averaging the depths at various locations. Mark off a known distance between two points to estimate stream velocity by timing how long it takes a floating object to travel the distance.

2. Multiply the cross-sectional area (20 square feet) by the velocity (83 fpm) to find the flow:

20 x 83 = 1,660 cfm

3. If you want your answer in gallons per second, multiply cfm by 7.48, then divide by 60:

1,660 x 7.48 ÷ 60 = 207 gps

ROUGH POWER ESTIMATION

Power output is a product of water flow, pressure, and the efficiency of the entire hydropower system. To roughly estimate the system's potential for power generation, multiply head (in feet) by flow (in gallons per minute) and divide by 10:

Head (feet) x Flow (gallons per minute)
÷ 10 = Watts

This simplified approach will give you a very rough expectation of the system's power output in watts; it is based on installer's experience but is not a substitute for thorough analysis of your project. In systems with lower component efficiencies, the "divide by" value may increase somewhat, indicating a loss in power.

As an example, if you have 40 feet of head and a flow of 75 gpm, this works out to 40 x 75 ÷ 10 = 300 watts of power that can be generated continuously, as long as the water is flowing at that rate. Multiply 300 watts by the number of hours of flow you can expect daily (or seasonally) and divide by 1,000 to arrive at the potential kilowatt-hours the site can produce for you:

$$300 \text{ watts} \times 24 \text{ hours} \div 1{,}000 = 7.2 \text{ kilowatt-hours per day}$$

Intake Site Selection

THE INTAKE IS where the water is diverted from the river and channeled into the penstock leading to your turbine. The intake's job is to provide a constant flow of clean water. This requires a debris screen (see Resources), or trash rack, to catch particles that can clog the penstock and damage the turbine. To prevent a plugged nozzle or equipment damage, the filter system must catch anything larger than the size of the nozzle orifice. This also means the debris screens must be cleaned frequently.

Sedimentation, caused by small particles (such as sand and dirt) washing down the stream, can quickly cover over your dam and block the intake. Sediment must be removed and the trash rack cleared regularly. To help avoid intake blockage, the intake can be positioned on the downstream lip of the dam, or upstream of the dam if there is a sizeable reservoir.

A quiet pool of water around your intake allows for smaller particles to settle out of the water before entering the penstock. The inlet pipe needs to be under water that is deep enough and replenished at a rate that is fast enough to keep the intake underwater, so that air does not enter the system. Air in the penstock not only reduces power but can also cause damaging shocks to the turbine as it's intermittently pounded by water. Consider durability at every step when assembling your hydropower system, keeping in mind that it will be exposed to the forces of nature year-round.

Penstock Pipe Selection

THE PENSTOCK PIPING must be sized to provide the required flow and pressure to the turbine without too much friction loss. Even though a pipe may be able to carry all the water to the turbine, if its diameter is too small, the water flow will be restricted and slowed on the way down, losing power along the way.

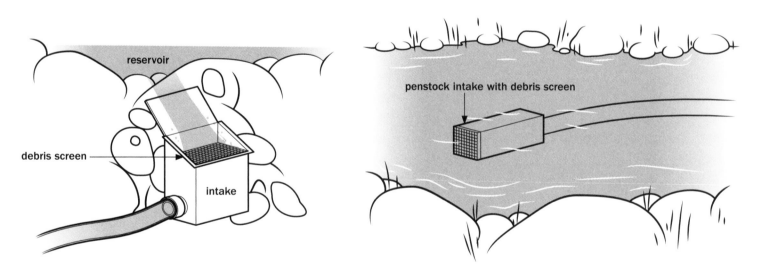

▲ The goal of the water intake at the top of the penstock is to provide a constant flow of clean water. Set up your intake system so that water flows through a trash rack or debris screen. The intake on the left requires a constant flow of water from a reservoir above. The intake on the right is submersed in the reservoir.

The variables to consider in penstock sizing are head, flow, generator capacity, and pipe length and layout.

A good design goal for your penstock is to limit total friction losses to 10 or 15 percent, so you can achieve a net head of 85 to 90 percent of gross head. Once again, pipe size and material will likely be a compromise between pressure drop, ability to work with a given material at your specific site, and cost. It will be important to discuss the finer details of your penstock design with the turbine supplier or manufacturer.

CALCULATING FRICTION LOSS

Most micro-hydro systems call for PVC pipe diameters between 2" and 8", depending upon head and flow. You probably don't want to exceed half a mile in total pipe length unless the value of the site in delivering power can justify the costs of pipes and wires.

The Hazen-Williams equation offers an accurate method for calculating friction loss based on pipe diameter and water flow. You can use the Hazen-Williams Friction Loss table (page 168) to find the head loss of several common pipe sizes. Or you can calculate the friction head losses in your specific piping configuration by applying the Hazen-Williams equation when you know:

1. Type of pipe and its roughness coefficient (c)

2. Inside pipe diameter (ID). Note that the ID is not the same as the pipe's nominal size. Measure or research on manufacturer's website.

3. Flow rate (in gpm)

For this exercise, we'll assume a 75 gpm flow, and the use of 2" (nominal) PVC pipe with a Hazen-Williams roughness coefficient (c) of 145. PVC is about as smooth as you can get, but pipes with a rougher interior will have lower values (corrugated steel has a coefficient of 60).

Friction loss in feet of head per 100 feet of pipe =

$$[0.2083 \times ((100 \div c)^{1.852})] \times (\text{flow in gpm }^{1.852})$$
$$\div (ID^{4.8655})$$

If you have 75 GPM flowing through a 2" ID PVC pipe, the formula looks like this:

$$[0.2083 \times ((100 \div 145)^{1.852})] \times (75^{1.852})$$
$$\div (2.067^{4.8655})$$

To break it down further looks like this:

$$0.105 \times 2969 \div 34.2 = 9.1 \text{ feet}$$

In this example, every 100 feet of penstock length results in a loss of 9.1 feet of head pressure. Therefore, for every 100 feet of pipe, you would subtract 9.1 percent from the gross head to find the net head value.

Accounting for Bends

The Hazen-Williams equation values are for straight pipe. You will also want to account for the effects of each elbow. It's best to use long-sweep 90-degree elbows or 45-degree elbows wherever possible to reduce friction losses due to bends and turns. The head loss for each long-sweep elbow is roughly equal to 3 times the nominal pipe diameter in inches, but expressed in feet of straight pipe. For example, a long-sweep 90-degree elbow on a 4" (nominal) pipe has a pressure drop equivalent to about 12 feet (3 x 4) of 4" straight pipe. A 45-degree elbow has a pressure loss of about half that of a long-sweep 90.

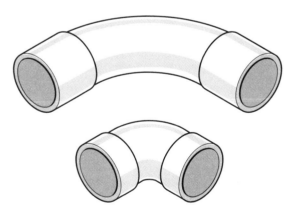

▲ Choose long, smooth curves to use in the penstock to reduce friction losses inside the pipe; the 90-degree long-sweep elbow above is better than the standard 90-degree elbow below.

PENSTOCK DESIGN EXAMPLE

Let's say you have 200 feet of head and 100 gpm of flow, and the total length of the penstock is 500 feet, including four long-sweep 90-degree elbows. With 200 feet of head, your penstock loss might range from 20 feet (10 percent loss) to 30 feet (15 percent loss). In this case, you don't want to exceed 6 feet of head loss for every 100 feet of penstock length (30 ÷ 500 x 100 = 6). Using the Hazen-Williams Friction Loss table (or the Hazen-Williams equation), you can see that at least a 3" pipe diameter is required to keep losses below 6 feet (or 15 percent). In this case, the losses amount to 2.26 feet per hundred, for a total head loss of 11.3 feet.

Now let's add in the elbow losses. Four 3" elbows are approximately equivalent to 48 feet (4 x 3 x 4) of straight 3" pipe, for a total effective penstock length of 548 feet. Fifteen percent head loss in this pipe is equal to 30 ÷ 548 x 100 = 5.5 feet of acceptable loss per 100 feet of pipe. Therefore, the 3" pipe selection is still a good choice.

Note the huge jump in loss between 2" and 3" pipe sizes. If you have a borderline situation, contact the turbine manufacturer for guidance on penstock pipe sizing. You want to be sure that the additional cost of larger pipe is worth it in terms of power output.

To calculate your net head, subtract the pipe loss from the total head:

Total head = 200 feet
Pipe loss with 548 feet of 3" pipe and a flow of 100 gpm = 12.6 feet (2.26 feet of loss per 100 feet x 548 feet ÷ 100)
Your net head is: 200 – 12.6 = 187.4 feet

Converting that to pressure available at the turbine:

187.4 x 0.433 = 81 psi

HAZEN-WILLIAMS FRICTION LOSS

The table below uses the Hazen-Williams equation to present friction head loss for various combinations of PVC pipe sizes and flow rates. Values are expressed as feet of head loss for every 100 feet of penstock length. The values can also be read directly as percentages. Find your penstock diameter along the top row, and your flow rate in the left column.

FRICTION HEAD LOSS FOR 100 FEET OF SCHEDULE 40 PVC PIPE

Pipe size, nominal inches

Flow in gpm	2	3	4	6	8
	Head loss (in feet) of water per 100 feet of pipe				
25	1.19	0.17	0.05	0.01	0.0
50	4.29	0.63	0.17	0.02	0.0
100	15.47	2.26	0.60	0.08	0.0
200	55.85	8.18	2.18	0.30	0.08
300	118.4	17.32	4.62	0.63	0.17
400	201.6	29.52	7.87	1.07	0.28
500	304.8	44.62	11.89	1.62	0.43
1000	1100.4	161.1	42.94	5.85	1.54

Pipe Maintenance

Moving water represents a potentially huge amount of energy. It's important to support the penstock well, especially at turns, where thrust blocks may be needed to counteract the water's force. When it comes time for maintenance, you must have a means to remove and/or divert the water from the penstock. Often this is provided by a valve at both the top and bottom of the penstock, with the top valve providing an air inlet (just below the shut-off valve) to allow the water to flow smoothly out the bottom. Always turn valves gently to avoid potentially damaging forces created by sudden changes in water pressure.

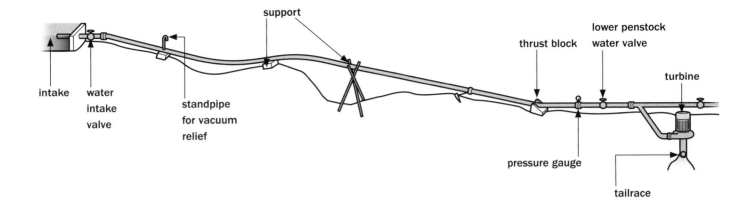

support

intake
water intake valve
standpipe for vacuum relief

thrust block
lower penstock water valve
turbine

pressure gauge

tailrace

▲ **Micro-hydro system components and layout.**

Turbines

NOW THAT YOU have determined net head and design flow of your stream, it's time to think about the kind of turbine that is best suited to your site. Specifically, the turbine is the piece of the hydropower system that collects the energy input; it's the interface between the water in the penstock and the shaft connected to the power generator.

Turbines are different from water wheels of old because water is delivered at high pressure and velocity through a nozzle that focuses a jet of water onto the runner. The runner rotates to transfer energy to the generator power shaft. Different turbines have different types of runners, and the turbine's housing design is integral with the runner in terms of how the water impacts and reacts with the runner and flows through the turbine and down into the tailrace.

CHOOSING A TURBINE

Choice of the best turbine design for your site depends upon the site's characteristics, as well as the head and flow. Most small hydro systems employ direct drive between the runner and the generator, meaning that there are no efficiency-robbing belts, gears, or pulleys. Belts and pulleys may be required in larger systems so that both the turbine and generator can run at their respective optimal speeds.

Hydropower Resources

There are many local, state, and federal guidelines and laws surrounding the use of water, water rights, and alteration of streams and wetlands. Due diligence is required to identify all of the legal issues that apply to your site. For larger, grid-tied hydro systems, you will need to look into licensing, permitting, design specifications, and fees that may be required. Here are some places to start your research (see Resources for websites):

- The Environmental Protection Agency's Clean Water Act Section 401 Water Quality Certificate provides guidance on water quality standards that may be required by your state for projects that impact streams or wetlands.

- The Federal Energy Regulatory Commission's responsibilities include licensing of new or existing power generation projects and oversight of all ongoing project operations, including dam safety inspections and environmental monitoring.

- The U. S. Army Corps of Engineers has jurisdiction over most U. S. waterways and wetlands, and offers some useful design guides.

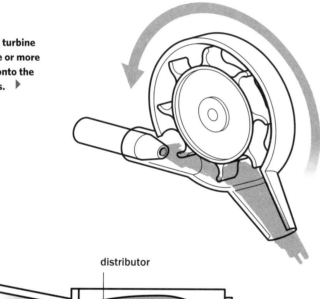

Alternator-based impulse turbine from Harris Hydro. A pipe from the intake feeds the four nozzles that spray water into the runner cups. The runner spins the generator, and the water spills down and away from the runner. ▶

An impulse turbine focuses one or more water jets onto the runner cups. ▶

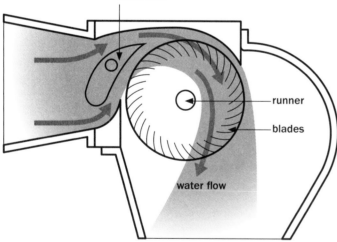

distributor

runner

blades

water flow

▲ **Cross-flow turbines allow water to flow across the runner blades, so that the water hits the blades at the top, falls through the runner, and hits the blades again adding energy. The distributor guides water through the turbine so that it hits the runner at the proper angle.**

There are two basic approaches to small-scale hydro turbines, categorized by how they capture the energy in water. You may have a very small stream with a very high head that creates lots of pressure but perhaps relatively little flow. Or you may have access to a river meandering through your property, powerful and high-flowing but without much head. Each situation requires a specific approach to capturing the available resource, and each turbine can be customized by the manufacturer, and by your control of the water flow hitting the turbine, for best efficiency at a specific site.

Impulse Turbines

Impulse turbines work well in situations with high water pressure (head) and low-to-medium water flow. Impulse turbines are not submersed in water; they operate in the air and receive water from an inlet pipe, sending it through one or more nozzles to create a focused, high-velocity spray. Additional nozzles may be used to increase the power output, but this can be achieved only when there is enough supporting water volume. The water jet(s) hits the runner's paddles, causing it to spin. Two common (and similar) turbine designs use Pelton and Turgo runners. These runners have a number of cups around the perimeter to catch the jet of water.

A good example of an impulse turbine is the alternator-based micro-hydro generator made by Harris Hydro. Beneath the housing is a Pelton-wheel runner connected to the shaft of the alternator. These micro-hydro systems can produce up to 2,000 watts of DC power, depending on the water flow rate, pressure, and output voltage. Useful power output starts at 25 feet of head and a water flow rate of 15 gpm, a configuration that will yield about 25 watts of power.

Crossflow Turbines

A crossflow turbine is a sort of modified impulse turbine where water flows in high volume and with low pressure through a large opening, rather than through small, high-pressure nozzles. Water passes over the runner blades in a somewhat similar fashion to a water wheel.

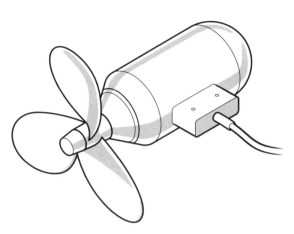
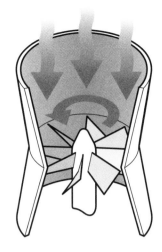

▲ **Two different versions of reaction turbines. They both take advantage of water flowing through the turbine blades in a stream or under a boat.**

Reaction Turbines

Reaction turbines are best suited for low-head and medium-to-high flow situations. A reaction turbine sits fully immersed in the water in a way that allows water to flow through it, causing it to spin. A wind generator is another good example of a reaction turbine in that it spins in reaction to a fluid force (air) moving through it. Large hydropower installations use reaction turbines, but there are a few small units on the market for home or recreational use.

Power Generators

IN MANY MICRO-HYDRO systems, the power generator is a specially designed alternator. Older-style alternators require brushes that need periodic replacement, while newer alternator designs use more efficient brushless, permanent-magnet technology. An alternator is used because it efficiently and affordably generates AC electricity. However, this is unregulated, variable-frequency electricity and is not the same as the AC power used in your home. The output of an alternator is ultimately **rectified**, or converted, into DC power.

Generating high-voltage AC power has a significant advantage over low-voltage DC generation. Low-voltage systems require very large (read: expensive) power cabling if there is significant distance between the generator and the battery bank, which typically is located in the home. High-voltage AC power can travel long distances in small wires without much power loss, and it can be transformed to higher or lower voltage, or it can be rectified to produce DC power that can be stored in batteries.

POWER MANAGEMENT

There are a few critical design factors and control features needed to ensure power quality and prevent damage from overspeed, in both grid-tied and off-grid systems.

Utility Power

Grid-connected hydro systems must deliver utility-quality power in terms of voltage, frequency, and power quality. The voltage and frequency of an AC generator depend upon its speed, and the speed depends on how much water is hitting the runner and how much load is on the turbine. Power conditioning is handled by the electronic components.

Preventing Overspeed

Whether grid-tied or off-grid, the generator cannot be allowed to spin too fast, or damage can occur. Overspeed can result from situations where you don't need the full power output at the same time the stream feels like delivering it.

Speed control is often accomplished by sending excess power to a **diversion**, **dump**, or **ballast load** (all the same thing) in order to maintain a constant load on the generator, resulting in a constant speed. This function is performed by an electronic controller. Diversion loads are usually electric air- or water-heating elements. In a battery charging system, once the batteries are fully charged, the controller sends the power to the dump load while providing just enough power to the batteries to keep them fully charged.

In a grid-tie system, the grid serves as the full load . . . until the grid goes down, and then the dump load takes on the burden.

Safety

Additional controls are required to ensure safe conditions for the generator and any people working on it. These include the following features:

- Electrical safety disconnect that cuts off the generator output power

- Shutoff valves to halt water flow

- Electrical brake (shorting the output to apply a maximum load to the generator), which prevents the generator from freewheeling or runaway in case the load is accidentally removed from the generator

POWERHOUSE DESIGN

You will probably want a box or small shed to mount and install hardware, make connections, and protect the generator and controls from the elements, including freezing temperatures, animals, and unauthorized personnel. Plan this powerhouse with turbine efficiency in mind: Don't make too many efficiency-robbing penstock twists and turns away from the ideal turbine site to get to the ideal powerhouse site. Be sure to provide a large enough water outlet at the turbine so that "backsplash" (when the water splashes and bounces around inside the turbine housing, impeding efficient runner operation) does not affect the turbine. The powerhouse also must include adequate space and access for servicing the equipment.

System Efficiency

THE KIND OF efficiency you can expect — in getting the energy from water to wire — from a micro-hydro system can range from 40 to 70 percent. Efficiency depends upon all the little details of each component in the system, with each piece affecting overall efficiency.

You will need to work with the turbine manufacturer for specific numbers to estimate efficiency and expected power production at your site and with the equipment you choose. The efficiency will vary depending on flow and head throughout the year.

System efficiency is calculated by multiplying the efficiencies of each component within that system. The Hydropower Component Efficiencies table (facing page) shows some approximate efficiency values you can expect for the various components in a well-designed micro-hydropower system. Turbine efficiency will vary quite a bit over the seasonal range of water power hitting the runner. As flow changes, the nozzle size can be changed (either by switching out fixed nozzles or using an adjustable nozzle) in order to maintain the optimum velocity for maximum efficiency, even though the power will drop off.

Electricity Transmission

Proper transmission cable selection, installation, and layout are as important as penstock design. Once you generate the electricity, you'll want to get it to the power control center as efficiently as possible. See chapter 10 for more information on power management, controls, and wire sizing.

HYDROPOWER SYSTEM COMPONENT EFFICIENCIES

Component	Efficiency
Penstock	85%
Runner	80%
Permanent-magnet alternator	90%
Wiring and controls	92%
Total System Efficiency	**56%**

MAINTENANCE CHECKLIST

Maintenance for a hydro system includes the following basic procedures:

☐ Lubricate generator bearings and replace brushes as needed.

☐ Inspect generator periodically for wear, vibration, damage, cracks, leaks, or corrosion.

☐ Keep the turbine's water intake clear of debris and ice.

☐ Adjust the water flow into the turbine according to seasonal fluctuation in water level.

☐ Check the penstock for leaks, and reset pipe anchors as needed.

☐ Inspect electrical connections periodically for tightness and corrosion.

10

Renewable Electricity Management

PREVIOUS CHAPTERS DISCUSSED a variety of resources for generating electricity from renewable resources. Each collection technology has its own set of specific equipment and site requirements, but there are some common components among home power systems. This chapter offers a brief introduction to those commonalities, including wiring considerations, controllers, inverters, batteries, battery chargers, generators, system monitoring, and other important "balance of system" components needed for a complete, safe renewable-energy system.

We'll go over some terms that are important to understand, and you'll get an idea of what's significant from an electron's point of view, as it travels through your system doing its job while avoiding trouble. By paying attention to the details, you can wrangle maximum benefit out of every last electron.

Grid-Tie vs. Off-Grid

AS DISCUSSED IN chapter 7, the primary distinction between various home-based electrical generation systems is whether they are interconnected with utility power (grid-tied) or are stand-alone (off-grid). Grid-tied systems come without the cost and hassle of storage batteries, but one big disadvantage is that if the grid fails, you are out of power, too. It is possible, though, to have a hybrid system that is grid-connected but includes battery backup. This approach can provide power for critical loads, or even the entire home, when the power goes out. However, such a system adds cost and complexity.

A grid-tied renewable energy system provides power (when available) to your home and diverts excess electricity to the local utility. In this way, the grid is used as a kind of storage facility for your power. For instance, if you have a solar-electric grid-tie system and the sun is shining, the photovoltaic (PV) modules contribute to the power needed by your home. If the PVs are generating more power than you need, the electrons actually flow into the power company's lines, spinning your electric meter backward as they go. In some cases, the amount of solar energy produced is tracked using a separate meter. At times when you need more than your own system is producing, the power company provides you with all the electrons you need from the grid.

The details of accounting for electrons sent back and forth vary by state and utility. At worst, you can expect to offset, or receive a credit on, your power bill through a system called **net metering**. If

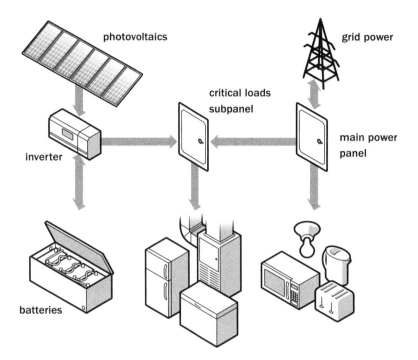

photovoltaics

grid power

critical loads
subpanel

inverter

main power
panel

batteries

Hybrid grid-tied PV system with battery backup for critical loads in the event of a power outage. In this case, the PV system is optional because the batteries can be charged with grid power when it returns.

you produce a surplus of electricity this month, and next month is cloudy, you can draw on your electrical "storage account" to meet your needs at no cost until you've used up your banked electrons. Of course, the power company will continue to provide all the power you need as you offset what you can with what you produce. At best, you might earn money for the power you produce by way of a **feed-in tariff**. This policy mechanism allows you to earn money as a small power producer.

OPTIONS FOR BACKUP POWER

You can use renewable energy on a scale that is appropriate for your needs. Many homeowners have gas-powered generators that can be used as a source of backup electricity in the event of a power outage. Another option is a

Distributed Generation

Having lots of small power generators, such as home PV systems, distributed throughout a utility's service territory, serves to decentralize power generation, a strategy appropriately called **distributed generation**. Relying on just a few large power plants makes everyone on the grid more vulnerable to power outages should one generator fail or the fuel supply be disrupted. Distributed generation offers a measure of diversity and resiliency to the power grid that will be better utilized as smart-grid technologies are used more widely. If you want to connect your renewable power generator to the grid, start by contacting your power company.

battery-and-inverter–based backup system, which can be used to provide electricity to critical loads, such as heat and refrigeration. One huge advantage to this approach is that the switch from utility grid to inverter power can be done automatically without any need for human management. This provides some peace of mind if you're away from home when the power goes out.

In this system, the critical load circuits are separated from the home's main circuit breaker box (service panel) and brought to a subpanel. The subpanel is fed by the inverter, and while the grid is active, the inverter merely allows grid power to pass through it and on to the subpanel. Meanwhile, the batteries are kept charged by either utility power or some renewable source. When utility power is not available, the inverter senses this and immediately draws power from the batteries, providing its own AC power to the subpanel. When utility power returns, the inverter automatically and instantaneously makes the switch back to grid power, and the batteries recharge.

Definition of Terms

WHEN DISCUSSING ENERGY generation and consumption, there are several terms that require clear definition.

Volts are a measure of electrical *pressure*, or how much force is pushing the electrons through the wire.

Amperes (amps) are a measure of the *flow* of electrons, or current. The way that electrical current is produced can be divided into two categories. Solar, wind, and hydroelectric generating systems usually produce direct current (DC). Batteries store and deliver electricity as DC. Alternating current (AC) is how electricity is supplied to your home by the power company. It is difficult to generate the high-quality AC power produced by the power company with residential-scale renewable sources because the inconsistency of the resource means that the voltage and frequency will vary too much to be used by your home's appliances. Additionally, AC power cannot be stored, while DC power can be easily stored in batteries and used later. DC power can be converted to AC with an electronic **inverter**, and AC power can be converted to DC with an electronic **rectifier**.

Resistance is the opposition to the flow of current and is measured in **ohms**. This will be important to understand when determining the wire size required to move electricity efficiently from one place to the other.

Power is the product of volts and amps and defines the *rate* at which work is performed. One horsepower is the amount of work required to move 550 pounds of weight a distance of 1 foot

THE OFF-GRID ROOM

I bought my first solar electric panel back in 1988 (a Solarex MSX53). I walked out of the store feeling like my own power company. It was as if the solar panel was a magic carpet that would take me to many exotic new places. And it did. That purchase eventually led to a job with the solar energy company I bought it from.

I used that first PV module (at the rental house I shared) to power an electric garden fence and to charge a small motorcycle battery that kept the fence operating at night. Soon, the battery had lots of wires hooked up to it, powering a clock, a radio, and DC lights in one room, creating an "off-grid" room in the

house where everything was solar-powered, even when the utility power failed. I learned a lot from that simple system — from battery care to the effects of tilt angle on the PV panel. Living with solar electric power makes you hyperaware of when the sun is out and how much power you're getting.

over a period of 1 second. When working with electricity, power is commonly expressed in units called **watts**. One horsepower is equal to approximately 746 watts.

Energy is a *quantity* of work performed. In terms of electricity, energy is expressed in kilowatt-hours (kWh). A 100-watt light bulb consumes power at a continuous *rate* of 100 watts. If that light bulb is left on for 10 hours, it will have consumed 1 kWh (100 watts x 10 hours) of energy.

British thermal unit (Btu) is another quantity of energy. When comparing fuels and their energy content, we need to find a common denominator for all energy measurement units, regardless of fuel, so we can compare apples to apples. The energy "apple" (at least in the United States and Britain) is the Btu. By definition, a Btu is the amount of energy required to raise the temperature of one pound of water (about a pint) by 1°F. For perspective on energy equivalents:

- 1 Btu is approximately the amount of energy released by completely burning a wooden kitchen match.

- There are about 125,000 Btus in a gallon of gasoline and 5.8 million Btus in a 42-gallon barrel of oil.

- One kilowatt-hour of electricity is equivalent to 3,413 Btus.

- A gallon of gasoline contains the equivalent of about 31,000 food calories. This is equivalent to the energy in more than 50 McDonald's Big Macs!

- It takes 320 pounds of lead-acid batteries to deliver the equivalent energy in one 6-pound gallon of gasoline.

Electrical Wiring

MOVING ELECTRICITY EFFICIENTLY requires the proper conductors for it to move through. The tasks of cutting, stripping, and connecting wires are fairly simple, and you may be able to work with your electrician to wire and connect your renewable energy system's components. However, do not attempt this on your own if you are not an electrician.

An experienced electrician will know all about the importance of the following:

- Wires need to be properly sized to carry a specified current over a certain distance.

- Fuses and circuit breakers must be carefully chosen to protect both conductors and equipment.

- Connectors and switches must be properly rated, sized, and placed within the system.

- Grounding and lightning protection are of utmost importance to ensure personal safety and prevent failure of electronic components due to power surges.

- A main disconnect switch is required to remove your renewable power generator from the system, for both service and safety needs.

Wind-, solar-, and battery-powered electrical systems now have their own sections in the National Electrical Code (NEC; see National Electrical Code and Safety on page 177). Specifically, NEC Article 690 relates to photovoltaics, Article 694 relates to wind, and Article 480 relates to batteries. Several additional sections of the NEC are also applicable to renewable systems. At present, there is a proposal to consolidate these articles, so these references may change in the future.

You can find all the specialty tools, suitable wires, connectors, and insulation at your local electrical supply store or renewable energy dealer. One excellent online resource for electrical cable, connectors, and tools is Del City Wire (see Resources).

WIRE SIZING

While not a substitute for code requirements or the skills and experience of a professional electrician, there are some basic guidelines for properly sizing wires. Wire size is important to get right because a wire that is too small will not deliver the full power of your renewable energy system; a wire that is too big will be more difficult to work with and add to the cost, sometimes significantly.

Wires are not perfect conductors because they offer some resistance to the flow of current. For example, a 10 AWG (American Wire Gauge) wire has a resistance of 1.29 ohms per thousand feet. A 2/0 ("2-aught") cable has a resistance of 0.1 ohms per thousand feet. These may sound like small numbers, but they have a big effect on the electricity flowing through the wires. Shorter wires, fatter wires, and higher voltages help to increase electrical transmission efficiency. The wire must be able to carry the required current at

a specific voltage over a distance with a minimum of the loss known as **voltage drop**.

Acceptable voltage drop for renewable energy systems is generally 3 percent or less. To ensure minimal voltage drop and maximum current transfer, the wire must be sized based on the supply voltage, the current (or amps) flowing through the wire, and the distance between the source (solar panels or wind generator, for example) and the storage (batteries or grid tie-in). The current-carrying capacity (ampacity) of wires can be found by using the Wire Size and Ampacity chart on page 179.

Calculating Wire Size

To estimate the wire size you need, look up the ampacity based on the maximum amps your power source will deliver. Next, calculate the **voltage drop index** (VDI) of the wire. To do this you need to know the one-way distance between the power supply and the battery bank, as well as the acceptable voltage drop percentage. Use the following formula:

VDI = (amps x one-way distance in feet)
÷ (% voltage drop x voltage)

Here's an example using a 24-volt solar electric system that delivers 50 amps of charging current to the batteries. The total distance of the actual wire run between the charge controller and the solar panels is 100 feet, and we are willing to accept a 3-percent voltage drop. If you look only at the ampacity of a wire as the sole sizing requirement, it would appear that an 8 AWG wire would be sufficient. The voltage drop index tells us something different:

VDI = (50 x 100) ÷ (3 x 24)

The calculated VDI is 69. Look up the nearest VDI from the chart and round up. Our example shows that we need a 3/0 copper wire. If you decide that this cable size is too big or expensive to work with, an alternative might be a higher-voltage system that allows for the use of smaller conductors. Many modern charge controllers manage the voltage and current so that

WIRE SIZE AND AMPACITY CHART

There is an important distinction to make between the ampacity of wires and the gauge required to keep power loss to a minimum. This chart shows the current-carrying capacity of both copper and aluminum wires, based on the wire gauge in both AWG and metric units. These ratings should be used only for direct current (DC)–carrying wires. Also listed in the table is a voltage drop index (VDI).

Wire Size		Copper		Aluminum	
AWG	Area mm^2	VDI	Ampacity	VDI	Ampacity
16	1.31	1	10		
14	2.08	2	15		
12	3.31	3	20		
10	5.26	5	30	DON'T USE	
8	8.37	8	55		
6	13.3	12	75		
4	21.1	20	95		
2	33.6	31	130	20	100
1/0	53.5	49	170	31	132
2/0	67.4	62	195	39	150
3/0	85	78	225	49	175
4/0	107	99	260	62	205

you can wire your PV panels at a higher voltage than the batteries require, thus minimizing wire size and cost.

Balance of System Components

MANY ADDITIONAL COMPONENTS are required to complete your renewable electric power system. Some are optional; some are not. Here is an overview of the main items.

INVERTER

An inverter converts the DC power — produced by the power source and stored in the batteries — into the AC power required by lights and appliances in your home. Most modern inverters deliver clean, pure **sine wave** AC power that allows you to operate everything you would normally use in a utility-powered house. Older or less expensive inverters may produce a "modified sine wave," which can be problematic with some types of appliances but fine for recreational use.

Power-handling capacity of the inverter can be increased by connecting two or more units together electronically, a configuration sometimes called **stacking**. Since inverters typically provide 120 volts of AC power output, another advantage to stacking is that the voltage can be doubled, providing 240 volts to high-power loads, such as well pumps.

Inverters used in **grid-tied systems** accept the renewable power source input and, through electronic manipulation, condition and synchronize that power with the grid to deliver the correct voltage, frequency, and phase. **Off-grid inverters** must match the battery voltage, and many

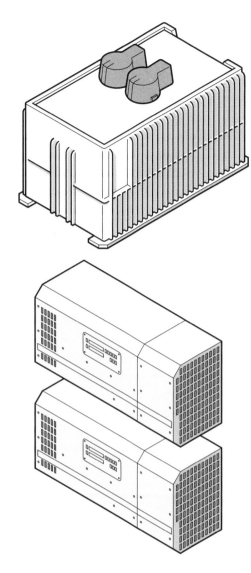

Inverters convert DC power produced by renewable electricity sources and stored in batteries, into 120 volt AC power ready for use in your home. Their capacity can range from a few tens of watts to several thousand watts. ▶

Inverters can be electrically connected to increase power delivery, and to double the voltage if 240 volt service is required. ▶

inverters also act as battery chargers when power from another source (utility or generator) is available. Some inverters can be used in both battery-based and grid-tied scenarios, allowing for a hybrid system that enables batteries to supply power to critical loads when grid power is not available. A few of the more popular brands of inverters are Outback, Xantrex, SMA, Fronius, Solectria, Apollo Solar, and Exeltech, to name a few (see Resources for websites).

CHARGE CONTROLLER

A charge controller is an electronic device that regulates the amount of power flowing from the power source into the batteries while preventing the electrons from flowing back to the power source. The primary function of the charge controller is to prevent overcharging, which can damage the batteries. To keep the batteries properly

charged, the controller can be adjusted to manage the charge current and hold the voltage at a specified level. It can also help to protect the batteries by disconnecting them from the load at a specified voltage. Choose a charge controller based on the system voltage and maximum amps delivered by the power source. All controllers have a display meter or lights indicating voltage, current, and/or charge status.

Many modern charge controllers use **maximum power point tracking** (MPPT). This technology optimizes charging by managing and manipulating the voltage and current available from the energy source to match the battery voltage and charging requirement. It's a bit like shifting gears in your car to deliver maximum power to the wheels from the same engine. One benefit of MPPT controllers is that you don't need to match PV voltage with battery voltage. Sending high-voltage electricity from the panels to the charge controller can increase efficiency by lowering line losses.

Not all charge controllers are alike, and not all energy-generating technologies are alike. The charge controller must be suitable for the generating source. Solar panels can be disconnected from the load, or the output wires can be connected together (shorted) without damage. Most wind and hydroelectric generators should not be allowed to "freewheel" in an open-circuit condition or they could be damaged from over-speed. These systems require a charge controller that can transfer the power to an alternate load (a dump, or diversion, load) when the batteries are full but the generator keeps producing power. Most wind and hydroelectric controllers are available from the manufacturer as part of a system package. Many inverter manufacturers also produce charge controllers, but some additional brands include Midnight Solar, Morningstar, Blue Sky Energy, and SCI (see Resources for websites).

SYSTEM MONITORING

A good system meter will show the battery voltage, the charge and discharge currents, and the state of charge of the batteries. A meter also facilitates

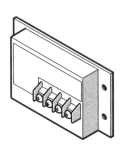

▲ **Charge controllers range from simple on/off switching to sophisticated power management with MPPT and metering.**

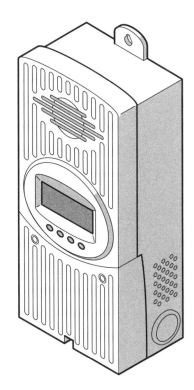

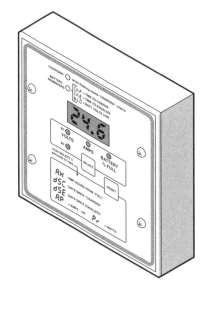

◀ **Power system monitor from Tri-Metric (see Resources)**

system troubleshooting, making it easier for you to know exactly what's happening throughout the system.

DIVERSION OR DUMP LOAD

This provides a place for excess power to be dumped or diverted away from the usual load. It can serve two purposes:

1. Once the batteries are charged in an off-grid system, any excess power generated has no place to go, so it's effectively wasted when the charge controller disconnects the power source from the batteries. If that power can be diverted to some "opportunity load" and put to good use, the renewable energy source can be more fully utilized.

2. A dump load can prevent a wind or hydro generator from spinning too fast once the batteries are charged and the controller disconnects the source from the load. In a grid-tied system, the load is the grid. If the grid goes down, the load disappears and the electrons need a place to go to keep the spinning generator under control and prevent it from tearing itself apart.

The most common diversion loads are electric resistance heating elements, such as those used for heating air or water. The load must be electrically matched to the power source to handle the voltage and current effectively. Renewable-energy equipment dealers and manufacturers offer products for specific applications. HotWatt (see Resources) offers a wide variety of air and water heaters suitable for diversion loads.

The electrical transfer from the charging load to the diversion load is normally handled through a charge controller that senses that the battery bank has reached a set voltage, or that grid power is not available. Power from the source is then diverted to the dump load.

LIGHTNING ARRESTOR

Also called *lightning protection devices* or *surge protectors*, lightning arrestors should be installed as close as possible to the power-generating source. On a PV system, it would be mounted on the combiner box. It is wired in parallel with the power cables, with one wire connected to a ground rod; if the voltage surges, the excess energy is quickly shunted to the ground.

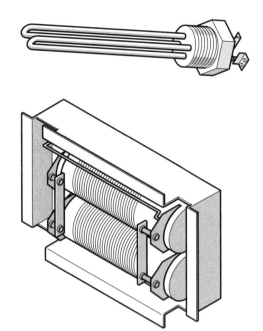

Typical dump loads: low-voltage water heater element (top) and electric resistance air heater (bottom) ▶

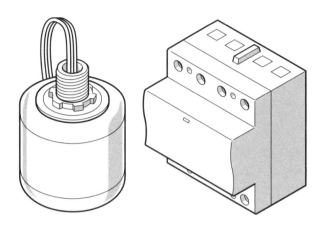

▲ Lightning protection devices. Two different kinds from Delta and ETI; see Resources

Lightning arrestors can handle surges of 10,000 amps or more in a matter of a few nanoseconds, but they should be replaced after taking a hit. Some devices provide visual indication that they need replacing, such as LED lights, and may also show visible signs of damage. Good grounding and a quality lightning protection device go a long way toward preventing damage to your power system from close lightning strikes. However, nothing in the world can withstand a direct lightning strike without damage or total destruction.

BACKUP GENERATOR

Having a backup generator in an off-grid system allows the use of high-power equipment without draining the batteries, or if the inverter doesn't have the power capacity to handle a large load. A generator also provides a source of power for battery charging during prolonged periods when the renewable resource may not provide enough power to meet your needs.

Generators can be powered by gasoline, diesel, kerosene, liquid propane gas, natural gas, wood gas, or biogas (the latter two with modifications to the engine). Lower-speed (1800 rpm) engines and generators are quieter and often last longer, though they are more expensive than the more common high-speed (3600 rpm) engine/

generators. Northern Lights and Kohler (see Resources) are just two manufacturers offering low-rpm, durable generators.

BATTERY CHARGER

Off-grid power systems are essentially battery chargers. However, when the renewable resource is not available, a separate battery charger can be operated by a source of AC power, such as a fossil fuel–powered generator. A hybrid power system (one that is grid-tied and includes battery backup) can use utility power to charge batteries. Battery charging also may be integrated with the inverter, as many inverters have built-in battery chargers. Before choosing a battery charger, read the following section on batteries and make sure to use a charger that is smart enough to charge batteries effectively. When using a fossil fuel–powered generator, be sure the generator can deliver the full power required by the charger.

ADDITIONAL SYSTEM COMPONENTS

A variety of additional components are required for your renewable energy system. Some are specific to the resource being harvested, many are common to power management. All must be properly designed and rated to handle the job at hand. This equipment includes:

- solar panel mounting racks
- wind towers
- water sluices
- dams
- electrical grounding
- disconnect boxes
- fuses
- circuit breakers
- weathertight cable connectors
- box to hold the batteries

Batteries

BATTERIES STORE ENERGY in the form of DC electricity, for use when the power source may not be producing power. When the sun shines, the water flows, or the wind blows, the batteries are being charged. The stored electrons are available to supply DC electricity during times of low or no power production, smoothing out an otherwise inconsistent supply of electricity. Sufficient energy storage allows for lower-capacity (less costly) energy-generating equipment because the energy stored over time in the batteries can be released at the desired rate, providing the required power to electrical loads.

BATTERY RATING AND TYPE

Batteries are classified by voltage, and their energy storage capacity is rated in **amp-hours**. Amp-hours are a measure of how many amps of current can be supplied to a load over a period of time. Mathematically, a 200-amp-hour battery should deliver 200 amps for one hour (or some combination of amps and time leading to the same number of amp-hours) before its charge is depleted.

In reality, the faster a battery is discharged, the less overall energy it will deliver due to its internal electrochemical efficiency. The amp-hour rating of batteries typically is based on a 20-hour discharge rate, meaning that the battery can deliver a certain number of amps over 20 hours, at which time the battery has been fully discharged.

Batteries used in a renewable energy system must be specifically designed for **deep-cycle duty**, meaning that they can be consistently and deeply discharged without significantly shortening their life span. Deep-cycle designation generally means that the battery can consistently withstand a depth of discharge (DOD) of up to 80 percent of its full rated capacity, meaning that 20 percent of the stored energy remains in the battery.

When a battery has delivered its rated amp-hours to a load (thus achieving its rated DOD) and is then recharged, it is said to have undergone one **charge cycle**. Cycle life drops dramatically with improper charging and poor maintenance, such as allowing the electrolyte level to drop too low or terminal connections to become corroded.

In addition to use in off-grid homes, deep-cycle batteries may be used in forklifts, golf carts, and wheelchairs. The most common technology used for home power storage is the **flooded lead-acid** battery, meaning that the lead-based electrodes are immersed in a liquid electrolyte (hydrochloric acid). Lead-acid batteries are used primarily because of their relatively low cost, availability, and familiarity. They are similar to the battery used in your car, but because conventional car batteries are designed for only about a 10-percent discharge before they require recharging, they would not perform well in a home power system.

CHOOSING BATTERIES

Batteries are likely to be the biggest maintenance item in your renewable energy system and represent a substantial part of the cost, so they're worth careful consideration. There are many battery brands and several technologies to choose from. When selecting batteries for renewable energy systems, compare the following performance factors so that you can determine their lifetime cost and maintenance requirements:

- storage capacity — rated in amp-hours
- depth of discharge (DOD) — 80 percent DOD is typical
- cycle life — number of charge cycles that can be endured before the battery must be replaced

- electrolyte reservoir capacity — a larger reservoir means less frequent watering is required

Lead-acid batteries have a low cost per amp-hour and generally survive through 500 to 1,000 charge cycles. Flooded lead-acid batteries have a liquid electrolyte that must be checked and filled periodically. As batteries charge, the electrolyte is depleted as gases are produced and released into the air. These gases are poisonous and must be removed from the house. Industrial lead-acid batteries are big, heavy, and costly, but when properly cared for they can last quite a bit longer than smaller batteries. Another consideration is how to dispose of the batteries at the end of their life. Lead-acid batteries are almost completely recyclable, while other types of batteries utilizing heavy metals may require disposal as hazardous waste.

Sealed lead-acid batteries utilize electrolyte that is gelled or absorbed into a substrate. They have no liquid electrolyte to fill or spill and do not off-gas while charging. This can be desirable in cases where low maintenance is desired, but care must be taken when charging so as not to boil off the gelled electrolyte, which cannot be replaced as with flooded batteries.

Nickel-cadmium (ni-cad) or other, advanced-technology batteries using exotic materials are quite a bit more expensive and often less forgiving in their charging requirements when compared to lead-acid batteries, but they hold great promise in terms of capacity and longevity.

Batteries need periodic inspections (every two or three months) to ensure adequate electrolyte level and clean, solid, corrosion-free cable connections. Other maintenance requirements include:

- Paying attention to the voltage and charge level on your system monitor to avoid heavy discharge or overcharge

- Keeping the batteries warm (70 to 90°F)

- Following a regular charging regime according to the manufacturer's charge profile, to maximize performance (see pages 186 to 189)

Battery life can range from 1 to 25 years, depending on the type of batteries you choose, how deeply they are discharged, how closely you follow the recommended charge regime, and whether you keep them filled with the proper amount and type of electrolyte (add only pure distilled water to lead-acid batteries).

Some battery brands made for use with renewable energy systems include Rolls, Surrette, Interstate, Trojan, and Deka (see Resources for websites).

SIZING YOUR BATTERY BANK

Once you've determined how many kilowatt-hours (kWh) are required to power your home each day, decide how many days you want your batteries to last between charging cycles. This depends primarily on the consistency of the charging resource. For example, if you need 3 kWh each day and you want the ability to go 3 days without the need for charging, then your batteries should have 9 kWh of storage capacity.

Batteries rated for 6 volts and 220 amp-hours are commonly used as a building block in small-to-medium-size **battery banks**, or groups of batteries. The amp-hour capacity rating typically reflects the battery's storage capacity over a period of 20 hours, during which time it is completely drained. If you have a 24-volt power system you will need to wire four 6-volt batteries in series (positive to negative) to achieve 24 volts (see pages 128 to 129 for information on series and parallel wiring).

Figuring Battery Capacity

Converting from the amp-hour ratings of batteries to kilowatt-hours of electric use requires a little math:

> volts x amps = watts
> 6 volts x 220 amp-hours = 1,320 watt-hours, or 1.3 kWh of energy storage

Because you don't want to discharge the battery all the way — only to no lower than 80 percent for a deep-cycle battery — adjust the rated storage value to find the battery's **useful capacity**:

1,320 watts x 0.80 = 1,056 watt-hours, or just over 1 kWh

Four of these 6-volt batteries (wired in series to produce 24 volts) will store approximately 4 kWh of electricity. Connecting two of these series-wired battery strings in a parallel configuration will store 8 kWh of electricity. However, keep in mind that mixing and matching batteries of different voltages, capacities, or ages is not recommended because the older batteries will waste energy by constantly draining the newer batteries (but without having the capacity to accept and hold the charge). Batteries of different capacities will recharge at different rates, leading to inefficiencies in charging.

Figuring Bank Capacity

The amp-hour capacity in a *series* string is the same as that of a single battery, but both the voltage and available energy (watt-hours) have increased. The amp-hour capacity in a *parallel* wiring arrangement is a multiple of the number of batteries in the battery bank. In this case, the voltage stays the same, but again, the available energy has increased. To determine the watt-hour capacity of the entire battery bank, multiply the watt-hours per battery by the number of batteries in the bank, regardless of wiring configuration.

Figuring Charge Rate

The greater the charging capacity, the faster your batteries will recharge. For example, if you have a 1 kW (1,000 watts) solar array (group of solar modules wired together) and a 10 kWh battery bank, here's how long it will take to recharge the dead batteries (assuming full sun reaching the solar array):

10,000 watt-hours ÷ 1,000 watts = 10 hours

As with all things, batteries are not 100 percent efficient. A typical lead-acid battery is about 75 percent efficient, meaning that 75 percent of the energy that went into it will be available for use. Therefore, you need to give a battery 25 percent more energy than what came out of it for it to be fully recharged.

BATTERY CHARGE PROFILE

Ensuring your batteries achieve a full charge and have a long life requires **charge profile optimization**. This is the optimal pairing of power generation to storage capacity, ensuring that the correct voltage and current are delivered to the batteries for the correct period of time. Manufacturers have recommendations specific to each battery so that it absorbs as many electrons as possible for a certain period of time, then is held at a set voltage until the charge current (in amps) reaches a certain level, after which the battery is held at a lower, float-charge voltage.

Battery Balancing Act

The size of the battery bank must be matched to your rate of power consumption and your need for periods of autonomy when no charging occurs. Equally as important is the electrical supply capacity of the charging system (i.e., how much power the PV system can deliver). It's important to recharge the battery bank fully in a reasonable amount of time when the charging resource is available. If there is not enough charging capacity, the batteries will never be fully charged. Too much charging capacity relative to battery size only results in wasted energy (and money) because batteries can accept a charge only as fast as their chemistry allows. A charge controller regulates the battery charge rate to prevent overcharging, but there is nothing that will compensate for too little charging capacity.

A battery has three basic charging stages, plus a periodic, controlled overcharging stage called **equalizing**. A high-quality charger automatically manages these stages and allows the user to adjust the parameters to meet the specific needs of the batteries. Voltages required are very precise values, and while this may seem nitpicky, these details are important if you want to maximize battery lifetime and efficiency. I offer some reference values below, but always follow your battery manufacturer's recommendations when setting up your charger.

The four charging stages are:

1. **Bulk** charging, during which the battery will accept as much current as it can take from the charging source, limited only by the battery's chemistry and the capacity of the charger. Empty batteries can accept a huge amount of current to fill them up, but more current than the chemistry can manage will lead to heat buildup in the battery and wasted charging energy. During bulk charge, the battery's voltage will rise until it has reached the voltage set by the charger. The bulk charge voltage is usually set at or slightly above the point at which the batteries cannot accept additional charge and the electrolyte begins to bubble and off-gas, as the electrolyte separates into hydrogen and oxygen. For flooded lead-acid batteries, the gassing voltage is around 2.37 volts per cell, or 14.2 volts for a 12-volt battery.

2. **Absorption** charge begins when the bulk-charge voltage set point is reached. The voltage is held constant for a period of time, and the charge current drops off as a result of the battery's increasing internal resistance as the state of charge rises. The absorption charge cycle can last for a preset period of time or until the current falls to a specified level.

3. **Float** charge starts after the absorption charge criteria have been met. Float charge can be considered a "maintenance" charge, where the voltage is held to a minimum and current is reduced to a trickle. Float voltage for a 12-volt flooded lead-acid battery is usually between 13 and 13.5 volts.

4. **Equalization** charging is a high-voltage, controlled overcharge that is performed periodically (perhaps monthly) and for a period of several hours. Equalize voltage for a 12-volt battery is over 15 volts. The purpose of an equalization charge is to break up sulfation, a buildup of lead sulfates on the electrodes. A battery left in a low state of charge for an extended period of time may develop excessive amounts of sulfation that cannot be removed. Sulfation reduces the amount of current a battery can deliver.

The Battery Charge Profile chart on the following page shows the recommended charge profile for certain Interstate brand, 12-volt flooded lead-acid batteries (see Resources). Note that it specifies a charge current of C/10, where "C" is the battery's 20-hour capacity in amp-hours. C/10 means that the capacity is divided by 10. For example, for a 220 amp-hour battery, the recommended charge current would be 22 amps. Faster charge rates result in excess heat buildup, and the battery may not absorb all the energy it needs to be "filled" even though it has met the voltage set point. Lower charge rates may not be able to overcome the cells' internal resistance. An ideal charge rate for renewable energy systems is between C/6 and C/12, or between 18 and 36 amps for our 220 amp-hour battery.

Adjusting Battery Temperature

Battery performance is rated at a specific temperature, usually 77°F. A cold battery will not accept or deliver its charge as readily as a warm battery, and a hot battery can suffer plate damage and a shortened life. A lead-acid battery, for example, does not want to be more than 95°F. To match the charge profile, the temperature must be compensated for by adjusting the charge voltage. For a lead-acid battery, charge voltage is increased by 0.028 volts per cell (0.17 volts for a 12-volt battery) for every 10 degrees below 77°F. If the battery temperature is greater than 77°F, the charge voltage is decreased by the same amount. Some battery chargers automatically adjust the charge voltage based on the battery temperature. Such

TESTING BATTERIES USING SPECIFIC GRAVITY

Testing the state of charge of a wet-electrolyte battery can be done by measuring voltage with an accurate voltmeter, or (more accurately) with a hydrometer to measure the specific gravity of the electrolyte. Voltage correlates with the specific gravity, but to be accurate in either case requires that the battery rest for several hours after being charged, and with no load, so that the chemistry stabilizes, and in turn so does the voltage.

To test the charge level with a hydrometer, draw electrolyte into the hydrometer and read the value on the floating scale inside the tube. The specific gravity readings between cells should not vary by more than 0.05 points. If they do, it's time to perform an equalizing charge. Specific gravity changes with temperature, so be sure to use a hydrometer with a temperature compensation scale. The table here shows the state of charge of both 6- and 12-volt batteries as indicated by a voltmeter and a hydrometer, at a temperature of 77°F.

BATTERY VOLTAGE RELATIVE TO SPECIFIC GRAVITY

State of Charge	Specific Gravity	6-volt Battery	12-volt Battery
100%	1.27	6.34	12.68
90%	1.25	6.30	12.60
80%	1.24	6.26	12.52
70%	1.23	6.22	12.44
60%	1.21	6.18	12.36
50%	1.19	6.14	12.28
40%	1.18	6.10	12.20
30%	1.16	6.05	12.10
20%	1.15	6.00	12.00
10%	1.13	5.92	11.85
0%	1.12	5.85	11.70

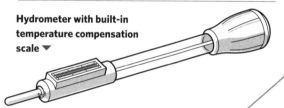

Hydrometer with built-in temperature compensation scale ▼

12-VOLT BATTERY CHARGE PROFILE (SAMPLE)

Battery Voltage	Charging Current (1)	Bulk Charge Voltage	Absorption Voltage	Absorption Time, hours	Float Voltage (2)	Equalization Voltage (3)	Equalization Time, hours
12	C / 10	14.4	15.3	2 to 4	13.4	15.6	2

1. C = 20-hour capacity in amp-hours

2. Float condition is for long-term storage (several weeks).

3. Equalize every 4 to 8 weeks.

temperature compensation requires a temperature sensor that's attached to the outside of a battery and connected to the charger.

BATTERY SAFETY

As useful as batteries are, they are also very dangerous. Always take great care when working with and around batteries. Keep the following information and guidelines in mind to prevent accidents:

• Batteries contain an enormous amount of energy. If you drop a wrench across the terminals of a battery, sparks will fly, the wrench will be welded to the terminals, the battery will heat up and may even explode.

• Remove all jewelry, wear insulating gloves, and use insulated tools.

• Cover all exposed battery connections and terminals that are not being worked on with

BATTERY BOX BASICS

A sturdy box made with acid-resistant interior surfaces, such as epoxy-coated plywood, offers many benefits to protect and house your battery bank, including:

- **Providing protection** against unauthorized or untrained access.

- **Keeping battery tops clean.** When you remove the batteries' vent caps to add water, you don't want dust and debris (or *anything* other than distilled water) falling into the batteries, which can wreak havoc with the electrolyte and shorten the battery life.

- **Controlling battery temperature.** Battery temperature rises while charging and discharging, but the batteries tend to cool off in places like unconditioned basements. An insulated box helps maintain the desired temperature: Sandwich a layer of 2" rigid foam board between ¾" plywood on the outside and ¼" epoxy-coated (or other acid-resistant treatment) plywood on the inside. Caulk the interior joints and install a hinged, tight-closing, insulated, and gasketed top.

- **Removing gases with an exhaust fan.** The box must be reasonably airtight with a suitable exhaust fan to move air through and out of the box to the outdoors. An air inlet is cut into the box near the bottom (the gases naturally rise up) and screened against debris and bugs. A PVC vent fan designed for battery boxes is available from Zephyr industries (see Resources). This unit has an internal damper to prevent backflow. Some charge controllers offer an auxiliary output that allows the fan to be automatically turned on and off based on battery voltage. This control is quite useful, as batteries only off-gas once they reach a certain voltage.

Tip: Make the box a bit larger to store your hydrometer, gloves, a funnel, and baking soda.

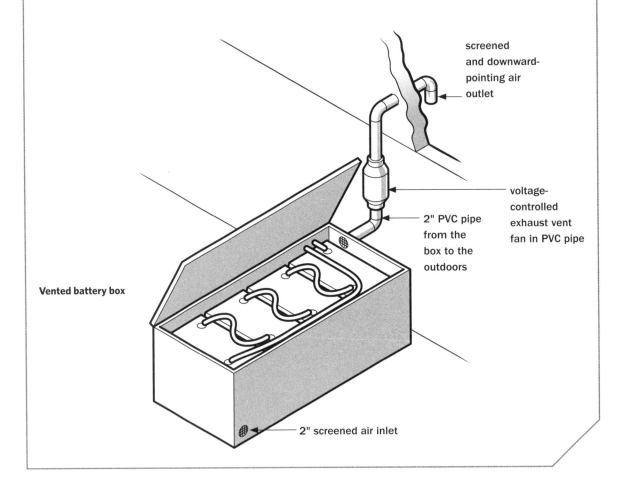

screened and downward-pointing air outlet

voltage-controlled exhaust vent fan in PVC pipe

2" PVC pipe from the box to the outdoors

Vented battery box

2" screened air inlet

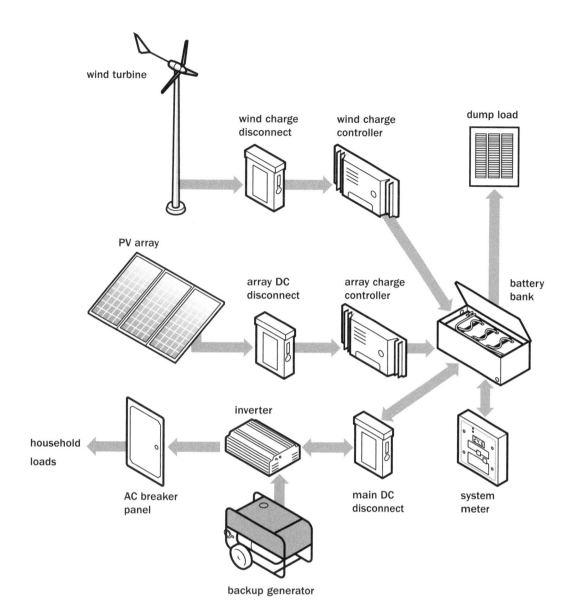

wind turbine

wind charge
disconnect

wind charge
controller

dump load

PV array

array DC
disconnect

array charge
controller

battery
bank

household
loads

inverter

AC breaker
panel

backup generator

main DC
disconnect

system
meter

**A hybrid wind and
solar electric power
system that is
grid-connected with
battery backup**

a nonconductive material, such as plywood, a rubber mat, or a durable PVC apron.

- Batteries contain deadly poison. Electrolyte in lead-acid batteries is hydrochloric acid. If you spill it on yourself or your clothes, it will burn them. If you get in your eyes, it could blind you. Keep water and baking soda handy while working around batteries. These can neutralize acid spills. Water (or a mix of water and baking soda) can be used to wash acid off skin or clothes; baking soda can be sprinkled on electrolyte spills.

- Batteries generate poisonous and explosive gases while being charged. Never smoke around batteries. Always ventilate the area where the batteries are located. Avoid sparks near batteries.

- Always disconnect the charging source(s) and the load(s) before working on batteries.

- Batteries are heavy, weighing anywhere from 60 to 120 pounds; be sure they reside on a sturdy, well-supported floor. Use straps to lift and carry batteries to avoid back injury.

- Build a sturdy, ventilated, insulated box to house your batteries (see Battery Box Basics on facing page).

Mapping Motivations ... and Watts

BECAME AN ENGINEER because I like to solve problems and make things work better. While getting my mechanical engineering degree, I learned how to calculate the efficiency of mechanical and electrical systems and how to design them so they operate as efficiently as possible.

CHRIS KAISER is an industrial and commercial building efficiency consultant, sustainability enthusiast, and cocreator and principal contributor to the Map-a-Watt blog (see Resources). Here's his perspective on efficiency, sustainability, pain, and gain.

I learned about companies taking steps to increase efficiency and operate more sustainably — and I thought all companies operated that way. Then I entered the real world. I saw inefficiency all around me.

My first job was selling automation and electrical control products to industrial customers. I wanted to sell them more efficient motors, new efficient lighting, energy monitors, and variable-speed drives that control motors better. I had a whole bundle of products to help with efficiency, but in most cases the customers would buy only the cheapest motor, not the most efficient.

Most purchasing managers I encountered cared about the up-front cost, not the lifetime operating costs of equipment. Most plant managers had no clue about how energy efficiency translates into financial gain and focused only on how many widgets can get out of the door with the least amount of financial pain. They didn't care about energy efficiency because the managers and accountants rewarded short-term economic goals instead of long-term savings. But now I work full-time selling energy efficiency solutions to companies that have realized that increasing efficiency improves their bottom line — and the beneficial environmental impacts are just icing on the cake.

One thing I've realized in sales is that people buy things that reduce their pain or offer a direct gain. If the purchasing agent or maintenance person isn't graded or doesn't receive some sort of incentive related to how much energy they save, they don't have any pain around energy costs and don't have any desire to save it. If energy is cheap, then costs haven't risen to the point where it causes management any pain. All companies love talking points on the environment or sustainability, but the majority of companies aren't going to act on operating more sustainably unless it solves a specific pain (such as with reduced operating costs) or increases gain (such as with more sales to environmentally conscious consumers).

It's not just companies that act this way; so do families and individuals. If we see no value or realize no gain in living more sustainably, we probably aren't going to do it — unless not doing it is so painful that we have no other choice. But if saving energy puts more money in your bank account each

---

If we see no value or realize no gain in living more sustainably, we probably aren't going to do it – unless not doing it is so painful that we have no other choice. But if saving energy puts more money in your bank account each month, it has a greater chance of happening.

---

month, it has a greater chance of happening. If your child has asthma that's triggered by poor air quality, you have pain around his or her suffering. If you tie that pain to realizing that smog is caused by air pollution from coal-fired power plants and gas-guzzling cars, you will be more likely to care about turning off the lights or consider buying a hybrid vehicle.

Living more efficiently and sustainably isn't a quick-and-easy, one-time deal or a gimmicky marketing campaign; it's a way of life. It's a way of operating, both personally and in business, that reduces the negative impacts of waste on our society. I try to live sustainably because it reduces the pain I feel when I see, smell, or feel pollution in my lungs, or I realize I'll have less money this month due to a high energy bill. Most importantly, living sustainably means living with less pain, which coincidentally provides the ultimate gain of living happier.

Living sustainably is all about happiness. I'm happy when I have more money. I'm happy when I'm hanging out with friends and family and helping my community. I'm happy when I'm enjoying the clean outdoors while hiking or biking. I'm happy when I'm eating good food from local farms and drinking good beer from my local brewery. I'm happy when I get a new car that has excellent gas mileage, which lowers my nation's dependence on foreign oil, saves me more money, and improves my local air quality. I'm happy when I learn something new and when I can share it with my extended online community. When people are happy, great things happen for everyone and everything.

Living more efficiently and sustainably isn't a quick-and-easy, one-time deal or a gimmicky marketing campaign; it's a way of life. It's a way of operating, both personally and in business, that reduces the negative impacts of waste on our society.

Biodiesel

T'S COMMON KNOWLEDGE to anyone who's spent time around a kitchen that vegetable oil is flammable. What's not so commonly known is that when Rudolf Diesel developed his engine back in the 1890s, it was designed to burn multiple fuels, including vegetable oil. He was under the impression that **someday** farmers would be providing both food *and* fuel. Now that the fossil fuel heyday is mostly behind us, his prediction is being fulfilled.

Somewhere around 2001, I caught the biodiesel bug. It seemed like the perfect way to offset my energy costs and reduce my reliance on fossil fuel. I had a diesel generator for backup power and was thrilled when my first batch of biodiesel burned with the smell of French fries in the exhaust, providing electricity for my house. All it took was a little time. Okay, it took a lot of time. But the fuel was nearly free and I could hardly resist.

Venturing into Biodiesel

I VISITED A dozen local restaurants and asked the managers what they did with their waste fryer oil. At first, I suspect they might've thought that I was the state health inspector. But when I explained what I was doing, they were very interested. The fact that I was offering to take their garbage away for free was an added bonus, since they usually paid to have this hauled away in barrels or waste bins.

Different restaurants have different qualities of grease. Some use the oil for so long that it's practically rancid and sometimes unusable even as a fuel. However, one high-end vegan restaurant in my area used its oil for only one day — it was practically new and made great fuel. I can't eat fried food now without thinking about biodiesel.

Collecting waste grease can be fairly clean and simple if the restaurant pours its waste oil into a 5-gallon container that you can carry away. Otherwise, it can be quite a disgusting job. You never know what you might find in the waste oil bin: rainwater, beer cans, cigarette butts, food waste — so wear old clothes when you go to collect grease.

Next I had to find a source of methanol and lye. To start, I used Red Devil brand lye from the supermarket. My propane gas supplier was able to part with a gallon of methanol, which is often used as an additive to keep propane lines from freezing in the winter. That was enough to experiment with. My father-in-law happens to be a chemist at a local university, and he was happy to join me in the experiment. This gave me a great excuse to buy my own set of chemistry tools, including a digital scale, glassware, hydrometer, and viscometer.

Back home, I set up a process that used an extended length of hose from the cooling system of the VW (see below) to heat the oil before mixing with it methanol and lye. This allowed me to use biodiesel burned in the car's engine to make more biodiesel. The electric mixing pump was powered from my home's solar electric system. After settling and filtering, the biodiesel worked great in both the generator and the VW. This gave me the confidence to use my homemade biodiesel in my daily driver, a newer VW Jetta TDI. Even at 40 miles per gallon, it takes a fair amount of time to gather and process the 375 gallons of fuel to travel 15,000 miles every year.

Every 50-gallon batch produced about 8 gallons of waste glycerin. The better oil produced a decent-quality glycerin, out of which I could make soap that proved to be an excellent degreaser. The lower-quality oil made a vile sort of goo that was best burned in a good, hot bonfire. I had read that glycerin could be composted and would readily biodegrade, but I didn't have much luck with that. Perhaps in an ideally mixed, hot compost pile it would have worked better. Over the years, I've ended up with many 5-gallon jugs filled with waste glycerin that have been disposed of in various ways, including compost, soapmaking, bonfires, and landfill.

Calamari Cruiser

I bought a beat-up 1985 VW diesel Golf so I could experiment with driving on vegetable oil. I modified the car by taking out the backseat and all the carpet, so it would be more like a pickup truck. It held a 55-gallon drum and a 12-volt fuel pump so I could transfer grease from the restaurants' grease bins to the barrel in the back of the VW. It worked great — and always drew stares and questions during the grease transfer. The car was dubbed the "Calamari Cruiser" because calamari was the main fried food of my favorite restaurant for collecting oil.

What Is Biodiesel?

BIODIESEL IS VEGETABLE oil that has been chemically modified to remove the heavy glycerin portion of the oil. This allows it to flow freely at temperatures down to around freezing, while straight (unmodified) vegetable oil must be heated to 120°F to flow freely through filters, injectors, and burners.

Both vegetable oil and biodiesel can be used in place of diesel fuel, home heating oil, and kerosene for use in diesel engines and oil-fired heating equipment. Using biodiesel requires processing the vegetable oil, but no changes to the vehicle or burner are needed. **Straight vegetable oil** (SVO) requires modification to the fuel system.

Biodiesel is chemically described as a **mono alkyl ester**. It can be used in its pure form or blended in any concentration with petroleum diesel. It can be made with vegetable oil, animal fats, or recycled (waste) fryer oil from restaurants. The oil is filtered and mixed with methanol (methyl or wood alcohol) with the aid of a lye (sodium hydroxide) catalyst to complete the chemical process known as **transesterification**. Transesterification is the chemical transformation of one type of ester to another.

Other alcohols, such as ethanol, can be used in place of methanol, and potassium hydroxide can be used in place of sodium hydroxide. These ingredients are less hazardous to work with, but the reactions are more sensitive, and the biodiesel yield is generally lower. Most commercial biodiesel makers and home brewers use methanol and sodium hydroxide.

DIY BIODIESEL

In the interest of ease of use for the biodiesel home brewer, this chapter describes the **base-catalyzed** production method using methanol and sodium hydroxide. The products of transesterification are methyl ester (used as fuel) and glycerin (soap). About 20 percent of vegetable oil is glycerin, which will be removed during the chemical process and replaced with alcohol. You'll need a plan to dispose of the waste so you don't end up with dozens of plastic pails full of glycerin scattered around the yard.

Benefits and Drawbacks

USING BIODIESEL INSTEAD of fossil diesel presents many benefits to both environment and engine, but it's not without quirks and peculiarities. All of these need to be addressed to achieve success in using biodiesel as a fuel.

BENEFITS

Biodiesel biodegrades about four times faster than fossil diesel. Its lack of sulfur eliminates the sulfur oxides (responsible for acid rain) produced when fossil diesel is burned. Because biodiesel is an oxygenated fuel containing 11 percent oxygen by weight, combustion is more complete and overall emissions are reduced by up to 90 percent when **B100** (100-percent biodiesel) is used instead of fossil diesel. These emissions include carbon monoxide, carbon dioxide, unburned hydrocarbons, and particulate matter (soot).

Soy Mileage

An average acre of soybeans grown in the U. S. produces about 42 bushels, at about 60 pounds per bushel. Each bushel of soybeans can produce about 1.5 gallons of soy oil, adding up to 63 gallons of oil per acre. In terms of heating energy value, if a gallon of soy oil contains 130,000 Btus, that's 8.2 million Btus per acre. A modern diesel passenger car gets about 45 miles per gallon, while a semitruck might get 6 miles per gallon. That works out to between 380 and 2,800 miles per acre.

Oxygen content in fuels effectively increases the volumetric efficiency of the engine as a result of more complete fuel combustion. This increase in combustion efficiency helps overcome the slightly lower energy content of biodiesel: You may not notice a significant difference in engine performance or fuel economy when using B100, but you can expect to experience some loss of each. Biodiesel's higher cetane number (a measure of the fuel's ability to ignite under pressure) translates to better fuel ignition. Increased lubricity means less wear and tear on engine and fuel system components, and it has a detergent effect on the entire fuel system.

Biodiesel is produced from recently grown biomass, rather than biomass that grew millions of years ago and has remained sequestered for that time (as is the case with fossil fuels). Therefore, no net increase in atmospheric carbon dioxide (CO_2) results from burning biodiesel because the same amount of carbon dioxide is released when that biomass decays naturally. Of course, in most cases, producing biodiesel requires the use of fossil diesel, but a net CO_2 reduction of over 75 percent can be realized when using B100. Finally, tailpipe emissions smell like fried food, which is considered a benefit by most people driving behind a biodiesel-powered vehicle.

DRAWBACKS

Shortcomings of biodiesel include a possible increase in nitrogen oxides (NOx) when burned, depending on the type of engine. Nitrogen oxides contribute to smog and ozone. Given the lower energy content of biodiesel, you may experience a 10-percent reduction in fuel economy if you're driving on B100. Biodiesel will dissolve any rubber in the fuel system, such as gaskets or seals, so all rubber parts that may come into contact with biodiesel must be replaced with synthetic materials.

Biodiesel has some cold-weather limitations. It begins to congeal (reaches its **gel point**) at around 45°F and must be mixed with fossil diesel when the temperature drops below this gel point; see Mixing Biodiesel with Other Fuels on page 204. Be aware that biodiesel acts as a solvent that will dissolve the fossil diesel deposits that have accumulated inside of your fuel tank and filter. Ideally, you should flush the fuel tank with a few gallons of biodiesel (let it sit for a day or so), drain, and repeat, until all of the gunk is gone. You should also replace the fuel filter after flushing the tank. If you don't flush the fuel tank, you'll quickly go through several fuel filters, as they will become clogged with sludge. Biodiesel will leave its own residue in the fuel tank that will build up over time. Should you switch back to fossil diesel after using biodiesel for a while, flushing the tank may again be needed.

Essential Ingredients

ONLY THREE INGREDIENTS are required to make biodiesel: vegetable oil, methanol, and a catalyst. Here's some information on each one and where best to find them.

VEGETABLE OIL

New, virgin vegetable oil is ideal for making biodiesel. You can also use oil left over from restaurant fryers. Waste fryer oil will yield a bit less biodiesel than fresh oil, but you can often find it for free at local restaurants, where they usually pay to dispose of it.

Restaurants, food manufacturers (salad dressing companies), and food processors are good sources for finding free or cheap vegetable oil. You can also buy it new in large quantities from restaurant food suppliers. Oil quality varies depending on how *used* it is. Older, more heavily cooked oil generally yields more waste in the biodiesel reaction than newer, less-used oil. Canola and soy oil are common, but you may find other types. Most cooking oils will make a suitable biodiesel.

Avoid using lard or anything else you can't pump at room temperature. Waste fryer oil often is available for free, but you may find that competition for the oil — from other biodiesel brewers like yourself — has made it a more valuable commodity.

METHANOL

Methanol (CH_3OH) technically is wood alcohol but generally is derived from fossil fuels. It is required for transesterification, the basis of converting vegetable oil to biodiesel. The amount of methanol required for a successful reaction varies a bit but typically is around 20 percent of the initial volume of oil in the recipe.

This fossil fuel is available in 55-gallon drums from chemical supply companies. You might be able to find smaller quantities at your local car racing track or from a propane gas supplier. For test batches, you can use "dry gas" gasoline additive that contains a high percentage of methanol. *Do not* use this type with ethanol or other alcohols.

CATALYST/LYE

Sodium hydroxide is also known as caustic soda or lye, or chemically as NaOH. Potassium hydroxide (KOH) also may be used, but 40 percent more is required in the recipe. The catalyst is required to chemically combine the oil and methanol to form biodiesel. The amount of catalyst varies based on the acidity of the oil.

Oil that has been heated, burned, or used for cooking is more acidic than new oil and will require more catalyst (a base) to neutralize the acid and create the chemical bonds. Liquid catalyst mixes better than dry or crystalized products. Sodium hydroxide generally is more widely available than potassium hydroxide, and the recipes in this chapter are based on using dry sodium hydroxide.

Lye is available in various quantities from chemical supply companies or, for small batches, use drain cleaner that is 100-percent sodium hydroxide, found at hardware and grocery stores. Keep the lye dry in a tightly sealed container.

Equipment Needs

TO MAKE YOUR own biodiesel, you will need equipment for collecting, storing, pumping, mixing, and heating the oil, as well as some basic chemistry equipment, and, of course, appropriate safety gear.

Collection. If you plan to collect waste fryer oil from local restaurants, first talk to the manager and find out how they currently dispose of it. It may go into a garbage bin in 5-gallon buckets, or it may go into 55-gallon drums or 300-gallon receptacles to be picked up by a waste oil hauler. If they dispose of it in 5-gallon jugs, your job is easy. If not, you'll need to transfer the oil from their storage vessel to a collection tank or buckets on your pickup truck. You can do this with a hand or motorized pump. If the oil normally is collected by a waste hauler, you may need to enter into a contractual agreement.

Pumping. If you're making only small quantities of biodiesel, an inexpensive, hand-cranked barrel pump works well. For moving a lot of liquid, a 12-volt DC-powered diesel-fuel transfer pump is a good solution. Just attach the power cable to your truck battery and run hoses between the collection vessel on your truck and the restaurant's oil storage bin. When you're back home, use the same pump to off-load the oil from your truck into your storage barrels.

The same pump can be used to:

- Fill the reactor, or mixing, tank with vegetable oil

- Pump the glycerin out of the bottom of the reactor and into a waste collection container (after mixing and settling)

- Remove wastewater mixture from the tank after washing

Tip: Draw from the Middle

Water and debris settle to the bottom of the waste oil bin. When using a transfer pump to collect waste fryer oil, adjust the length of the intake tube so that it won't sit on the bottom of the barrel. Debris and scum water accumulate on top of the grease, so also avoid pumping off the top. Water in the oil will ruin a batch of biodiesel.

- Transfer the biodiesel from the reactor to the fuel storage or settling tank
- Pump biodiesel through a filter and into your car's fuel tank

You can get by with a single pump, but it's handy to have two. Diesel fuel pumps are not designed for use with water, but you should be able to avoid problems by making sure the last liquid through the pump is oil or biodiesel. This keeps the seals lubricated and prevents rust from forming inside the pump. Good sources for fuel-handling equipment include large auto supply shops and the online retailer Northern Tool and Equipment (see Resources).

Mixing tank. A suitable mixing tank for the home brewer can be made from a 50- to 200-gallon plastic barrel with a conical bottom, which facilitates pumping out the glycerin after processing. Such tanks are available from specialty agricultural supply stores, such as PolyDome. or industrial plastic manufacturers, such as U. S. Plastic Corporation (see Resources).

Heat. You will need to devise a way to heat the vegetable oil to 120°F for mixing. This can be done in an old water heater, in a barrel with a submersible electric heater, or using another heat source and heat exchanger of your own design. Good places to find heaters suitable for this task include farm stores; agriculture supply outlets; auto parts stores; and online resources, such as Diesel-Therm and Jeffers Pet (see Resources).

Mixing motor. If you're making only a few gallons of biodiesel, you can get by with a portable drill and a paint-mixing attachment. A ⅜" drill will do fine for starter batches, but you may soon find yourself upgrading to a ½" drill with a larger motor. For mixing larger quantities, you'll want to use a ½-horsepower (hp), 1800-rpm motor and a good mixing blade attachment to provide vigorous

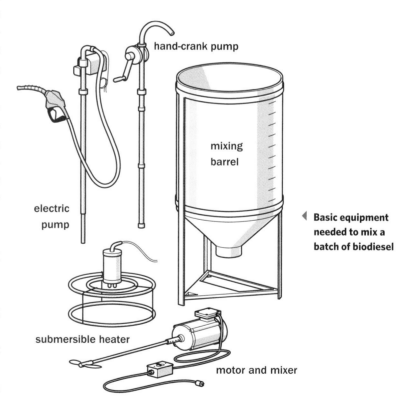

hand-crank pump

mixing barrel

electric pump

submersible heater

motor and mixer

◀ **Basic equipment needed to mix a batch of biodiesel**

Warning!

Never use a diesel fuel pump for pumping methanol or gasoline! Methanol forms explosive vapors at a temperature of 52°F, and sparks from a pump motor could easily ignite the fumes.

agitation. Remember that methanol is flammable. Sparks from motor brushes can ignite methanol vapors, causing an explosion. Use only Class 1-rated "explosion-proof" motors around flammable vapors. A Class 1 motor is constructed to contain an explosion within itself without rupturing. A suitable motor-mixer combination, designed for use with drums and barrels, is available from Neptune Chemical Pump Company (model F-3.1; see Resources). These are costly but will pay for themselves the first time they prevent an emergency-room visit.

Lab ware. Some basic lab equipment is required for measuring and weighing the ingredients and determining the proper amount of lye to use for the biodiesel reaction. The more often and longer the grease was heated for cooking, the greater amount of lye is needed for a complete reaction. The process for determining this is called **titration** (see Making a Test Batch with Used Vegetable Oil, page 203).

For lab ware, Frey Scientific is a good source (see Resources). Order extra; glass breaks. A basic lab kit to get started converting waste vegetable oil to biodiesel includes:

- Electronic scale for weighing quantities of lye up to 2 kilograms (kg), with a minimum resolution of 1 gram (g)

- Calibrated, 1.5-milliliter (mL) pipettes for measuring small quantities of oil and alcohol for titration when waste vegetable oil is used

- Hydrometer to measure the specific gravity of the biodiesel

- Two 10-mL graduated cylinders

- Glass beaker or wide-mouth jar

- 2-liter (L) beaker to measure your lye

- 1-L bottle to store your 1-percent solution of sodium hydroxide

- pH paper

- Funnels for pouring various wet and dry ingredients

- Scoop for measuring out lye

- Stirring rods

Filtering. Biodiesel must be filtered before it goes into your fuel tank. You can use a "sock"-type filter (a filter bag that looks like a big sock) or standard automotive fuel filters in line with the fuel pump to filter out particles greater than 10 microns.

Storage. Biodiesel can be stored for over a year in a clean, dark, dry environment. Use only containers suitable for liquid storage, made from black mild steel, stainless steel, fluorinated polyethylene, or fluorinated polypropylene. Keeping the biodiesel cool helps it last, but longer-term storage may result in fungal growth within the biodiesel (especially if any water is introduced into the fuel), requiring treatment with a chemical biocide. Keep in mind that there may be state or local regulations governing storage of large quantities of fuel or vegetable oil, so check with your city or local health department for more information.

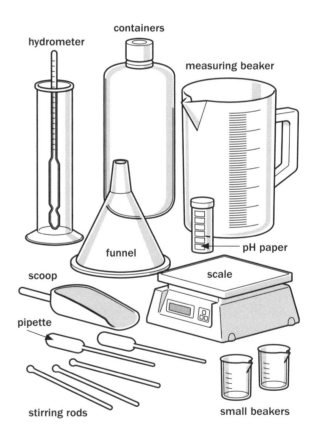

▲ **Lab ware required for titration, measuring, weighing, mixing, and testing biodiesel**

Safety

BIODIESEL IS FLAMMABLE, but much less so than diesel fuel. It has a minimum **flash point** (the temperature at which vapors will ignite when exposed to a spark or flame) of 266°F, compared to about 165°F for diesel fuel. Biodiesel is relatively safe and less toxic than fossil fuels, and it's not difficult to make and use, but it does require close attention to detail and respect for the materials and equipment you will be working with. *Don't get lazy with safety.*

It is extremely important to use personal safety equipment when working with methanol and lye. Obtain, read, and understand the **material safety data sheet** (MSDS) from the supplier of any chemicals you use. Handle biodiesel as you would any other fuel, and always take steps to prevent personal and environmental contamination. Biodiesel must be made outdoors or in a very well-ventilated area, using great care to protect yourself from hazardous materials and conditions. At a minimum, you will need:

- Chemical-resistant goggles
- Organic vapor respirator
- Nitrile gloves
- Chemical-resistant apron
- Clothing that completely covers all of your skin
- Ground-fault circuit interrupter (GFCI) — use this type of electrical outlet (receptacle or extension cord) for plugging in any electrical devices used in your processing operation
- Vinegar and water to neutralize spills of the corrosive lye and sodium methoxide

Many of these supplies are available at hardware stores or online through such places as Industrial

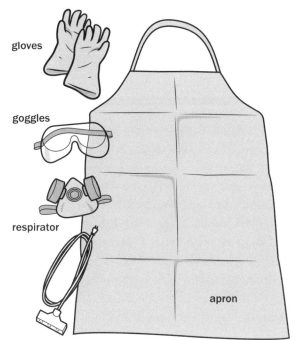

gloves

goggles

respirator

◀ **Safety equipment**

apron

GFCI extension cord

Safety Company, Grainger, and Direct Safety (see Resources).

Methanol is an alcohol, and the vapors are flammable when mixed with air in concentrations between 6 and 36.5 percent. It has a flash point (the temperature at which it will begin to evaporate, forming explosive vapors) of 52°F. Methanol is poisonous and can cause blindness and internal bodily damage; a quantity of only 2 ounces is enough to kill a human if swallowed. It is readily absorbed into blood through the skin, with exposure reactions similar to those from ingestion. Inhaling the fumes can make you sick in many ways: *Always wear a respirator.*

Sodium hydroxide, or lye, has a pH of 12 and will burn your skin, eyes, mouth, lungs, and clothing. It can be fatal if swallowed.

Sodium methoxide (methanol-lye mixture) is highly toxic, flammable, and explosive. Exposure

A Cautionary Tale

A farmer friend of mine made several successful batches of biodiesel during the summer months so that he could heat his greenhouses during the winter. When the weather turned cool in the fall, he moved the mixing operation into a greenhouse. After starting a batch by mixing the methanol and lye, he returned a short time later to find all his tomato plants dead; they had been burned by sodium methoxide fumes. When he entered, not yet knowing what had happened, his eyes, nose, and lungs were instantly burned. It took weeks for the burns to heal. Making biodiesel is an outdoor activity!

to this extremely strong base will burn your lungs, skin, and eyes.

Glycerin contains both unreacted methanol and catalyst, and therefore is somewhat caustic. (This is why it makes such a good soap.) Depending on what you want to use it for, and how complete your biodiesel reaction was, you may need to clean and neutralize your glycerin.

Biodiesel and vegetable oil can spontaneously combust under the right conditions. Do not leave oil-soaked rags in enclosed containers. Also, vegetable oil is slippery! If you spill it, clean it up immediately. This is especially true if you spill it at the restaurant where you are picking up the oil (ask me how I know).

Basic Steps for Making Biodiesel

IF YOU'RE FORTUNATE enough to be using fresh, new vegetable oil (as opposed to waste fryer oil from a restaurant), you will be reacting a quantity of carbohydrates in the form of vegetable oil with 20 percent (by volume) methanol and a small amount (0.35 percent) of sodium hydroxide as a catalyst.

I highly recommend that you start your biodiesel experiments with a small blender batch using fresh oil, and work your way up to more substantial volumes and converting waste oil. This will give you some practice with the process and equipment before moving on to larger batches.

During the process, the bonds of the triglycerides that make up vegetable oil will be broken by the catalyst. Glycerin is replaced with alcohol to form methyl ester, and the catalyst then bonds with the glycerin, which will settle out as waste. The result will be 80 to 90 percent fatty acid methyl esters (biodiesel) and 10 to 20 percent glycerin (a carbohydrate). As a point of reference, new canola oil weighs about 7.6 pounds per gallon, but used oil will weigh a bit more. Methanol weighs 6.63 pounds per gallon. Your finished biodiesel will weigh about 7.4 pounds per gallon.

THE BASIC PROCESS

First we'll cover the basic steps of making biodiesel, followed by the specific processes for making test batches with new oil and waste oil. Make sure to do all of your mixing with mechanical mixers or pumps suitable for use with caustic and flammable chemicals.

1. Gather your oil. If using waste oil, strain it to remove bits of food and other debris. Try not to use oil that has been burned or has gone rancid. You can make decent biodiesel from some pretty nasty oil, but higher-quality oil means a better reaction and more usable fuel for your efforts. The oil quality varies depending on the source and on whether it is fresh-pressed or waste from restaurant cooking. Do not attempt to use oil that has any water in it — it will ruin the entire batch (again, ask me how I know). If it smells like something that died and rotted, find another source (you can ask me about that, too).

2. Perform a **titration** (see Making a Test Batch with Used Vegetable Oil, page 203) if you are using waste fryer oil, to determine how much lye to use for a complete reaction.

3. Mix methanol with the correct amount of lye for about 15 minutes to form sodium methoxide. This is an exothermic chemical reaction, meaning heat will be created.

Collecting oil from a restaurant with the Calamari Cruiser ▶

Managing Waste Glycerin

The waste glycerin resulting from making biodiesel can be cleaned and made into soap, or it can be composted as with other organic materials. Be aware that the pH of the waste glycerin will be quite high due to the lye used in the mix. The pH will neutralize when combined with other organic matter. There will also be unreacted methanol in the glycerin, which you don't want on your skin, but this will evaporate over time.

4. Heat the oil to 120°F. While you can have a successful reaction at temperatures above 70°F, the mixing time will be longer and you may end up with more unreacted byproducts, along with lower-quality biodiesel.

5. Add the vegetable oil to the sodium methoxide mixture, and mix for about an hour. Avoid splashing the methoxide.

6. After mixing, allow the glycerin to settle to the bottom of the mixing tank (this should take 4 to 8 hours).

7. Pump or drain the glycerin into a waste container after it has settled. If you wait longer than 12 hours, the glycerin will begin to solidify and cleanup will be more difficult because the glycerin will be too thick to be pumped.

8. Pump the biodiesel immediately into a holding tank and let it settle for a week or so, during which time excess methanol will evaporate and smaller particles will fall to the bottom. If necessary, you can "wash" the biodiesel before storing it by mixing it with water to remove impurities, such as small glycerin particles and any excess methanol (see Washing Biodiesel on page 205). If you're making large amounts of biodiesel, it's a good idea to recapture the methanol by boiling it off the biodiesel and capturing it through distillation for reuse. This is prudent from both an economic and environmental standpoint.

MAKING A TEST BATCH WITH NEW VEGETABLE OIL

Starting off with a small blender batch using new vegetable oil is a good way to get your feet wet with making biodiesel. First, get a cheap or used blender; the plastic and rubber parts will likely disintegrate after a few batches, and you won't want to use the same blender for food. If you're working indoors, be sure to protect yourself, your

---

BIODIESEL QUALITY AND MEETING ASTM STANDARDS

Professionally made and commercially sold biodiesel needs to meet the American Society for Testing and Materials (ASTM) D6751 specification. It must be registered with the U. S. Environmental Protection Agency under 40 CFR Part 79. Homebrew biodiesel often does not meet those high quality standards but is useful as a fuel if properly made and thoroughly reacted. Commercial-grade biodiesel meets the following standards:

- Glycerin content of less than 0.24 percent

- Methanol content of less than 0.2 percent

- Water content of less than 0.05 percent

- pH of 7

- Specific gravity between 0.86 and 0.90

Unless you have a friend who works in a chemistry lab, most of these tests (other than pH and specific gravity) can be quite expensive if you want the accuracy required by ASTM.

clothing, and all work surfaces from splashing or spilled chemicals. Heed all of the advice in the safety section (pages 199 and 200), regardless of batch size. Basic solutions will burn skin, clothes, and finished surfaces just as quickly as acidic solutions.

This recipe is for 1 quart of biodiesel (with metric units for making 1 liter shown in parentheses) but the process and proportions are the same regardless of how much you'll be making. You will need:

1 quart (1 L) of NEW vegetable oil

6.4 ounces (200 mL) of methanol (20 percent of the amount of vegetable oil)

0.11 ounces (3.5 g) of sodium hydroxide (0.35 percent of the amount of vegetable oil) from a fresh, unopened container

Safety gear (see pages 199–200)

Pot and thermometer for heating oil

1.5 quart (or larger) glass container

Here's what you do:

1. Put on your safety gear.

2. Measure all ingredients.

3. Heat the vegetable oil to 120°F.

4. Carefully pour the methanol into the blender.

5. Gently pour the lye into the methanol and close the top tightly. Blend at low speed until the lye is completely dissolved, a few minutes, to make sodium methoxide.

6. Stop the blender and — keeping your face away from it — take off the top.

◀ **Sample of mixed biodiesel showing separation of glycerin (bottom) and oil esters (top)**

7. Slowly pour the vegetable oil into the sodium methoxide and blend at medium speed for 15 minutes.

8. Pour the biodiesel into a glass container and let it settle.

If all has gone well, you should see two layers start to form in the container after about 10 minutes. Depending on the type and quality of oil and the completeness of the reaction, anywhere from 75 to 95 percent of the product will be biodiesel. After 8 hours or so, much of the settling has occurred and you can pour the biodiesel off the top, filter it, and use it as fuel. Once the biodiesel is drawn off, you can dry out the bottom layer of glycerin by leaving the open jar outdoors in a protected area for a week or two, so that the alcohol has a chance to evaporate, then use the glycerin as soap. To purify the glycerin for other uses, further processing may be required.

Testing the Mix

To determine if your mix is a success, dip a jar or beaker into the reactor immediately after the mixing is completed (step 5, page 201), and draw off a pint or so of the liquid. Within 15 minutes, you should see two layers begin to form in the mixture. The top layer is biodiesel; the bottom layer is a slurry of glycerin, lye, and particles that may have been in the oil. After 8 hours, most of the settling has occurred. If you use a beaker, you can easily see how much glycerin was produced compared to biodiesel. Cleaner, less-used oil will produce less (and cleaner) glycerin.

MAKING A TEST BATCH WITH USED VEGETABLE OIL

The only difference between making biodiesel with used oil instead of new oil is the amount of lye added. All other processes are the same. To determine the quantity of lye required, you will need to perform a titration — a process to determine the concentration of an acid or a base — of the oil. This is necessary because oil that has been heated will be more acidic than fresh oil and require more catalyst (a base) to neutralize. Over time, you may find that oil from a single source has a fairly consistent quality, and experience may teach you how much lye to add without performing a titration, but otherwise a titration is always required for best results.

You'll need the following materials for the titration:

Sample of the vegetable oil you wish to use

Fresh bottle of isopropyl (rubbing) alcohol, at least 90 percent pure

1 L distilled water

1 g sodium hydroxide

1 L bottle (glass or chemical-resistant plastic) with tightly closing cap

Pipettes to measure 1-mL increments of liquid

pH paper

Glass mixing containers

Mixing rod

To begin, make a 1-percent sodium hydroxide solution by adding exactly one gram of lye to the 1-liter bottle, fill it with 1 liter of distilled water, and mix thoroughly. This solution becomes the basis for many titrations. Complete the following steps for a titration of used vegetable oil:

1. Pour 10 mL of rubbing alcohol into a small beaker.

2. Add 1 mL oil, using a pipette.

3. Mix the oil and alcohol thoroughly (they mix more easily when warm). A well-mixed solution will be cloudy and will not separate quickly.

4. Using a clean pipette, add 1 mL of the 1-percent sodium hydroxide solution to the oil/alcohol solution, and stir.

5. Test the pH of the solution by dipping the pH paper into it.

6. Repeat steps 4 and 5, as needed, until the pH reaches 9; this completes the titration.

To apply the titration result, note how many mL of sodium hydroxide solution were added to reach a pH of 9, then add 3.5 to that number; this is how many grams of lye are required for each liter of vegetable oil in your biodiesel reactor. For example, if you add 3 mL of sodium hydroxide solution to reach a pH of 9, you would need 6.5 g/L of catalyst for your biodiesel recipe. Use the Lye Quantity Table below to determine the total amount of lye to add to your recipe.

For the record, the more lye that's required, the greater the percentage of waste: If you need to add 7 mL of NaOH to your titration, the waste glycerin will be about 25 percent of the total batch quantity.

LYE QUANTITY TABLE

If you have to add this many milliliters of NaOH solution for a successful titration	You'll need this many grams of lye for a 10-gallon batch	Or this many ounces
1	170	6.0
1.5	189	6.6
2	208	7.4
2.5	227	8.0
3	246	8.6
3.5	265	9.4
4	284	10.0
4.5	303	10.6
5	322	11.4
5.5	341	12.0
6	360	12.6
6.5	379	13.4
7	397	14.0

Mixing Biodiesel with Other Fuels

HOME HEATING OIL, kerosene, and diesel fuel essentially are the same thing. Kerosene is almost identical to heating oil and diesel fuel, but it's slightly more refined, has a lower gel temperature, is slightly lighter in weight, and is slightly lower in energy content. There are also differences in the federal standards for allowable sulfur content of each fuel, and they are taxed differently. Kerosene and heating oil are considered "off-road" fuels and are not subject to the same taxes as diesel engine fuel.

Due to the highway tax imposed on motor fuels, it is illegal to use untaxed biodiesel or kerosene in place of diesel fuel for highway vehicles, though you may be able to apply for a research permit to have taxes waived for personal use. Kerosene is dyed red so that it is readily identifiable and distinguishable from diesel fuel and home heating oil. When using biodiesel on the highway, off-road, on a farm, or in your generator or heating system, check with your appropriate state agencies regarding taxes, use, storage, waste disposal, and fire codes.

Mixing fuels requires no special process; simply filling the fuel tank with the desired amount of each fuel allows the liquids to mix and stay mixed. Any amount of biodiesel mixed into fossil diesel, kerosene, or heating oil will perform well in your diesel engine or oil burner, provided there are no rubber gaskets or seals or other materials that will be degraded by biodiesel (check with the engine or burner manufacturer to ensure the suitability of biodiesel for your equipment).

Be aware that you may void all warranties if you use mixed fuel or straight biodiesel (B100). Some engine manufacturers offer specific warranties when fuels are used that contain small amounts of biodiesel. Diesel engines and oil burners have been optimized for best performance with fossil fuels.

COLD CLIMATE CONSIDERATIONS

Below a certain temperature, diesel fuel begins to congeal, or gel, and the particles become too big and the fluid too thick to flow through the fuel system. Pure fossil diesel has a gel point of around 20°F, while biodiesel has a gel point of around 45°F. In cold climates, refineries mix anti-gelling agents into diesel engine fuel. Once the temperature drops below the gel point, biodiesel should be mixed with fossil diesel. A mix of 20 percent biodiesel with 80 percent fossil diesel — called a **B20 mix** — will raise the gel temperature of the fossil diesel by 3 to 5°F, and should work well at temperatures down to –20°F (in areas where anti-gelling additives are used in the fossil diesel fuel). With B20, you should not notice any difference in equipment performance.

Washing Biodiesel

WASHING BIODIESEL IS a way to remove any unreacted methanol, lye, and glycerin after mixing. Unless you plan on selling commercial-grade fuel, it is not mandatory to wash biodiesel before using it, though you will find varying opinions regarding washing. Impurities in your fuel may lead to fuel system problems in the long run.

Allowing the biodiesel to sit for a week or two is generally enough time for most of the unreacted ingredients to settle, leaving a clear biodiesel fuel with a neutral pH. Before washing, you may want to run a few simple tests. Put a sample of your biodiesel in a clear jar so that you can observe it over time. The particles that settle to the bottom of the jar can be seen, and the pH can be easily measured with pH paper.

Testing for methanol content is a bit more difficult but is important because if too much methanol remains in the biodiesel it can damage an engine, burner, or fuel system. The easiest method to test for methanol content is to weigh a sample of biodiesel, heat it to above methanol's boiling point of 148°F (aim for 165 to 200°F), then weigh the sample again. The difference in weight is the amount of methanol that has boiled off. Methanol weighs 6.63 pounds per gallon. You can use the same process to determine water content, by heating the biodiesel to above the boiling point of water (212˚F).

How to Wash Biodiesel

You can wash your biodiesel after mixing and the initial settling by adding water to it so that you have about a 50/50 mix of water and biodiesel. Operate the mixer for about 2 minutes, after which time the whole slurry should be a milky yellow color. After settling (24 to 48 hours), there will be three layers in the tank. A milky water mixture will settle to the bottom, a soapy layer may form in the middle, and biodiesel will be on top. Pump or drain the bottom layers and discard them. Repeat this process 2 or 3 times until the water at the bottom of the tank is clear, and the biodiesel is a clear, amber color.

ENVIRONMENTAL CARE

Biodiesel is fairly nonreactive and will biodegrade over time. However, the ingredients used to make biodiesel can create serious environmental impacts if not handled properly. Both methanol and lye are toxic to all living things. *Avoid spills and fumes*. Better biodiesel kits and larger-scale manufacturing plants use sealed reactors that keep vapors contained and are able to recover methanol from the biodiesel. They can also recover glycerin through the process of distillation.

Veggie Oil Conversion

'VE BEEN DRIVING a diesel pickup on straight vegetable oil (SVO) for a few years now. I decided I'd rather convert the vehicle once than convert the fuel forever. Vegetable oil is much thicker than diesel fuel, and so won't flow through fuel filters or injectors unless it is heated to reduce its viscosity.

HILTON DIER III, my friend and renewable energy consultant, has been a renewable energy wrench-twister, consultant, and advocate for the many years that I've known him. Together, we began an electric car conversion business back in 1990, and today we continue to practice what we preach, with a twist: We traded electricity for waste vegetable oil as a transportation fuel. I prefer biodiesel because it's less worry for me about engine damage, I don't have to hack my car, and I can use the fuel I make in both my car and diesel generator. He runs his diesel pickup truck on straight vegetable oil. Here's what he has to say about it.

An SVO vehicle has an extra fuel tank just for the vegetable oil, plus a series of heaters to get the oil up to a temperature (150 to 190°F) where it flows as well as petroleum diesel. An SVO system also has valves and controls for switching between diesel and SVO. This can be done manually with dashboard controls or automatically with microprocessors under the hood. It's a good feeling, driving down the road fueled by an organic waste product rather than a petroleum product, but it isn't a perfectly convenient thing to do. Here are a few important things to think about before buying or converting an SVO vehicle.

- Figure out your vegetable oil supply first. You'll be using waste vegetable oil (WVO, generally interchangeable with SVO), probably from a restaurant, maybe from a food processing business. Get an idea of the type of diesel vehicle you'll want, its fuel efficiency, and the miles per month you'll want to drive it. This will give you a monthly vegetable oil budget. The beauty of a dual fuel (SVO/diesel)-system is that you can always fall back on diesel, but it's pointless to spend the money on a conversion if you can only run it on SVO now and then.

- In some areas where SVO vehicles are popular, you'll find out just how limited supplies of WVO are. In other places you'll be competing with professional companies that earn a living picking up and recycling WVO from restaurants and food processors. Make the situation convenient for your supplier and don't step on any toes.

- Not all WVO is created equal. Fast-food chains and burger joints generally use hydrogenated oil, which turns to sludge at room temperature. You can't use that. Some restaurants use and reuse the oil until it is brownish-black and full of impurities. This will be more acidic, which will abuse your injection pump, and it will be full of particles that will clog your filters. High-quality restaurants generally have high-quality oil.

- Make sure you have an appropriate, heated space for processing and storing your WVO. It is a messy process, and the oil won't flow well in cold temperatures. Your kitchen is not a good place. Oh, and mice love the stuff. They will leave little souvenirs in your bag filters, and they will drown themselves in any container of oil you leave open. Make sure you can close everything up. Also make sure that you are willing to handle a sticky, slightly pungent liquid on a regular basis.

- Set up your filtering and storage to avoid lifting heavy containers. I get my WVO in 5-gallon jugs, which weigh about 35 pounds each. Holding one carefully in midair to pour into a funnel gets old. I built a hinged jug holder with a long handle so I could decant WVO into the bag filter with one hand. The filter, set up next to and halfway up a flight of open stairs, gravity-feeds into a repurposed fuel oil tank, which has a hand pump installed to fill my truck. I never have to lift a jug above knee height.

- Let your WVO sit at room temperature for a couple of weeks so that the water and food waste can settle out. Then decant the good stuff through stacked sock filters: 100 micron, 25 to 50 micron, then 5 micron. McMaster-Carr is a good source for filters and other hardware (see Resources). The stacked arrangement means that you'll end up replacing the 5-micron filter less often. I made my filter holder by cutting circular holes in the tops of two tall, five-gallon buckets, bolting them together top to bottom, and sealing the seam between the two. I cut the holes slightly smaller than the 8" diameter of the ring in the top of the filter. The lower bucket has a hose fitting at the bottom. An old pot lid keeps the mice out when I'm not using it.

- If you buy a vehicle that has already been converted, be extra cautious. A diesel engine can run well on SVO for a long time, *if* the owner has done a proper conversion. If not, the injection pump can get clogged and etched, the piston rings can get carbonized and stuck, and much mayhem can be inflicted on the fuel system in general. Have a competent mechanic check out the vehicle thoroughly.

- If you are doing the conversion yourself or are having it converted, spend the extra money on a flat plate heat exchanger. These are identical to those used as hot water heat exchangers (described in chapter 6). They are available for around $100 online. The standard conversion kits have a fuel tank heater and a fuel filter heater. Add the finishing touch of the heat exchanger, just before the injector pump, and you'll have hotter oil and better combustion. This is especially important in northern climates. Insulate every part of the SVO system.

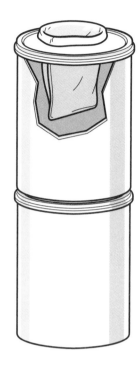

▲ **Veggie oil bag filter using two 5-gallon buckets. The top bucket holds the filter in place over the bottom bucket, which holds the filtered oil.**

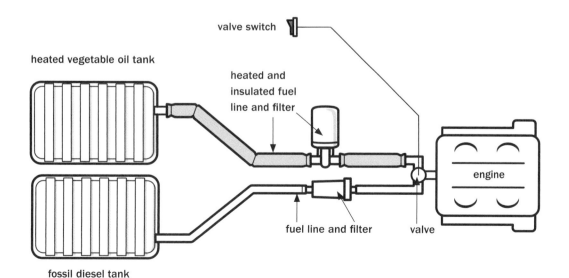

◀ **Basic diagram of a dual-fuel system for a diesel vehicle. The heat for the fuel filter and fuel lines can be provided by engine coolant, or they can be electrically heated.**

valve switch

heated vegetable oil tank

heated and insulated fuel line and filter

engine

fuel line and filter

valve

fossil diesel tank

Veggie Oil Conversion

- With an SVO system, you start out driving on diesel and then switch to SVO when the engine (and SVO heating system) is up to temperature. During a below-zero winter cold snap, the oil may never get warm enough. Cold oil will lead to injector pump damage. When you are near your destination you will hit a switch that flushes out the SVO system with diesel. This prevents clogging and leaves the whole fuel system full of diesel for starting. Don't push your luck. Flush the system early.

- Clean your injectors. I neglected this at first, and it cost me a set of injectors. This means getting a couple of cans of diesel injector cleaner and running it through the system every 3,000 miles or so. Don't just use the kind you pour into the fuel tank but rather something like Lubro Moly Diesel Purge that can be used undiluted to clean injectors. You can do this by emptying the SVO filter and filling it with the cleaner, or by rigging a temporary fuel "tank" made from a quart-size plastic bottle with a couple of pieces of clear tubing connected to the fuel filter's *send* and *return* lines, with the other ends sunk in the bottle of cleaner. Then, sit in your driveway revving the engine until the cleaner is almost used up. It takes maybe 15 minutes to make your diesel engine run much cleaner and smoother.

Running on SVO isn't the most convenient way to drive, but for someone willing to put in the occasional bit of work, it's a cheap and eco-friendly alternative. (See Resources for SVO conversion kit manufacturers.)

Create a Biodiesel Kit

Once you achieve consistent success with small batches of biodiesel, you may be inclined to start making enough to power your diesel vehicle or oil-fired heating equipment. You can have some success with a five-gallon bucket and drill mixer attachment, but when you're ready to make batches of 40 gallons or more, it's time to get serious about automation, durability, and safety. The processor described here has served me well in making dozens of batches of biodiesel suitable for use in diesel engines. This is a fairly expensive kit (my total was over $2,000) but will last a very long time when properly cared for, and it could very well be a good investment as fuel prices continue to rise.

Before we get to the steps for building the biodiesel kit, let's look at each of the main components:

Motor. At over $1,000, the most costly item by far is the explosion-proof mixing motor. This Class 1, Group D motor is required because it operates in an environment that may include explosive concentrations of methanol vapors, and a standard motor could easily ignite these vapors and cause an explosion (see page 198).

Mixing tank and stand. My conical tank and lid from Polydome (see Resources) work quite well for mixing. This tank is made with $\frac{3}{16}$"-thick, medium-density polyethylene and is designed for acid and caustic chemical storage. It is UV-stabilized and has a useful working temperature limitation of 140°F. The tank normally comes with a discharge pump on the bottom, which may work well for pumping out sludge and biodiesel; however, I have not employed this feature and ordered the barrel without the pump. You may find tanks of other sizes and materials through U.S. Plastics (see Resources).

The tank stand is welded steel, with a white, baked-enamel finish and integral motor mount. The mixing paddle (included with the motor) is a two-bladed propeller with a $\frac{5}{8}$"-diameter stainless steel shaft that's long enough to extend into the 8 gallons of methanol/lye mixture — a typical quantity for the first part of the batch mixing.

My tank has a 55-gallon capacity. Larger tanks and stands are available at additional cost. All other kit parts remain the same with the larger tanks. Keep in mind that the tank, stand, and motor may involve significant shipping costs due to weight and size.

Fuel pump. The 12-volt DC fuel pump has a telescoping suction tube with strainer, a 10-foot hose, and a nozzle. It's a magnetic-drive pump that can be used to transfer vegetable oil and biodiesel at up to 10 gallons per minute (gpm). I needed a longer hose, so eventually replaced it along with a nozzle with automatic shutoff. *Do NOT use the pump for pumping methanol!*

I use the same pump for moving vegetable oil from the barrel at the restaurant into a transfer tank on the pickup and back into larger storage tanks at the shop. Then I use it to fill the reactor tank with vegetable oil, and finally to pump the liquid glycerin out of the bottom of the reactor and into a waste collection tank or jug after mixing and settling. It can also be used to pump waste water from the tank after washing, but always be sure to pump biodiesel or oil through it before storage to prevent the internal parts from rusting. You may want a second pump that can stay put, along with an in-line filter to fill your car's fuel tank.

Heater. The thermostatically controlled heater in the parts list on page 210 is safe for use in plastic

---

tanks and turns itself off when the oil reaches 100°F. You should strain the vegetable oil to remove larger bits of food and other debris before preheating. While you can have a successful reaction at temperatures above about 70°F, lower temperatures yield unpredictable results, longer mixing times, and possibly more unreacted byproducts — meaning lower-quality biodiesel. It's best to heat the vegetable oil to at least 100°F before adding it to the mix.

Lab ware. The lab equipment is used for titration (see page 203). Remember, this is required only when using waste fryer oil. If you plan to use virgin vegetable oil, you need only a scale to weigh the lye, as the quantity will not vary from batch to batch.

Parts note: Manufacturers change prices, part numbers, designs, and availability without notice, so no prices are given in the parts lists, and there is no guarantee of availability. However, similar products to those listed should be available locally or online. You can download motor/mixer specifications from Neptune Chemical Pump Company. See Resources for suppliers and products noted in these plans.

MATERIALS
Mixer Parts

One ½ hp Class 1, Group D motor/mixer, 1750 rpm (Neptune F-3.1; includes 2-bladed folding, 316SS mixer)
One tank stand (Polydome PT-304S; specify stand that includes motor mount)
One 55-gallon conical tank, graduated (Polydome PT-304; graduated tank with gallon markers; specify without the optional pump motor)
One tank cover (Polydome PT-304C; specify split hinged inside cover with motor slot cut out)

Electrical Supplies

Three waterproof, strain-relief cable connectors (Del City Wire #2612)
One weatherproof switch box with weatherproof cover
One 10-foot length 14-3 SJOOW electrical cable (oil-, acid-, abrasion-, and flame-resistant)
Two 14 AWG, #8 stud spade connectors
Three 14 AWG, #8 stud ring connectors
Four yellow wire nuts
One 120-volt, 15-amp AC plug (NEMA 5266 style)
2" length ¼" heat shrink tubing
One 120-volt, 15-amp switch
One 14 AWG (minimum) portable GFCI extension cord (use if your AC outlet is not GFCI-protected)

Hardware

Two ⅜" x 1" #16 hex head bolts
Two ⅜" lock washers
Two ⅜" flat washers
One ⅝" loom clamp
Thread-lock

Safety Equipment

Nitrile gloves
Splash-proof chemical goggles (preferably antifog; it's worth it!)
PVC apron
Full-face respirator with carbon filter cartridge for organic chemical vapor filtering

Lab Equipment
(for titration; see step on page 213)

NOTE: Much of this equipment is available from school science or lab supply outlets, such as Frey Scientific (see Resources).

One dozen 1.5-mL pipettes (to measure fluids for titration)
pH paper
One 1-L narrow-mouth HDPE bottle
Two 10-mL graduated glass cylinders (to measure alcohol and lye solution for titration)
Two 30-mL glass beakers (to measure and mix for titration)
One 1-L polypropylene beaker (for measuring and pouring lye)
One hydrometer and jar (to test specific gravity of your biodiesel; available through brewing suppliers)
One 1-g resolution, 2 kg max. scale (Ohaus CS2000)

Miscellaneous Equipment

One barrel-mount pump with telescoping suction pipe, hand crank or electric; use to pump oil and biodiesel (Northern Tool)
One 1,000-watt submersible heater with thermostat control (Jeffers W-449; must be safe for plastic tanks)
One thermometer

ASSEMBLING THE BIODIESEL KIT

1. Wire the motor, switch, and plug.

Remove the electrical connection plate from the motor. Screw one waterproof, strain-relief cable connector into the threaded hole in the motor's electrical connection box. Screw two more of the connectors into the threaded holes on each end of the weatherproof switch box.

Cut a 7-foot length of the 14-3 cable. Strip about 3" of the cable jacket and ¼" of insulation on each individual wire, at both ends of the cable. Insert one end of the cable through the connector on the motor's electrical connection box. Crimp a ring connector onto the green wire and screw it to the hole inside the motor's electrical connection box. Connect the black and white wires as shown on the wiring diagram on the inside of the motor's electrical connection box cover, using wire nuts. Attach the box cover. Insert the other end of the cable through one of the waterproof connectors on the switch box.

Using the remaining 3-foot length of the 14-3 cable, strip the jacket and wires at both ends, as before. Attach one end of the cable to the 5266 AC plug. Insert the other end of the cable through the remaining open waterproof connector on the switch box. Slip about 1" of heat-shrink tubing over the black wire ends, leaving the tubing loose. Crimp a spade connector onto each black wire in the switch box, making sure it's secure. Heat-shrink the connection, then attach the connectors to the switch.

Crimp a ring connector onto each green wire in the switch box, then screw them to the box. Mount the switch and install the box cover.

Double-check all your work, making sure there are no frayed wires or loose crimps. With the motor on a workbench or floor, block it so that it will not roll either way. Turn the switch to the off position, plug in the power, and test the motor.

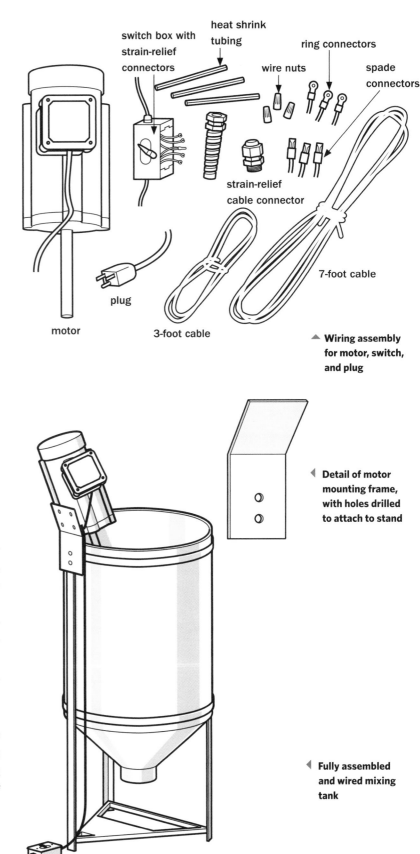

▲ **Wiring assembly for motor, switch, and plug**

◀ **Detail of motor mounting frame, with holes drilled to attach to stand**

◀ **Fully assembled and wired mixing tank**

2. Mount the motor to the tank stand.

NOTE: You will need a helper for mounting the motor to the tank stand.

If you're using the Polydome PT-304S tank stand, cut off the two horizontal supports and remove the threaded clamp from the Neptune motor mount frame. If you're using a standard barrel or drum for the mixing tank, you'll need these pieces to secure the motor to the barrel.

Drill two ½" holes into the motor mount frame to match the mounting holes on the tank stand as shown. Secure the motor mount frame to the motor using the four $\frac{7}{16}$" bolts, nuts, and washers included with the motor/mixer kit and following the manufacturer's instructions.

Stabilize the mixing tank in its stand and attach the mixing motor using two $\frac{3}{8}$" x 1" bolts, washers, and lock washers, orienting the lock washers so they face the bolt head. Have a helper hold the motor in place while you tighten the bolts securely into the captive nuts on the mixing

tank frame (torque to at least 25 foot-pounds). Attach the loom clamp to one of the motor frame mounting bolts; the clamp holds the power cord out of the way and provides strain relief. Make sure the mixing tank is stable and reasonably level.

3. Install the motor shaft and mixing paddle.

Attach the mixing paddle to one end of the mixing shaft, using an Allen wrench. Secure the connecting barrel to the other end of the mixing shaft, then attach the open end of the connecting barrel to the motor shaft (follow the instructions included with the motor mount). Be sure to make these connections very tight so they won't loosen while in operation. Add a little thread-lock to be safe.

NOTE: Retighten all fasteners after mixing your first batch of biodiesel.

4. Set up the transfer pump.

Attach the suction tube or hose to the transfer pump as directed by the manufacturer. Be sure to install a screen filter to prevent large debris from clogging the pump.

If you're using the DC pump, set up the pump and nozzle as directed by the manufacturer. Attach the red clip to the positive terminal of a 12-volt car battery, and attach the black clip to the negative battery terminal; be sure the power switch is in the OFF position before connecting to the battery to avoid sparks and unintended pumping. You may want to use an auxiliary battery to avoid draining the starting battery while you are pumping vegetable oil or biodiesel.

Motor, motor mount, mixing shaft and paddle, power cable and switch assembly ▷

Warning!

Dress in full safety gear and wear a respirator when handling or mixing methanol and lye, and during the biodiesel mixing process. Be sure to use the appropriate filter cartridges, and follow the manufacturer's instructions to ensure the respirator fits properly.

MIXING A BATCH OF BIODIESEL

1. Heat the oil.

NOTE: The mixing tank holds a total of 55 gallons. It is recommended that you mix 48 to 50 gallons (including methanol) at a time for best results. Using too little means the mixing paddle won't extend far enough into the methoxide to give a good mix; using too much could result in some spillage over the side once the mixer is turned on.

Heat the oil in an open-top storage barrel before pumping it into the mixing tank: Drop the submersible heater into the storage barrel and plug the heater into a GFCI-protected extension cord or wall outlet. It will take about 2 hours to heat 50 gallons of vegetable oil from 60 to 100°F. The heater will turn itself off once it reaches 100°F, then back on again if the oil cools below 80°F. Hotter oil helps improve mixing, yielding a better biodiesel. An insulated barrel speeds heating. Use a thermometer to be sure the oil is hot enough.

2. Perform a titration (for used oil).

Perform a titration to determine the proper amount of lye to use; see page 203. It's also a good idea to mix a small test batch in a blender to be sure you have a good recipe; see page 202. Generally, you will use 8 gallons of methanol, 40 gallons of vegetable oil, and a quantity of lye (sodium hydroxide). If you're using new vegetable oil, you'll need 530 grams (18.7 ounces) of lye. If you're using waste fryer oil, the quantity of lye is based on your titration results.

NOTE: To increase success and reduce waste, avoid using fryer oil that requires more than 3 grams of NaOH for a successful titration.

3. Mix the biodiesel.

When mixing a batch of biodiesel, do it straight through with no breaks between steps. Lay out all of the equipment and measured ingredients you will need to make sure everything is ready to go. You don't want to be fumbling around for things with a pound of lye in one hand. You are ready to mix a batch of biodiesel when:

- Titration is complete, and the test batch is successful

- Oil temperature is 100°F (37.8°C)

- Mixer is plugged into the GFCI outlet and tested to be sure it works

- Methanol and lye are scaled up to your batch size and ready to add

- Transfer pump is ready to pump the heated oil into the mixing barrel

- You are dressed in old clothes and protective gear (respirator, goggles, gloves, apron).

To begin the mixing process, place the cover on the mixing tank and open the lid. Be sure the mixer is OFF. Carefully and slowly pour methanol into the mixing barrel; avoid splashing! (If you're transferring the methanol from a 55-gallon tank, use a hand-crank or other suitable pump.) Close the tank cover and turn on the mixing motor.

◀ **Pouring lye into methanol**

Once the methanol is done with its initial splashing around inside the tank, open the hinged cover and *very slowly and carefully* pour the lye into the methanol. Close the lid, turn on the mixer, and let it run for 15 minutes. Turn off the mixer and let the methoxide stand for a few minutes until the fumes dissipate.

Pump the heated vegetable oil into the mixing barrel. Turn on the mixer and let it run for at least 45 minutes. Depending on your recipe, the temperature, and the quality of your ingredients, you may find that you need to mix for longer or shorter times. Actual mixing time may vary from 30 to 60 minutes.

Turn off the mixer and immediately take a sample of your biodiesel in a clear jar. If the mix is successful, it should begin to separate within 15 minutes.

Allow the mix to settle for at least 4 hours (but no more than 12) in the mixing tank. There's no need to disassemble the mixer or remove the paddle. Once the mix has settled, use the

you wait too long to pump off the glycerin, it will solidify and you will not be able to pump it. If this happens, pump the biodiesel off the top and scoop out the sludge from the mixing tank.

NOTE: The biodiesel will continue to settle for several weeks. If you pump fresh biodiesel into a storage tank, pump fuel for use from the middle of the storage tank rather than from the bottom. Periodically drain or pump the settled glycerin off the bottom of the tank and dispose of it.

4. Clean up the equipment.

Remove any unreacted ingredients, sludge, or other debris from the mixing tank before mixing another batch of biodiesel. The sooner you clean up, the easier it will be. Clean all lab equipment used for titration and measuring.

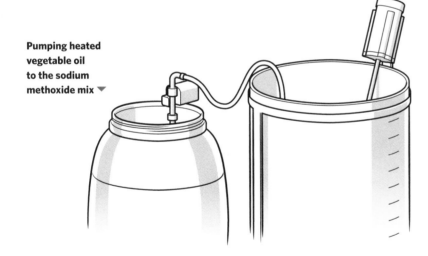

Pumping heated vegetable oil to the sodium methoxide mix ▼

transfer pump to pump the glycerin layer from the bottom of the mixing tank: Extend the suction tube or hose all the way to the bottom and pump the sludge into a waste container. When all the sludge is removed, stop pumping. You can let the biodiesel settle further in the mixing tank, if desired, or pump it right into a storage tank. If

Wood Gas

BURNING WOOD IN an open fire produces light, some heat, and lots and lots of smoke — as you know if you've ever forgotten to open the fireplace flue. The process isn't terribly efficient, and it obviously doesn't do wonders for our air quality. But when you burn that same wood in a very hot, oxygen-restricted environment, you break down the compounds of the wood into clean-burning combustible gases, along with ash or charcoal. Those gases — wood gas — can be used to heat a cook stove or even power a car, and their emissions consist of carbon dioxide and water vapor. There's no smoke.

Gas produced from wood and other carbon-based materials (primarily coal) has been used since the beginning of the industrial revolution. It's had many names — synthesis gas, syngas, producer gas, gengas, town gas — and has been used variously to provide heat, light, and transportation fuel. During World War II, over one million European cars and trucks traveled with onboard gasifiers that provided fuel for their modified gasoline engines. In this chapter, we'll look specifically at making and using wood gas.

Wood Doesn't Burn

STRIKE A MATCH and look at the flame. Notice that the flame is not actually in contact with the match but rather *surrounds* it. Likewise, wood in a fire does not actually burn; it is being heated to the point where combustible gases are released.

When the hot gases combine with oxygen in the air, they **oxidize** (combine chemically with oxygen) and burn. It is the liberated gases that are burning while the wood becomes carbonized, turning to charcoal, and eventually the charcoal burns to ashes. Charcoal is wood that has been reduced mostly to carbon. Ashes are minerals that are left over after all the carbon has been burned off the charcoal.

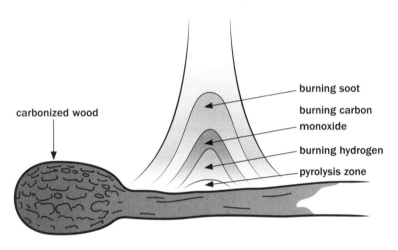

carbonized wood

burning soot

burning carbon monoxide

burning hydrogen

pyrolysis zone

▲ **Stages in the combustion process**

The blue part of the flame is burning hydrogen gas (oxidizing to form water vapor); the orange flame is carbon monoxide oxidizing to produce carbon dioxide (CO_2); and burning soot and tars produce a yellowish flame. With the addition of more oxygen to support further combustion, the carbon is reduced to ash. The ashes cannot be further oxidized. If you choked off the air from hot wood coals and stopped the burning, you would be left with charcoal suitable for use in a charcoal barbecue grill. When making charcoal, combustion is controlled so that no additional oxygen will further oxidize the carbon coals.

How Wood Gas Generators Work

GASIFICATION WORKS BY heating the biomass fuel — wood or almost any other carbonaceous material — to nearly 500°F, releasing flammable gases from the fuel without immediately burning them. The combustible gases produced are hydrogen (H_2) and carbon monoxide (CO). When burned, they create only carbon dioxide and water vapor (H_2O). The gases can either be burned within the unit (as with a gasification cook stove) or pulled off for external use as a gaseous fuel.

A wood gas generator consists of a chamber for holding and heating the fuel to create flammable gases in a controlled environment, a system for cooling and filtering the gases, and a system for distributing the gases to where they will be burned. Each of these pieces must be carefully designed to create an integrated system for managing the thermochemical decomposition of biomass.

NOT A WOOD STOVE

A wood gas generator is different from a wood stove. A wood stove burns chunks of wood, using lots of air to allow for complete combustion. It creates heat, coals, and ashes and lets the smoke go up the chimney. Most stoves are not very efficient at capturing all the heat in the wood. However, some modern wood stoves increase their efficiency by preheating incoming air to 400°F or more and injecting it into the exhaust stream, supporting more complete combustion.

A typical wood gas generator uses smaller chunks of biomass (or most any other material containing carbon, hydrogen, and oxygen) as a fuel. Such materials include:

- coal
- wood chips
- wood pellets
- pinecones
- corncobs
- rice husks
- nuts
- seed shells
- dry animal dung
- agricultural waste
- almost any other material that contains carbon

These fuels are burned in an oxygen-restricted environment to produce flammable gases through the staged combustion process of gasification.

A wood gas generator can be as simple as a small cook stove utilizing the efficient wood gas production and combustion process. You can build a simple gasifier cook stove (see plans later in this chapter) or buy one ready-to-use from Spenton LLC (see Resources). In a gasification stove, wood is not burned directly, but rather the wood gas is contained after it's liberated from superheated wood, then it's burned separately.

GASIFIER COOK STOVES

Here's how a gasifier cook stove works: A starting fire is lit on top of the wood fuel, and a controlled amount of "primary" air is forced through the fuel by the unit's integral fan. The heat liberates gases from the incandescent (but not flaming) fuel, and they rise to the top of the stove where additional, preheated "secondary" air is introduced by the fan through a series of holes at the top of the stove; here, the hot gases are oxidized by the incoming oxygen and ignite, producing a flame. The combustion zone moves downward toward the unburned fuel, leaving charcoal and ash on top.

These gasification stoves burn wood very efficiently, with little or no smoke, and produce very high temperatures. Heat output can be controlled by adjusting the airflow through the stove. Wood gas cook stoves produce charcoal, which can be further burned in the stove or removed to use in a charcoal grill or as **biochar** (see Biochar on page 218).

CHALLENGES OF WOOD GAS GENERATION

At the other end of the scale from cook stove units are large, combined-heat-and-power gasifiers that provide efficient, renewable heat and electricity for homes and businesses. This may sound like a simple engineering and marketing task, but significant technical challenges remain in the design of a turnkey wood gas generator. Despite its long history of use, gasification technology that can support a wide variety of fuel qualities and user savvy is still in its early stages. Making and using wood gas is a very hands-on approach to energy generation.

Unlike biogas and compost, wood gas doesn't just want to happen. It's not waiting around for you to take advantage of its hidden bounty. Managing combustion and capturing the resulting gases is difficult to manage in a practical sense. Two small companies that are making a big difference in this area are Victory Gasworks and All Power Labs; see Resources.

The basic challenges in developing a successful wood gas generator and harnessing wood gas involve:

- Creating an airtight gas system with a pressure relief function that manages and moves air, fuel, gases, coals, and ashes to the right places at the right time, and for the correct amount of time

- Creating a suitable combustion chamber and reduction zone

- Controlling the oxygen supply to the combustion zone

- Capturing and filtering the combustible gases released during pyrolysis

- Delivering the gas to where it will be burned

- Modifying equipment to burn the wood gas efficiently

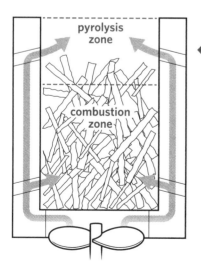

◀ **In this gasifier cook stove, the fire is lit from the top and primary combustion air is supplied from underneath the fuel load. Pyrolysis gases are released from the hot wood and burned with the addition of secondary air supplied through the top holes.**

Biochar

Biochar is charcoal that's used as a soil amendment. Because charcoal is primarily carbon, when it burns it releases lots of carbon dioxide, a potent greenhouse gas. But instead of burning the charcoal, it can be crushed and put back into the soil. This process is one way to sequester carbon in the ground, while helping to improve soil, rather than release it into the atmosphere as CO_2. See Resources for more information on biochar, including a source for NASA's *Biochar Activity Kit:* instructions, plans, and a teaching guide for building and using a three-can, top-loading, updraft (TLUD) gasifier cook stove.

Four Stages of Gasification

IN ORDER FOR a combustible material to burn, it needs to be hot enough, and there must be enough oxygen to feed the combustion process. The intent of gasification is to create combustible gases in an oxygen-restricted environment and capture the gases before they burn. They can then be shunted or piped away to be used as fuel.

Think of wood gas generation as a controlled, multistage combustion process. In fact, what you are doing is taking apart the combustion process and controlling each phase within that process to your advantage. The amount of air allowed into the combustion zone is less than what is required for complete combustion, and this is the key to controlling each phase of the process.

The following four stages are presented as a way to understand the activity inside a gasifier. Managing these processes is the goal of successful wood gas generator design. While each stage occurs in a specific place in the gasifier, the processes don't happen in isolation but rather in equilibrium with one another in a complex yet elegant thermochemical process.

Stage One: Drying. The fuel must be as dry as possible before it's loaded into the generator, ideally with a water content of less than 20 percent. Further drying will occur when the fuel warms in the gasifier and as air or fuel gases move through it. Drying consumes energy and keeps things cool until enough moisture is removed to allow the required **pyrolysis** (see stage 2) temperature to be achieved. If the fuel is too wet, it will take more energy to evaporate the

water before combustion processes can occur. Excessive water vapor also increases the formation of organic acids, which can lead to corrosion within the gasifier and poor-quality gas. All water vapor must be removed from the fuel before moving on to Stage Two.

Stage Two: Pyrolysis. An igniting fire is lit to heat the fuel. Once the fuel is heated to around 500°F in the absence of oxygen, the volatile solids decompose and release gases and liquids. The exact composition of these gases and liquids is related to the fuel being burned, but they will contain some combination of hydrogen, oxygen, and carbon. When biomass is used as a gasification feedstock, an acidic, tar-like pyrolytic oil is also produced. Pyrolytic oil is the result of moisture content and resins contained in woody materials. In a well-designed and properly managed gasifier, it will be burned as fuel. If the oil escapes into the fuel gas stream, it must be removed because it can gum up gas plumbing and engine parts.

Stage Three: Combustion. When carbon and hydrogen (contained in most **feedstocks** or organic material) combine with oxygen, the result is heat, carbon dioxide, and water vapor. After the igniting fire dies out, combustion in a wood gas generator is perpetuated by the biomass fuel, tarry oils, and hot carbon coals, along with just the right amount of air. The heat of combustion (approximately 2000°F) drives further pyrolysis and sets the stage for the final phase of wood gas generation, named for its chemical action of reduction.

Stage Four: Reduction. In chemical terms, reduction is the opposite of combustion. Combustion occurs when a material is oxidized by the addition of an oxygen atom, which releases heat. Reduction occurs when an oxygen atom is removed from a molecule by adding heat. In a gasifier, the fuel is oxidized (burned during combustion), forming (in part) CO_2 and H_2O, which come into contact with the hot carbon coals in the gasifier. The oxygen in the gases is strongly attracted to the carbon in the charcoal, and the heat gives it the energy needed to create new chemical bonds.

This is the point where the thermochemical magic happens in a gasifier. Through reduction, CO_2 gives up one oxygen molecule to the carbon in the coals to become carbon monoxide (CO). Water vapor (H_2O) gives up one oxygen molecule to become hydrogen gas (H_2), and the remaining oxygen is free to further oxidize the carbon. Carbon and hydrogen also react to form methane (CH_4). The processes of oxidation and reduction continue in equilibrium until there is no fuel left.

Gasifier Operation

A WOOD GAS generator must be tightly sealed so that any air entering the system is intentional and controlled. In many designs, fuel storage, combustion, and gas production all can take place within the same container.

The drawing on page 217 shows a gasifier in which the dry biomass is loaded into the fuel chamber from the top of the container and covered with an airtight lid. The fuel slowly dries and moves downward toward the combustion zone, then to the reduction zone, where it is consumed. Finally, ashes collect at the bottom of the barrel. The fuel moves by gravity and/or mechanical action to feed the carefully controlled combustion zone.

Some fuel material can get stuck in the barrel, causing combustion to stop for lack of fuel. This also can increase the possibility of channels developing within the fuel bed that can allow enough

4 PROCESSES IN GASIFICATION

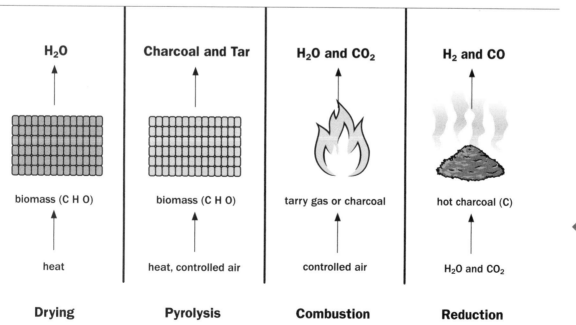

H₂O	Charcoal and Tar	H₂O and CO₂	H₂ and CO
biomass (C H O)	biomass (C H O)	tarry gas or charcoal	hot charcoal (C)
heat	heat, controlled air	controlled air	H₂O and CO₂
Drying	**Pyrolysis**	**Combustion**	**Reduction**

◀ The four processes in gasification can be broken out for illustration, but all processes are happening simultaneously.

oxygen into the unit to reach flammable or explosive levels. To reduce the potential for this effect, the fuel storage must be agitated to keep it moving. Small fuel materials (such as pellets, chips, or shells) may need to have recirculated gases blown through them to prevent the fuel from settling into a solid mass that is too dense for air to move through. This approach is called a **fluid bed** — as opposed to a **fixed bed,** where there is no intentional agitation. Gasifiers mounted on vehicles actually benefit from the bouncing, which keeps the fuel moving down to the combustion zone.

Firing Up

The combustion zone is at the bottom of the generator container and is accessed through a small airtight door for lighting and cleanout. A starting fire is lit in the combustion zone to ignite the fuel. Once started, air supply is choked off. As the fuel moves from storage to the combustion zone, it burns on a cone-shaped metal hearth with lots of holes in it, which allows ashes to drop down to the cleanout area. Air supply to the combustion zone is limited so that combustible gases released from the fuel do not burn. There's only enough air to keep the coals hot enough for pyrolysis to occur.

The movement of gases through the generator, including the amount of air drawn into the combustion zone, can be facilitated by the vacuum produced within an internal combustion engine, or with a fan that pulls gases through the gasifier and blows them into a burner. The induced flow forces the combustion gases through the hot coals, where the reduction of those gases occurs to make the wood fuel gases. The flow of air into the generator, and the flow of gases out of it, keeps the entire system under a slight negative pressure. This is important: If there is a leak, allowing too much air into the generator, the system will cease to function or, worse, all of the fuel will burn uncontrollably.

Cleaning and Filtering Wood Gas

GAS PURIFICATION REQUIREMENTS vary based on how the wood gas will be used. Wood gas produced in a gasifier includes ash, tar, and water vapor. These impurities may cause problems in a simple gas burner, but they all *must* be removed if you plan on using the gas in an engine; engines are more sensitive than gas burners and require cleaner fuel.

The gas can be initially filtered using a centrifugal canister, or **cyclone separator**. This is a cone-shaped chamber through which the gas swirls. In the process, ashes and larger particulates drop

---

WOOD GAS COMPOSITION

In addition to hydrogen gas and carbon monoxide, there also are small amounts of carbon dioxide and methane present in the wood gas. Nitrogen gas (N_2) amounts to about one-half of the gas volume and does not add to its energy content. The actual components and amount of each gas depend upon the feedstock and the completeness of each stage in the gasification process. A typical makeup of gas produced from wood might be:

- 50 percent nitrogen gas (N_2) — air is used as a source of oxygen, and nitrogen is the main ingredient in air. Nitrogen passes through the gasifier with no effect other than to take up space and reduce the overall energy content per volume of wood gas.

- 20 percent carbon monoxide (CO) — a flammable gas

- 20 percent hydrogen (H_2) — a flammable gas

- 5 percent methane (CH_4) — a flammable gas

- 5 percent carbon dioxide (CO_2) — a nonflammable byproduct of combustion

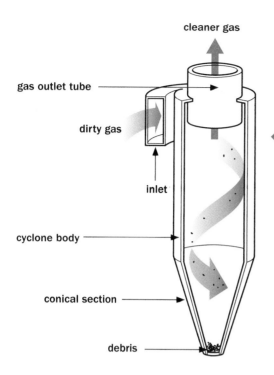

cleaner gas

gas outlet tube

dirty gas

inlet

cyclone body

conical section

debris

◀ Inside a cyclone separator. "Dirty" gas enters through the side inlet and swirls around as particulates fall to the bottom. Cleaner gas exits through the top.

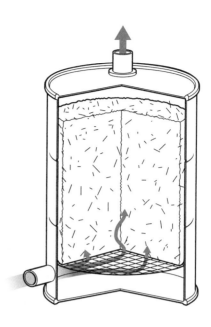

◀ Cutaway view of secondary particulate filter using drum filled with fine wood chips or straw

to the bottom and, as the gas cools, some of the water vapor will condense out of the gas stream.

After the cyclone separator, the gas can be passed through a low-tech filtering container of wood chips or even hay for further filtering, assuming the temperature can be reduced enough to prevent the filter medium from burning. This simple but effective filtering system can last for several years in a reasonably well-used gas generating system. Some commercial filter units use glass-fiber filters capable of withstanding high temperatures.

The hot gas is then cooled in a heat exchanger. Reducing the temperature causes the water vapor, tar, and oils to condense out of the gas stream, increasing the volumetric energy density of the gas. The gas can then be further filtered through a paper or cloth fine-particulate air filter before going into the engine or burner.

Using Wood Gas

AFTER FILTERING, WOOD gas can be used in place of, or mixed with, natural gas, liquid propane gas, gasoline, fuel oil, or diesel fuel. It can be used for heating or in gasoline- or diesel-powered engines.

When replacing liquid fuels (which are generally delivered to engines in a gaseous or vaporized state, by way of fuel injectors or a carburetor), wood gas is supplied through the engine's air intake. Using an automotive throttle body to open and close the wood gas supply, and a fuel shutoff valve to open and close the original fossil fuel supply, each system can be isolated and used independently.

On average, wood gas has an energy content of about 150 Btus per cubic foot. This is a relatively low value because one-half of the fuel volume is noncombustible nitrogen. For comparison, that's about 15 percent of the heating energy in a cubic foot of natural gas. In order to burn, wood gas is delivered to an engine or burner using an air-to-fuel ratio (measured in terms of mass, or weight, not volume) of approximately 1:1, in contrast to gasoline combustion, which requires an air-to-fuel ratio of 14.7:1.

Wood gas–powered engines tend to be a bit more sluggish than their gasoline counterparts and lose about 20 to 30 percent of their power due to the lower energy density (heating value per unit of fuel by weight) of wood gas, along with a greater pressure drop in fuel delivery created by the gas generating system.

With a flame temperature of around 3,600°F, wood gas can be used in a heating or cook stove

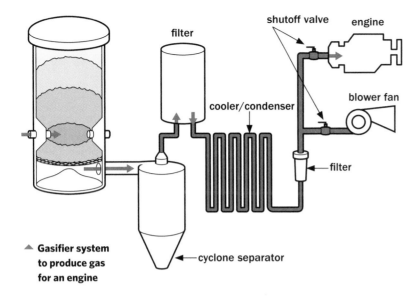

shutoff valve — engine

filter

cooler/condenser

blower fan

filter

cyclone separator

▲ **Gasifier system to produce gas for an engine**

burner, but the gas must be under a small amount of pressure. A high-temperature centrifugal blower can be used to pull the wood gas out of the generator to the burner, and the air-to-fuel ratio will need to be adjusted at the burner (see chapter 13 for more discussion about burners).

POWERING GASOLINE VEHICLES

To use wood gas in a spark-ignition (gasoline) engine, the engine typically is started on gasoline. The wood gas generator is ignited using an external blower to start the combustion. Once the fire is burning and wood gas is flowing, the gasifier's blower is turned off; the gasoline flow is reduced while the wood gas flow is increased, and the vacuum from the engine pulls air into the gasifier, driving the gasification process.

In addition to achieving the proper air-to-fuel ratio, using wood gas as an engine fuel requires advancing the ignition timing to compensate for its slower flame. Carbureted engines have the advantage of being able to use both wood fuel and gasoline simultaneously. Modern fuel-injected engines will likely have a computer control that must be modified or reprogrammed to adjust the timing and turn off the injectors. Combustion can be controlled manually with cable controls from throttle-body valves, or electronically using an oxygen sensor as a feedback control in the exhaust stream; this in turn adjusts the air-to-fuel ratio.

POWERING DIESEL VEHICLES

Compression-ignition (diesel) engines typically are operated in a dual-fuel mode. This is because the wood gas will not ignite under the normal diesel compression ratio. In a typical scenario, the engine is started on diesel fuel, then the flow of wood gas into the engine's air intake is increased, which has the effect of automatically decreasing the diesel fuel flow as the energy requirements are met with the wood gas to maintain the desired rpm.

Diesel fuel flow can be reduced by about 80 percent while increasing the wood gas flow. Injector timing must be advanced, and the injector pump may need to be adjusted for a lower fuel-flow rate to accommodate the additional energy input of the wood gas. In some cases, fuel injectors may need to be changed or modified to avoid overheating and fouling during dual-fuel mode. Power output is typically reduced by 15 to 20 percent.

Improving Acceleration

As a wood-powered vehicle accelerates, more fuel is required by the engine and more wood gas is drawn through the system using the vacuum generated by the engine. As more air is drawn through the gas generator, the rate of combustion increases in the gasifier. However, it takes some time for the combustion process to respond and deliver the required amount of fuel. The result is poor acceleration. This can be overcome by installing a *small* volume of expandable gas storage: When you're stopped at a red light, the storage container fills with wood gas. When you accelerate on the green light, the fuel storage is drawn down while the combustion builds once again.

Storing Gas

IN A PRACTICAL sense, wood gas cannot be stored and should be burned as it is produced. Its low volumetric energy content makes any storage scheme quite bulky. Compressing the gas requires energy and can change the chemistry of the gas — carbon monoxide is not very stable and may devolve into a potentially explosive combination of carbon and oxygen.

The best way to store wood gas is in wood! The next best ways are to generate electricity and store it in batteries or to use it to heat water. Some World War II-era vehicles used expandable gas bags for storage, but the obvious dangers of fire or explosion make this an impractical solution.

HOW MUCH GAS CAN YOU MAKE?

In terms of energy output, 1 pound of perfectly dry wood releases approximately 8,500 Btus when completely combusted. (The actual energy content released when burning wood in the real world is based on the wood's density, moisture content, and combustion management.) By comparison, 1 gallon of gasoline (just over 6 pounds) contains about 125,000 Btus. So in reality, about 18 pounds of wood with 20 percent moisture content contains approximately the same heating energy potential as one gallon of gasoline. This gives dry wood an energy density of about one-third that of gasoline.

The wood gasification process is in the range of 60- to 75-percent efficient at converting energy stored in biomass to the energy released by the gas produced on combustion. The exact makeup and amount of gas produced, and the energy contained in the gas, vary according to the fuel's characteristics. For example, coconut shells and charcoal produce relatively high-energy gas compared to rice hulls and wheat straw.

On average, one pound of wood converted to gas in a gasifier produces approximately 40 cubic feet of gas. Each cubic foot contains about 150 Btus, or about 6,000 Btus of energy produced for each pound of wood fuel consumed. A general fuel consumption rule of thumb for driving

a car on wood gas is about 1 mile per pound of wood. That's over 4,000 miles on the equivalent of a cord of hardwood. If a cord of wood costs $300, your fuel cost is about 7.5 cents per mile — about the same cost per mile if you paid $2 per gallon for gasoline and your car gets 25 mpg. Of course, gasoline is more convenient, but there are many more value propositions to consider than cost and convenience.

Quantifying Your Gas Needs

When designing a wood gas generator, start by determining how much gas you need to produce, and at what rate. Assuming you want to power a gasoline engine with wood gas, here's an example of how much gas you'll need for each horsepower (hp) of engine rating:

$$1 \text{ hp} = 746 \text{ watts (or 2,546 Btus)}$$

With a typical internal-combustion engine efficiency of 20 percent, you'll need to increase the Btu output by a factor of 5:

$$2,546 \times 5 = 12,730$$

Therefore, you'll need to produce 12,730 Btus of wood gas each hour.

At 150 Btus per cubic foot, that's about 85 cubic feet of gas production per hour. Since each pound of wood yields about 40 cubic feet of gas, that works out to just over 2 pounds of wood burned for each horsepower-hour. However, this doesn't mean that if you have a 200-hp engine you'll need 400 pounds of wood to drive for an hour, but you would if you were pushing the engine to produce maximum power for that hour.

This rule of thumb can help you think about maximum required gas production rate. Your gasifier will produce only as much as you ask it to produce, based on how much air is moving through it. But keep in mind that oversizing can create a set of problems. Therefore, it's best to determine how much gas you require under the most typical situation and adjust the design accordingly to meet short-term peak-load requirements.

The Importance of Proper Sizing

To ensure maximum efficiency, the gas generator must draw in air at the correct volume and rate to support all gasification processes that meet the demand of the load. More air increases the rate of combustion, but if too much air is moved through the system, the pyrolysis gases won't spend enough time in the reduction zone to be fully converted, or the reduction zone could cool off. The result is poor-quality gas and increased amounts of tar. Proper sizing and good design are essential to good performance and quality gas production.

Types of Gasifiers

THERE ARE THREE basic gasifier designs, named according to how air and fuel gases are moved through them. The design choice depends primarily on the type of fuel to be used, according to its energy value, moisture and ash content, density, and charring properties.

Downdraft gasifiers bring air into the combustion zone that resides in the lower-middle half of the unit. Air enters into the combustion zone and is drawn downward, away from the fuel into the reduction zone, then is channeled out. Heat radiating upward evaporates moisture from the feedstock, and as charcoal beneath is consumed, the feedstock sinks closer to the combustion zone.

Downdraft gasifiers generally are the most common design for fueling engines on wood gas because they typically have relatively low tar production and respond well to changes in gas requirements where the load varies. Downdraft gasifiers are best suited for fuels with low (less than 25 percent) moisture content.

The biomass cook stove described on page 217 is an example of an *inverted downdraft* gasifier, which has a container of biomass that is lit from the top with tinder. The pyrolysis reaction moves from top to bottom, where the gas is drawn off and burned.

Updraft gasifiers admit air into the bottom of the unit at the combustion zone, drawing the heat and gases up through the fuel (preheating and drying it) and out the top. This design is relatively efficient but generally produces more moisture and tar, so it's best suited for low-moisture, low-tar fuels, such as charcoal, rather than resinous biomass, like wood.

Crossdraft gasifiers bring air into the combustion zone toward the bottom of the gasifier to feed the gasification process, with the fuel gas outlet on the opposite side of the fuel load. These are often used for gasification of charcoal, in which very high temperatures are generated.

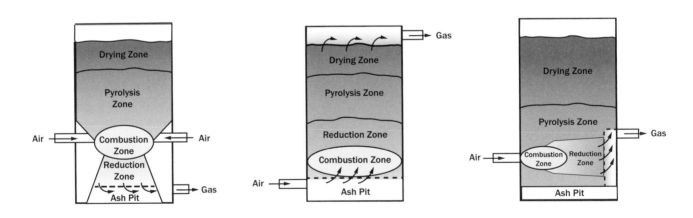

▲ Schematic of downdraft gasifer   ▲ Schematic of updraft gasifier   ▲ Schematic of crossdraft gasifer

Working Safely around Wood Gas

GAS SYNTHESIZED FROM carbonaceous fuels contains a high proportion of carbon monoxide. This colorless, odorless gas is harmful or fatal in very small doses. Any experiments you do must be done outdoors, and all gas produced must be captured and channeled away from living creatures.

Do not pipe the gas indoors for use with a cook stove, heating appliance, or any other purpose. Natural gas and propane gas have the odorant ethyl mercaptan added for leak detection. The pungent smell alerts occupants of a gas leak or if a pilot light has gone out. Wood gasifiers may smell a bit smoky, but pure wood gas is odorless, so leaks and gas accumulation are not easily detectable, creating a potentially dangerous situation.

If you experience symptoms such as drowsiness, headache, or nausea while working around wood gas, you are likely being poisoned by carbon monoxide and should move immediately to fresh air and seek medical attention. Wherever you are using wood gas, you should have a personal CO meter and alarm, as well as a stationary alarm for your shop. Spend a little extra on these safety devices and get models that read low concentrations, such as those available from CO Experts or Pro-Tech Safety (see Resources).

The main combustible gases produced by a wood gas generator are hydrogen and carbon monoxide. Each is flammable when mixed with air in a very wide range of concentrations. For hydrogen, the range is between 4 and 75 percent; for carbon monoxide, the range is between 12 and 75 percent.

Always take great care when working around such flammable gases. Air leakage into a gasifier could cause a fire or explosion when oxygen comes into contact with hot fuel. Remember: Generating wood gas involves high temperatures and combustion. Take suitable precautions against fire and have a fire extinguisher available near the gasifier.

ENVIRONMENTAL IMPACTS OF WOOD GAS

Both carbon monoxide and hydrogen gas are clean-burning fuels. When burned, they each combine with oxygen to form carbon dioxide (CO_2) and water (H_2O). Engines powered by wood gas are much cleaner in terms of emissions than fossil-fuel engines. This is partly due to catalytic converters in the exhaust systems of wood gas vehicles that control nitrogen oxides produced when the nitrogen present in wood gas is oxidized in the engine.

When biomass from the current generation of feedstock is used to produce wood gas, there is no net increase in global warming potential of the carbon dioxide released as exhaust. The CO_2 produced on combustion is equal to the amount that would be released by the biomass decomposing naturally over time. Ideally, that CO_2 is absorbed by the next generation of biomass. Of course, the process is accelerated with combustion; it takes many years for a tree to break down naturally and completely release the equivalent CO_2.

Tar produced by wood gas can be problematic to dispose of because it is caustic. It's best to use dry fuel and design your generator so that tars and acids are burned and reduced to ash. Ash production and disposal generally are not a problem. Volume ranges from 1 to 20 percent depending on the fuel material used.

Cooking and human health. When using the gasification process for cooking, there is almost no smoke and a much hotter flame than that of a conventional wood fire. This means fewer particulates, faster cooking, and, importantly, healthier humans. This last point has been the focus of several organizations bringing clean, efficient cooking options to developing nations. Considering that nearly one-half of the world's population cooks daily meals on open fires, efforts of organizations such as the Aprovecho Research Center and the Global Alliance for Clean Cookstoves (see Resources) have greatly increased the quality of life for many families throughout the world.

Two Men and a Truck

JOHN IS NOT an energy geek, car fanatic, or science expert. He is a cut-from-the-cloth DIY man with practical and relevant skills to make things work. He was looking for a hands-on project and admits that he didn't do as much research as he could have before building the gasifier. This led to a few mistakes but resulted in a useful, working gasifier that powered his truck on wood chips for many satisfying, gasifying miles.

JOHN DUNHAM got the alternative energy bug in college. He looked into converting a diesel car or truck to burn waste vegetable oil or biodiesel but could not find a suitable and affordable diesel vehicle. When John was scratching his head for senior project ideas, it was his father who first stumbled onto the thought of running an engine on wood gas.

The idea intrigued both of them. Using plans dating from the 1970s — from *Mother Earth News* — along with information gathered from online forums such as those hosted by All Power Labs (see Resources), they embarked together on a project to build a stratified, downdraft wood gas generator and use it to power a pickup truck.

They scavenged barrels, pipes, and scrap metal. With their plans, welding skills, and ingenuity, the project started to take shape: Two stacked 55-gallon drums make up the fuel hopper (on top) and combustion chamber housing (on the bottom). The fire tube is 12"-diameter snow-making pipe (scavenged from a local ski area) that drops from the bottom of the top barrel, through and to the bottom of the bottom barrel. The combustion zone is ideally in the middle of the fire tube (it can be in the wrong place if not properly designed), and the reduction zone is at the bottom, where hot coals drop onto a grate made from a stainless steel colander. A small door cut in the bottom barrel provides access to light the fire in the combustion zone and also serves as an ash cleanout.

Once the fuel in the gasifier is burning, the lighting door is closed, and air is drawn through the entire system via a blower motor on the outlet of the gasifier that pulls air through it. This creates negative pressure inside the gasifier and provides enough draft to get the fire going. After about 15 minutes, wood gas is generated and can be burned in the engine. The blower also serves for testing whether the gas is ready to burn: If a fire can be lit at the outlet of the blower, the gas is ready. Once the flame has stabilized, the blower is turned off and wood gas is diverted to the engine.

Here's how John describes the system:

1. Air enters through the top of the top barrel. Its quantity is determined by the draft created by either the blower or engine vacuum and ultimately is restricted by the size of the air entry port.

2. The air moves through the stored chips and down into the fire tube, along with chips, which fall into the tube by gravity to the combustion zone.

3. The combustion gases are pulled down the tube, through the reduction zone, and through the open space in the barrel surrounding the tube, which cools them as they exit the barrel and enter the "cyclone" particle separator.

4. The gas moves into a barrel filled with wood chips that act as a filter and further cool the gas. (After a period of time, the filter chips can be replaced with fresh chips and the old chips used for fuel.)

5. The gas flows through pipes (made from car exhaust pipe) to the front of the truck and into a series of metal tubes, where it cools, releasing moisture and tar through condensation.

6. Finally, the wood gas enters the engine through the air intake. Individual throttle bodies are used to control both the wood gas intake and air intake by way of cables operated from the cab of the truck. This allows John to manually adjust the proportion of wood

> There does come a time when you really need to just go out, get your hands dirty, and try it for yourself.

gas and air for best engine performance. A carbon monoxide detector lives in the truck at all times.

John uses a plug-in electronic tuning module (a unit made by Tweecer; see Resources) to access the truck engine's computer and adjust spark timing to compensate for the slower burn of the wood gas, and to turn off the fuel injectors to stop the flow of gasoline when wood gas is flowing. This approach allows the truck to operate on either wood gas or gasoline with the turn of a dial.

The switchover can't be done on the fly, however, because there's a manually operated valve on the outlet of the condenser that must be opened for wood gas operation and closed for gasoline operation. With additional controls for independently regulating gasoline and wood gas flow, the truck could run on a combination of both wood gas and gasoline. At some point, John will hook up two hot-wire anemometers — one located in the wood gas intake, the other in the air intake — to receive visual feedback on the quantity of both air and gas flowing into the engine. This feedback can allow for more precise control of the fuel-to-air mixture when running on wood gas.

As mentioned, many lessons were learned, including discovering that the combustion zone in the fire tube can move up and down the tube depending on the draw of air through the system. This led to a problem with lighting the fuel because the draft wanted to pull the igniting fire away from the fuel. A

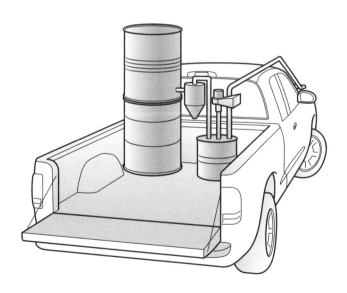

▲ John's gasifier truck

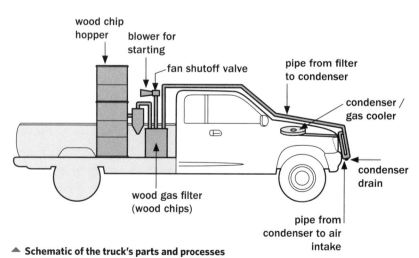

wood chip hopper

blower for starting

fan shutoff valve

pipe from filter to condenser

condenser / gas cooler

condenser drain

wood gas filter (wood chips)

pipe from condenser to air intake

▲ Schematic of the truck's parts and processes

Two Men and a Truck

carefully designed combustion zone helps to control this situation. As for preferred fuels, John has used both hardwood and softwood chips and feels that the former provided better performance.

The initial gasifier design was oversized, which means that the combustion and reduction processes were not ideal. This led to excessive tar in the fuel, which can damage or ruin an engine. Since the ideal air-to-wood gas ratio is 1:1, the gasifier needs to be sized to provide about half the displacement volume that the engine requires at the most typical (or desired) rpm. John remedied this by reducing the outlet below the combustion zone from the full 12"-diameter pipe to a 6" opening. This constricted the airflow and increased the air velocity, making a much hotter combustion zone.

After learning this the hard way, John discovered that others have covered the same ground and have published tables online that could have helped with the initial sizing (there are some excellent Internet forums for wood gas users and experimenters, and this DIY community is happy to share information). With his system designed to operate best at an engine speed of 3,000 rpm, the truck will stall if it idles for long periods.

John is already planning his next gasifier project, complete with better combustion zone design, an insulated combustion tube, improved gas filtering, and preheated combustion air. He also plans to incorporate mass airflow sensors to provide feedback for automatic mixing of air and wood gas for optimum ratio, as well as temperature monitoring of both the combustion zone and the downstream gas temperature, to be sure that it's cool enough to condense the tar out of the gas. John shares the following thoughts for anyone thinking of taking on a similar project:

"Despite the mistakes I made and problems I encountered in this project, I had an enormous amount of fun while learning many things that I would not have been exposed to otherwise. While I would encourage anyone interested to read the articles and books on the subject (certainly more of them than I did before I started), there does come a time when you really need to just go out, get your hands dirty, and try it for yourself. Be safe, and keep that CO detector close."

Despite the mistakes I made and problems I encountered in this project, I had an enormous amount of fun while learning many things that I would not have been exposed to otherwise.

Build a Simple Wood Gas Cook Stove

Cooking with wood gas allows you to move away from old-school, smoky wood fire to modern, clean-burning pyrolysis fire. This project shows you how to build a top-loading, updraft (TLUD) stove that not only helps you understand biomass gasification, it's also useful for cooking. All you need is a coffee can (or any other tin can, preferably unlined; the size does not matter much), some hardware cloth (metal mesh), and some fuel.

The fuel can be almost any dry biomass — from twigs or wood pellets to cherry pits or dry corncobs. Try different fuels and experiment with can sizes and air-hole diameters. It may take some trial and error to find the optimum design, but the materials are inexpensive, and the project takes only about 15 minutes.

This simple cook stove design belies the full potential of clean cooking. Larger stoves employing this basic design can be used to build a complete cooktop and oven arrangement with a single fire.

MATERIALS

One tin can (unlined)
One piece hardware cloth
    (wire mesh) with ¼" grid
Handful of dry biomass:
    twigs, nutshells,
    wood pellets, etc.

1. Make the bottom air inlets.

Remove any paper from the can, then if you haven't already, remove the lid from one end of the can with a can opener and empty the can of its contents. Turn the can open-side down and drill or punch eight evenly spaced ⅛" holes around the perimeter of the bottom lid, about halfway in from the edge, then make one hole in the center; these are the primary air inlets to support fuel combustion.

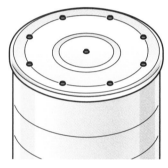

▲ **Holes in bottom lid of can**

2. Make the side air inlets.

Position the can with the open end up and mark a line around the perimeter of the side, about one-third of the way down from the top. Drill eight evenly spaced ¼" inch holes along this line; these are the secondary air inlets and will provide oxygen to burn the pyrolysis gases before they leave the stove.

Can with side inlet holes ▶

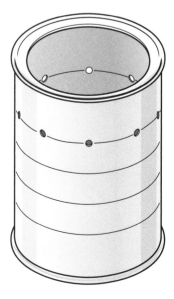

3. Add the fuel screen.

Cut a square piece of ¼" hardware cloth or similar metal mesh the same width as the interior diameter of the can. Bend down the edges of the mesh to create a shelf that will keep the fuel about ½" above the bottom air inlet holes.

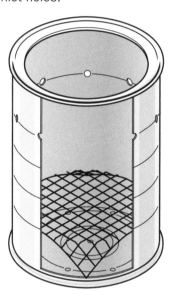

▲ **Fuel shelf made with ¼" hardware cloth**

4. Fire up the stove.

Set the stove on top of a fireproof surface that allows air to enter the primary air inlets on the bottom of the can; an open barbecue grill works well for this. Fill the can about one-third full with dry biomass. Put some paper or tinder on top of the biomass and light it from the top. A little fuel soaked in alcohol makes a good starter, too.

After a few minutes the biomass should start to burn from the top down as combustion air is drawn upward due to the draft created by the hot fuel on top. The fire may be a bit smoky at the start and again at the end of the burn, but once it settles down the smoke will subside, and you'll see blue flames appear at the inside of the secondary air inlets; this is the pyrolysis gas being burned with the addition of more oxygen.

To use the stove for cooking, balance two pieces of angle iron (so that the angles are over the center of the can) on top of the stove to rest a pot on for cooking. A more stable design would be to cut a notch halfway through the center of two 1"-tall pieces of thin metal bar stock. Slide the bars together at the notch and lay it across the top of the can. Allow 1" or 2" of space between the top of the can and the bottom of the pot.

If desired, you can improve the stove's cooking performance by adding a metal wind guard made from a larger can or a piece of metal roof flashing; cut this to the same height as the stove and encircle it around the stove can, leaving a few inches of space between the two.

A chimney added on top of the stove will improve draft and create a hotter fire. You can make a suitable chimney from another can that rests on top of the stove can, or from roof flashing encircled tightly around the stove can (above the secondary air inlet holes) and secured with a metal hose clamp.

Experiment with different hole sizes and locations. Different fuels behave in different ways.

Cutaway view of completed cook stove ▶

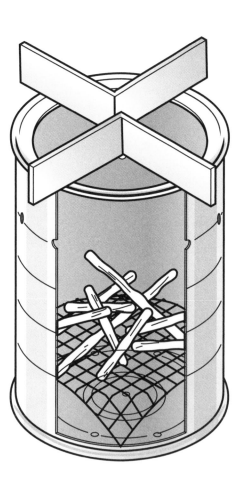

Biogas

BIOGAS IS A mixture of gases formed anywhere organic material decomposes in the absence of oxygen, such as underwater, deep in a landfill, bubbling out of municipal solid waste, or in the guts of animals (including you). Sometimes called swamp gas, biogas is produced through the biological and chemical process of **anaerobic digestion** (AD). This is a natural process that happens without any assistance from you or me.

Simply put, anaerobic digestion is the microbial decomposition (digestion) of carbohydrates in an oxygen-free (anaerobic) environment. It begins with a process similar to the fermentation of alcohol, but AD occurs in the absence of oxygen and continues past fermentation. In fact, oxygen is toxic to the process, in that it inhibits the growth of methane-producing microbes, also known as **methanogens**, which are ultimately what we want to encourage for the production of biogas.

The Basics

THE MAIN INGREDIENT of biogas made in a controlled environment is methane. Methane (chemically known as CH_4) is a hydrocarbon made up of one molecule of carbon and four molecules of hydrogen, and is lighter than air. Methane is also the primary component of natural gas, commonly used for cooking and heating, although biogas is not as energy-dense as natural gas. The methane content of the biogas you make will probably range from 50 to 80 percent, compared to about 70 to 90 percent with utility-supplied natural gas. Natural gas also contains up to 20 percent other combustible gases, such as propane, butane, and ethane, while biogas does not.

The exact makeup of biogas depends in part on the source of the gas, which is based on what is fed to the digester, and in turn what was fed to the producers of those ingredients. Noncombustible components of biogas can be considered impurities. These will be primarily carbon dioxide (CO_2), along with small amounts of water vapor, nitrogen (N_2), and possibly trace amounts of hydrogen sulphide (H_2S). If air contaminates the process, nitrogen can dilute the biogas. Other trace impurities may be formed as well. You can remove these impurities if desired, but depending on how you intend to use the biogas, you may not need to.

PRODUCING BIOGAS

To produce biogas, you first mix water with organic material (often called **feedstock**) such as animal manure or vegetable material, add a starting culture, then close it all up in an airtight container. You maintain a temperature within the container that is close to the temperature inside an animal (around 100°F) and, in about a week, you should be generating biogas.

The airtight container where this process is captured and controlled is called an **anaerobic digester** or **methane generator**. I prefer the term *generator* for the system in general, because it implies the intention of producing something, while anaerobic digestion is a process that happens with or without our intention or intervention.

While design specifics can vary, a methane generator usually contains a filler tube for feeding the digester vessel; an effluent outlet to remove digested solids and liquids (also called **digestate**); and a gas outlet. You can make a small generator from a single 55-gallon barrel, but any digester vessel smaller than 200 gallons should be considered experimental because it will not make enough biogas to be useful for any practical purpose.

Keeping Things Simple

The biological and chemical processes of AD, along with all the nuances of feedstock variables, are complex. However, if you dwell on the complexity of the science, you may never get started.

Organic materials mixed into a slurry and put in an airtight container produce combustible gases, nitrogen-rich liquids, and compostable solids. ▶

gas storage

biogas

organic material (feedstock)

slurry

digestate

liquids

solids

Talkin' Biogas

Throughout this chapter, **biogas** refers to the gas produced by the generator, impurities and all, while **methane** refers specifically to the chemical compound methane, which is the combustible component of both biogas and natural gas. They are not the same.

Save that step for when you turn professional. Anaerobic digestion is a natural process of decay that wants to happen by itself — any encouragement you offer can only be helpful.

In fact, you could probably ignore the rest of this chapter and find a sealed container, put a home brewer's airlock on top, fill it halfway with water and halfway with any sort of organic material you can find, and have some success in making biogas within a week. But if you want to understand the process and be reasonably efficient about it, read on.

HOW IT WORKS

Once your digester is filled with organic material and water, biochemical processes begin to happen. First, the ingredients will break down and ferment, then acids will begin to form, followed by the desired methane production. There are four stages in the breakdown of organic material within a biogas generator. These four stages can be separated into two phases: acid formation and methane formation. The waste of one stage feeds the next. Once a generator is operating and producing gas, these processes happen simultaneously rather than as discrete sets of chemical reactions.

Hydrolysis starts when water is mixed with organic material. Hydrolysis is the enzymatic breakdown of complex proteins, carbohydrates, fats, and oils into amino acids, simple sugars, and fatty acids. The broken-down (depolymerized) material is in a chemically accessible form and ready to be fermented by acid-forming bacteria.

Acidogenesis, or fermentation, happens when acid-forming bacteria oxidize the simple compounds formed during hydrolysis to create carbon dioxide, hydrogen, ammonia, and organic acids.

Acetogenesis is the conversion of organic acids into acetic acid. Acetic acid is the main ingredient in vinegar and is the food for the final stage of decomposition within the generator. Acid-forming bacteria are fast-breeding and hearty, producing lots of CO_2.

Methanogenesis is the creation of methane-producing microbes, or methanogens

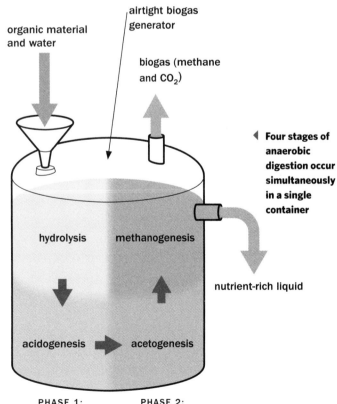

organic material and water

airtight biogas generator

biogas (methane and CO_2)

hydrolysis methanogenesis

acidogenesis acetogenesis

nutrient-rich liquid

PHASE 1:
ACID FORMATION

PHASE 2:
METHANE FORMATION

◀ Four stages of anaerobic digestion occur simultaneously in a single container

(single-celled, nonbacterial microorganisms from the group *Archaea*). Methanogens combine hydrogen and CO_2 produced during the acid-forming phases to create methane. In contrast to the acid formers, methanogens are slow to reproduce and extremely sensitive to temperature, pH, and the presence of oxygen.

HOW MUCH BIOGAS CAN YOU MAKE?

There are many variables in all processes of generating biogas. These include the type and quality of feedstock, the type of generator and how well it is maintained and fed, and other factors that you will read about later. I point this out because it is almost impossible to calculate exactly how much gas you can produce from any given "recipe." There are, however, rules of thumb that offer enough guidance to point us in a generally useful direction.

Rule of thumb for biogas production: A well-managed generator may produce approximately its own volume of biogas each day. To put this in terms of energy production, a bit of math is required:

- A 55-gallon drum has a volume of about 7.35 cubic feet.

- One cubic foot of methane contains 1,000 Btus of energy.

- Biogas containing 60 percent methane offers 600 Btus of energy for each cubic foot.

- 7.35 cubic feet x 600 Btus per cubic foot = 4,410 Btus.

A typical gas cook stove burner might burn through 15,000 Btus of fuel per hour on maximum heat. At this rate, a 55-gallon methane generator can potentially produce enough gas in a day to supply the burner for about 18 minutes, allowing you to boil about 2 gallons of water (assuming a 60-percent transfer efficiency between the energy in the flame and the water in your pot).

This might be enough in some cases, but in a practical sense, a small family with modest daily cooking needs will require the output of a warm, well-fed, 200-gallon (27-cubic-foot) methane generator at a minimum. This much biogas represents about 16,000 Btus and offers about one hour of cooking time, or enough energy to boil around 8 gallons of water.

Lots of Variables

The quantity and quality of methane you make depends on the nutrient value of the feedstock and how well the microbes convert the available nutrients into methane. For practical purposes, biogas production and quality are functions of your specific recipe and generator management. Important things to understand about making biogas are:

- Recipe development and the carbon-to-nitrogen ratio of ingredients

- Solids, liquids, and digestible quality

- Temperature

- Feeding rate

- Retention time

- pH

- Mixing

We'll cover all of these topics, but as you can see, any estimate of methane yield for each unit of digestible material has quite a few variables. That means any lists you come across indicating specific values for any of the variables are only estimates. Your actual production *will* vary. I encourage you to not get lost in the numbers or details, and simply experiment. You can learn at least as much by doing as by studying!

Recipe for Making Gas

ALMOST ANY COMBINATION of organic materials can be digested, including vegetables, food scraps, grass clippings, animal manure, meat, slaughterhouse waste, and fats — almost anything as long as it contains carbon and/or nitrogen. Avoid using too many woody products, like wood chips and straw, which contain large amounts of lignin (which is resistant to microbial breakdown and tends to clog up the digestion process). Also avoid material that may be contaminated with heavy metals or other toxins, and materials with large amounts of ammonia or sulphur.

The ideal ingredients are those materials you have a plentiful, convenient, and consistent supply of, so you can make consistent and useful quantities of biogas. If you have experience with mixing compost, you already have a good idea of what the recipe needs to be: If you can compost it, you can digest it.

THE RIGHT RATIO

Organic material can be classified by how much carbon and nitrogen are in its makeup, and the ratio between those two elements. This ratio is expressed as **C:N**. Carbon is a source of energy for microbes; nitrogen is needed for protein and used to build cell structure. A C:N ratio of between 20:1 and 30:1 is suitable, with the higher C:N range being ideal. Too much carbon slows the decomposition, while too little means organisms don't grow. Too much nitrogen produces ammonia, but with too little you don't get enough of the right kind of microorganism growth.

Don't worry too much about getting the C:N ratio perfect at first, since a wide range of C:N will produce something you can use. Experiment and find a recipe that works well with ingredients you have available, then perfect the recipe for maximum gas production.

Brown vs. Green

In general, things that are brown are high in carbon. This includes cardboard, wood chips, cornstalks, dry leaves, pine needles, and straw. Green things are generally higher in nitrogen, including food and garden wastes, grasses, seaweed, and manure (an obvious exception to the general color rule). The chart, Evaluating Raw Materials, on pages 238 and 239, lists approximate C:N ratios, moisture content, weight, and estimated volatile solids content of common organic materials. Any combination of these can be mixed to yield the ideal ratio.

Keep in mind that the C:N ratio describes the chemistry of the material — it does not mean that you need 30 times more brown material than green material. The actual amounts of carbon and nitrogen in any material will vary depending upon the specific makeup and age of that material, and the data should be taken as a general approximation of C:N ratios. Note: The terms "TS," "VS," and "FS" are discussed on page 237.

Calculating C:N Ratio

You can determine the C:N ratio of any combination of ingredients — and the relative quantity of each ingredient required in your mix — if you know the C:N ratio and the approximate moisture content of the material. Note that C:N ratios are given in dry weight, not wet weight. This is because moisture content varies greatly, even with quantities of the same material.

Factoring in moisture content requires an extra step in the calculation, but it's a fairly simple step when weighing ingredients to add to your recipe. You can get away with using volume measurements only if the water content of the ingredients is similar.

Using the C:N ratios and moisture content data from the Evaluating Raw Materials chart, here's how to calculate the C:N ratio of your entire recipe:

1. Make a list of each of the ingredients you wish to add, along with their carbon values (the C side of their C:N ratios).

2. Weigh the total amount of each ingredient. (You can use any weight unit you want — pounds, kilograms, ounces — just be consistent.)

3. Find the moisture content percentage of each ingredient in the chart.

4. Figure the dry weight of each ingredient using this formula:

Wet weight x (1 − moisture content percentage) = dry weight.

5. Multiply the carbon value of the ingredient by its dry weight to find the carbon units.

6. Add up all of the dry weights (step 4) of the ingredients.

7. Add up all of the carbon units (step 5) of the ingredients.

8. Divide the total carbon units by the total dry weight. The number you get is the carbon value (C) of the ratio, where the nitrogen value (N) is 1.

Here's an example using ingredients you might have around your homestead (see chart below). The quantities are relative and can be scaled up or down as needed. We want to develop a recipe using chicken droppings, grass clippings, kitchen scraps, and paper. Chicken manure has a C:N ratio of 6:1, and you have 2 pounds (wet) to add to your recipe.

Chicken manure has a 70-percent moisture content, so to figure the dry weight:

$$2 \times (1 - 0.70) = 0.6$$

Round off any decimal places (let's not worry about precision as it is not obtainable, or required, outside of a lab).

Next, multiply the carbon value and dry weight to find the carbon units:

$$6 \times 0.6 = 3.6$$

Entering the numbers for each item onto a list or chart, you can easily add up all the dry weights and carbon units to complete the calculation:

$$45 \text{ (total carbon units)} \div$$
$$1.6 \text{ (total dry weight unit)} = 28$$

Your C:N ratio for this recipe is approximately 28:1.

Consider this example a place to start and not the final word on the available carbon and nitrogen in any given recipe. Experiment to find out what works best for you using your specific ingredients. Notice the very small amount of newspaper required, due to its high carbon content. Small changes in the addition of woody material will make large differences in the C:N ratio.

Solids, Liquids, and Volatile Solids

MOST MATERIALS CONTAIN some amount of water, and many contain lots of water. Water is an aid to digestion, but it cannot be digested. If you evaporate all the water from a material, you're left with only the solid portion. Organic materials

RECIPE EVALUATION — C:N RATIO IN YOUR RECIPE

Ingredient	Carbon value	Wet weight	Moisture content	Dry weight	Carbon units (carbon value x dry weight)
Chicken manure	6	2	70%	0.6	3.6
Grass clippings	17	2	82%	0.4	6.1
Kitchen scraps	20	2	69%	0.6	12.4
Shredded newspaper	500	0.05	10%	0.05	22.5
Total dry weight		1.6			
Total carbon units		45			
C:N ratio of recipe		28:1			

▲ Breaking down organics:
Food scraps = 25% total solids and 75% water
Total solids = 90% volatile solids and 10% fixed solids

◀ Smaller pieces of biomass will break down more quickly and completely.

might be 75 percent water by weight, more or less, leaving 25 percent total solids.

Not all solid material is susceptible to the bacterial breakdown required for anaerobic digestion. The portion of the **total solids** (TS) available for AD are called **volatile solids** (VS), which might be 80 or 90 percent of TS, more or less. The rest of the solids are known as **fixed solids** (FS), and these are unavailable as food for the digesting microbes.

If you evaporated all the water from a material and then burned the solids, only the volatile solids would burn up, and you'd be left with a pile of ash (the fixed solids). Chemically speaking, this approach is a bit rough in determining exactly how much of which part of the material gets digested, but it's the best we can do at this point.

VS AND GAS PRODUCTION

Knowing the VS is important for determining how much gas can be produced for any given material.

One pound of volatile solids can theoretically yield a maximum of 30.5 cubic feet of biogas. In reality, anywhere from 10 to 60 percent of the VS will be converted in the digesting process, so the practical result is that you can expect anywhere between about 3 and 18 cubic feet of biogas production per pound of volatile solids. The gas will likely be somewhere between 55 and 80 percent methane, so the methane yield for each pound of VS consumed in the generator can range from 1.7 to 14 cubic feet.

Studies of biogas yields from various feed stocks are generally represented in terms of cubic feet of methane produced for each pound of VS converted under the specific conditions of that test. Expect that your results will be different from anything you see published. The specific composition of VS within organic wastes, along with the environment within the digester and how you manage the whole process will all affect gas production.

Ballparking Water

The commonly accepted range of total solids in the mix for optimal biogas generation is between 2 and 10 percent, meaning that 90 to 98 percent of the material inside your generator, including the moisture in the material itself, can be water. Other considerations in determining how much water to add are material handling and digester volume.

EVALUATING RAW MATERIALS

Refer to this chart as you read through this chapter and work through the examples. It compiles all of the useful information you'll need about typical ingredients that might be available to you. Look up the material you're considering for your recipe, then read across the chart to get an idea of its value to your mix. For example, manure from laying hens has a nitrogen content range from 4 to 10 percent by weight, with an average of 8 percent. It has an average C:N ratio of 6 to 1, average moisture content of 69 percent, and weighs about 1,479 pounds per cubic yard. The important value of volatile solids (VS) will vary within the range of most animal manures, lying somewhere between 70 and 85 percent of total solids (TS).

CHARACTERISTICS OF RAW MATERIALS

Material	Type of value	%N (dry weight)	C:N ratio (weight to weight)	Moisture content % (wet weight)	Bulk density (pounds per cubic yard)	Average % VS of TS
Crop Residues and Fruit/Vegetable-processing Wastes						
Corncobs	Range	0.4–0.8	56–123	9–18	-	98
	Average	0.6	98	15	557	
Cornstalks	Typical	0.6–0.8	60–73 a	12	32	95
Fruit wastes	Range	0.9–2.6	20–49	62–88	-	75
Vegetable wastes	Typical	2.5–4	11–13	-	-	90
Animal Manures						
Broiler litter	Range	1.6–3.9	12–15 a	22–46	756–1,026	
	Average	2.7	14 a	37	864	
Cattle	Range	1.5–4.2	11–30	67–87	1,323–1,674	
	Average	2.4	19	81	1458	
Dairy tie stall	Typical	2.7	18	79	-	
Dairy free stall	Typical	3.7	13	83	-	
Horse, general	Range	1.4–2.3	22–50	59–79	1,215–1,620	
	Average	1.6	30	72	1379	
Horse, race track	Range	0.8–1.7	29–56	52–67	-	
	Average	1.2	41	63	-	
Laying hens	Range	4–10	3–10	62–75	1,377–1,620	
	Average	8	6	69	1479	
Sheep	Range	1.3–3.9	13–20	60–75	-	
	Average	2.7	16	69	-	
Pigs	Range	1.9–4.3	9–19	65–91	-	
	Average	3.1	14	80	-	
Turkey litter	Average	2.6	16 a	26	783	

Estimated average between 70 and 85% VS for most manures

Material	Type of value	%N (dry weight)	C:N ratio (weight to weight)	Moisture content % (wet weight)	Bulk density (pounds per cubic yard)	Average % VS of TS
Domestic Wastes						
Garbage (food waste)	Typical	1.9–2.9	14–16	69	-	90
Night soil (humanure)	Typical	5.5–6.5	6–10	-	-	85
Paper from domestic refuse	Typical	0.2–0.25	127–178	18–20	-	97
Refuse (mixed food, paper, and so on)	Typical	0.6–1.3	34–80	-	-	90
Sewage sludge	Range	2–6.9	5–16	72–84	1,075–1,750	87
Agricultural Products						
Corn silage	Typical	1.2–1.4	38–43 a	65–68	-	
Hay, general	Range	0.7–3.6	15–32	8–10	-	
	Average	2.1	-	-	-	
Hay, legume	Range	1.8–3.6	15–19	-	-	
	Average	2.5	16	-	-	most grass products average 90-95% VS
Hay, non-legume	Range	0.7–2.5	-	-	-	
	Average	1.3	32	-	-	
Straw, general	Range	0.3–1.1	48–150	4–27	58–378	
	Average	0.7	80	12	227	
Straw, oat	Range	0.6–1.1	48–98	-	-	
	Average	0.9	60	-	-	
Straw, wheat	Range	0.3–0.5	100–150	-	-	
	Average	0.4	127	-	-	

Note: Data was compiled from many references. Where several values are available, the range and average of the values found in the literature are listed. These should not be considered as the actual ranges or averages, but rather as representative values.

Table Credit: NRAES and the Cornell Waste Management Institute grants permission to reprint "Characteristics of Raw Materials" (see Resources for website), taken from the On-Farm Composting Handbook, NRAES-54. Estimated VS content added by the author.

SETTING UP A WASTE MANAGEMENT SYSTEM

Having a system for processing your organic waste material will save time, minimize mess, and generally make an easy task of the whole process. When things are easy, they get done!

We set up a fairly simple system in a corner of our garden hoop house, using an old sink and a ¾-horsepower under-sink food grinder. A heavy-duty extension cord brings power to the disposal, a 5-gallon bucket catches the ground-up food and grass cuttings, and a water hose flushes material through the grinder. With some rubber gloves at the ready, we've got a relatively easy way to manage digester feeding.

What once went straight to the compost pile now takes a detour through the food grinder and biogas generator. The effluent from the generator spills over into a 5-gallon bucket as fresh material is added, and finally makes its way to the compost pile.

◀ Sequence of preparing food scraps

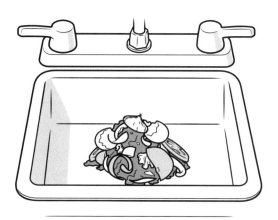

1. compostable raw material placed in sink

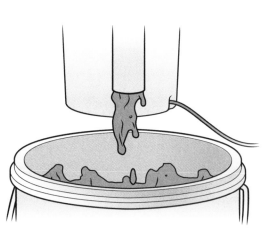

2. waste disposal under sink drains slurry into a bucket

3. slurry poured into digester

FEEDING YOUR BIOGAS GENERATOR

Now that you've developed a recipe, you can begin to feed the generator. Solid material should be chopped or shredded into 1" or smaller bits. Having more surface area available to the microbes promotes better digestion of organic material, yielding more efficient production of biogas.

Add enough water to make a pourable slurry, then mix it all up and pour the slurry into the digester. The general rule of thumb is to add the same amount of water as solid material, but this ratio will vary according to how much water is already in the organic material in your recipe, as well as how much fiber is in the organic matter. More fiber requires more water to break it down, but less water means more room for digestible solids, and thus more gas.

A power drill with a paint mixer attachment works well for mixing a manure slurry, but for food and plant scraps, you may find that a blender, garbage disposer (see Setting Up a Waste Management System, facing page), or a yard waste chipper/mulcher is a convenient way to chop up material before adding water. Fibrous material may digest more readily if it has been allowed to age (allowing fungi to begin breaking down the fiber) for a few days before putting it into the generator. Just don't age it for too long or energy will be lost.

INOCULATION

When you first load the digester, you will need to inoculate it with a culture of methane-producing organisms (methanogens). These microbes exist naturally in animal dung, so if you're using manure you don't need to worry about adding them. But if you want to digest only food scraps or grass clippings, you'll need to inoculate it initially to get the biological processes going.

A good culture can come from farm animal manure (ideally cow or pig manure), slurry from another operating digester, pond muck, or a shovelful from the bottom of a compost pile that has not been turned in a long time. Once added to the slurry at the digester's initial startup, the methanogens will reproduce on their own so you don't need to continue to feed manure or other source of methanogens.

Depending upon your location, finding a good source of inoculating manure or pond muck may be the toughest ingredient to find. With proper management, though, you may need to do this only one time, even with a batch digester that is periodically emptied and refilled. Much like keeping sourdough bread culture alive, some of the last batch is added to the next batch, inoculating the fresh material with the bacteria generated from the previous batch. Try not to expose the inoculant to air for too long, as oxygen will kill the methanogens.

The Process

Fill the digester about 80 percent full with slurry. More air space is okay if you're starting with just a small batch of material, but it will take a bit longer for your gas quality to improve so that it will burn. At first the generator will produce a large amount of CO_2, as the available oxygen is consumed and the methane producers catch up to the acid producers. This means that the first batch of gas will be mostly CO_2 and will not burn. Once the oxygen in the generator is used up, the process stabilizes and shifts from aerobic (microbial breakdown with oxygen) to anaerobic digestion.

Temperature

TEMPERATURE IS PROBABLY the most critical detail in the generation of biogas. Different groups of methanogens have been identified that respond to different temperature regimes.

You must choose which regime you wish to work with, and design your system accordingly:

- **Psychrophilic** methanogens survive at cooler temperatures, above 32°F (0°C).

- **Mesophilic** methanogens are active at temperatures between 70 and 105°F (21–41°C).

- **Thermophilic** methanogens will dominate at temperatures between 105 and 140°F (41–60°C).

Not much information is available about digestion in the psychrophilic range, but the gas production rate will be quite low. Mesophilic methanogens are much more tolerant to changes in their environment than thermophilic and are faster producers than psychrophilic. Mesophilic methanogens may produce gas at 70°F, but the rate at this temperature will be very slow.

One advantage to the thermophilic methane producers is that they are much faster at digesting, requiring only about half the material retention time to achieve similar gas production compared to the mesophilic range, thus requiring a smaller-volume digesting vessel. However, to achieve this efficiency you will pay a penalty in energy consumed to provide the required heat. For our purposes, we'll focus on the safer and more easily managed mesophilic temperature range.

FINE-TUNING TEMPERATURE

The mesophilic conditions you are trying to mimic within the digester are similar to those inside an animal's gut — that is to say, oxygen-free and a temperature of around 98°F (37°C), plus or minus a few degrees, for maximum gas production. Biological activity within the digester will produce some heat, but depending on your climate you may need to supply heat to your digester.

To reduce the amount of external heat required, place the generator in the sun or inside a greenhouse. An insulated wrap can be made with thin, flexible foam insulation, or even bubble wrap, covered with black, UV-resistant, 6-mil polyethylene plastic.

Cold Climates

To produce gas during the winter in cold climates you'll need to provide an additional source of heat. A larger generator may produce enough gas for some of it to heat water, which can be circulated via closed piping acting as a heat exchanger inside the digester. Or, you can wrap the outside of the barrel with flexible tubing (covered with insulation) and pump hot water through it (a good use for the solar batch collector project in chapter 6). Another option is a submersible, thermostatically controlled electric water heater designed for heating plastic animal waterers (see Resources).

In any case, you will need to weigh the costs of providing heat against the benefits of gas production. Alternatively, if you live in a hot climate you may want to provide some shade so that the temperature inside the digester does not rise much above 105°F.

Retention Time and Loading Rate

IN MOST CASES, material you put into a well-maintained methane generator operating in the mesophilic range will be fairly well digested in about a month. You might get more gas from the same material over a longer period, but the production rate will fall off over time.

Handling more material for a longer period also requires a larger generator. Retention time (also called **hydraulic retention time**, or HRT) is the amount of time material stays in the digester vessel, where it is broken down over time. Retention time is determined by how quickly the material breaks down, the gas output requirement, and the volume of the digester vessel. The desired retention time and the volume of the digester determine the rate at which you feed the digester so that material stays in the vessel for the optimum amount of time. Material should stay in the digester long enough to produce most of the gas that it can, but not too long because without enough nutrients the methanogens will die.

For example, if you have a 55-gallon drum (with a volume of 7.35 cubic feet) and you need a retention time of 30 days to produce the optimum amount of gas, you want to add 55 gallons of material over a period of 30 days, or just under 2 gallons (0.27 cubic feet) per day. Every time you add 2 gallons of new material, nearly 2 gallons of digestate will flow out the effluent tube. Over half of what you add will be water, so your average daily gas production is limited to the digestible content (VS) of the added organic matter.

Optimum retention time will vary with recipe and temperature, so experiment and find out what works best in your situation. For maximum efficiency, you want to match retention time with gas production and the rate at which you feed material, known as the **loading rate**.

LOADING RATE

Loading rate refers to the amount of volatile solids added to the generator with each feeding, as well as the frequency of those feedings. Loading rate is generally expressed in terms of pounds of VS per cubic foot of generator volume and is managed according to the properties of the material being used. If the material put into a digester is fairly consistent, you don't need to go through the exercise of calculating VS every time. After a while, you will get a feel for how much and how often to feed, based on your recipe and the observed results of gas production.

The rate at which you feed the generator is a function of how large your digesting vessel is and how well the methane-forming microbes keep up with the acid-forming bacteria. The rate will vary somewhat depending upon the digestibility of the feedstock you're using, since not all VS behave in the same way. Assume that only about one-half of the VS added will be digested, and expect that amount to vary for different feedstock — and even among the same feedstock, depending on its exact makeup. You will get different amounts of VS and gas from the same cow at different times of the year if that cow is on pasture for the summer and grain and hay for the winter.

Where to Start

As a general rule, start the generator on a small amount of VS — say, 0.1 pound VS for every cubic foot of digester capacity. As the methanogens become established, you can add more, perhaps up to 0.25 pounds of VS for each cubic foot of *filled* digester capacity. Some sources of VS will convert to gas at a higher rate and more completely, and you'll need to adjust the loading rate for best results. If gas production is fast and efficient, you can decrease retention time and move material through more quickly by increasing your loading rate.

Calculating VS and Gas Production

To understand just how much VS is in your recipe, and therefore how much gas you can expect to produce, refer to the table Evaluating Raw Materials (pages 238 and 239) and take a look at the column "Average %VS of TS." We'll pull values from that column and expand on the Recipe Evaluation chart (page 236) to calculate the VS content in the sample recipe.

Of the 6 pounds of organic material fed to the generator, only about 1.4 pounds is VS. If you put all this material into a 55-gallon drum (7.35 cubic feet of volume) that is 80-percent full (including 5.9 cubic feet of additional water), your loading rate works out to be:

$$1.4 \text{ pounds} \div 5.9 \text{ cubic feet}$$
$$= 0.24 \text{ pounds per cubic foot}$$

This is a good loading rate for an operational 55-gallon drum digester. Assuming a 50-percent

RECIPE EVALUATION: VOLATILE SOLIDS

Ingredient	Approximate wet weight	Moisture content	Dry weight	% of TS that are VS	VS weight
Chicken manure	2	70%	0.6	77%	0.46
Grass clippings	2	82%	0.4	80%	0.29
Kitchen scraps	2	69%	0.6	90%	0.56
Woodchips/newspaper	0.05	10%	0.05	97%	0.04

conversion of VS to biogas, total production over time might be:

$$(1.4 \text{ pounds VS}) \times (30.5 \text{ cubic feet of biogas per pound VS}) \times (50\% \text{ conversion rate})$$
$$= 21 \text{ cubic feet}$$

These are very general estimates, and your biogas production may be quite a bit less. The retention time (the amount of time it takes to convert the VS to biogas) will be determined through direct observation by you. Once your digester is loaded and operating, keep track of the rate of gas production by observing the displacement of the gas collection tank. When the production rate starts to drop off, it's time to feed. You may need to feed every day or once a week, depending on the conditions inside the digester and your recipe.

MIXING

Some material fed to your generator will form a crusty layer of scum on the surface. This scum can prevent gas formation, so daily mixing is recommended. You can mix by shaking, stirring, or otherwise agitating the slurry as long as no air is introduced into the generator. A barrel-sized generator can simply be rocked back and forth to mix, but larger vessels often require an internal mixing system. More agitation means less scum formation, and better mixing means more efficient gas production. You can help keep scum to a minimum by avoiding large amounts of lignin and by chopping ingredients into small pieces.

Types of Methane Generators

THERE ARE TWO general approaches to methane generator construction: **batch** and **continuous-flow**. I've also developed a third type of construction, which is sort of a hybrid of the two other processes.

BATCH GENERATORS

The simplest design is a batch processor where you fill the digester once, collect the gas until there isn't any left, then empty the digester

---

TWEAKING YOUR GENERATOR'S DIET

The methanogenesis (methane-forming) process is the limiting factor in how much to feed your generator. If you feed too much, the acid-forming process overtakes the relatively slow methane-forming process. This leads to a low (acidic) pH level that inhibits methane-forming activity, poisons the process, and ultimately will shut down gas production. If you're feeding the generator regularly and gas flow stops, you can check for this imbalance by testing the pH of the slurry. The pH should normally be right around 7 or 7.5. Perform this simple test by dipping pH paper into a sample of liquid digestate. Here's what to do with the results:

1. If the pH is below 6.5, the digester has gone "sour." Stop feeding for a few days so that the methanogens can catch up, then test the pH again.

2. If things are really bad and stay bad for more than a few days, you may need to add some baking soda to help bring the pH back up. Add only ¼ to ½ cup at a time for a 55-gallon barrel generator, wait a day, and check the pH again.

3. When the pH has returned to normal, gas should start bubbling out again. Resume feeding and adjust your recipe and/or the rate at which you feed, while monitoring pH as you make adjustments.

The ecosystem within the generator evolves to suit the recipe. If you change recipes in a sudden and dramatic way, the chemistry will react and readjust. Think of how you feel after you try a new food or travel to a foreign country and indulge in local cuisine. Make small, incremental changes to your generator's diet to avoid a bellyache.

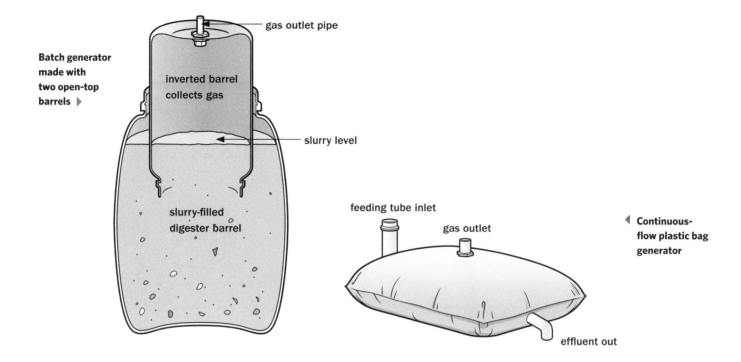

Batch generator made with two open-top barrels ▶

gas outlet pipe

inverted barrel collects gas

slurry level

slurry-filled digester barrel

feeding tube inlet

gas outlet

◀ **Continuous-flow plastic bag generator**

effluent out

vessel, compost the digestate, and start all over. A batch processor is time-intensive and involves the dirty job (at least once a month) of emptying the digester and starting a new batch. The advantage is that it's simple, cheap, and works great for small, easily managed volumes of material.

You can build a fairly simple batch processor using an open-top 55-gallon barrel for the digester vessel, and a second, smaller barrel (also open-top) for gas collection. Drill a hole into the bottom of the smaller barrel and insert a gas collection tube. Then, put the smaller barrel into the larger barrel with the open end down and the gas outlet tube up. Push the top barrel all the way down into the slurry, allowing air to escape from the gas outlet. Close the gas outlet to seal the barrel.

As gas is produced, the top barrel will rise up, and the gas can be drawn out through a pipe to a burner. This is a good way to get started quickly and to gauge gas production as the barrel rises. You can determine the gas quantity by measuring how much of the barrel rises above the slurry and calculating the displaced volume.

The simplicity of a batch generator ends when it comes time to empty and refill the barrel: 55 gallons of effluent is not a pretty sight to behold and can be difficult to manage. This is not such a big drawback with a small, 5-gallon-bucket batch digester like the project on page 253.

CONTINUOUS-FLOW GENERATORS

As you move beyond experimenting with test batches of buckets and barrels in your biogas hobby, you may find that a continuous-flow process is more practical than a batch setup. Continuous-flow is also called *plug-flow*, because a "plug" of material is introduced to the generator periodically.

A continuous-flow generator has a feeding inlet on one end, an effluent outlet on the other end, and a gas outlet pipe in between. It's sized according to the daily input volume and desired retention time. As you put new material in, it pushes older material through the generator until it gets to the effluent outlet, where the digestate it is removed. You can build a very simple continuous-flow generator with a UV-resistant polyethylene plastic bag (such as those used for bagging hay) and PVC piping for the inlet, outlet, and gas connection.

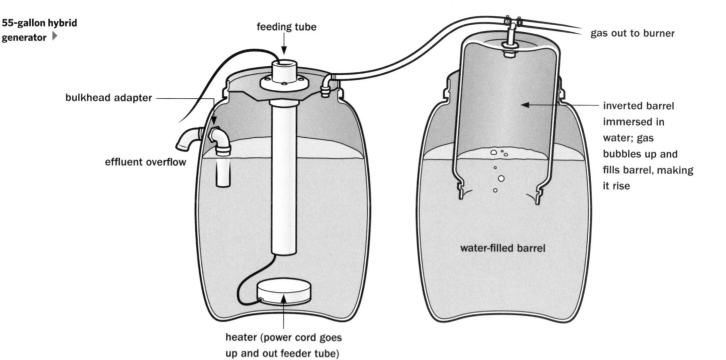

feeding tube

gas out to burner

bulkhead adapter

inverted barrel immersed in water; gas bubbles up and fills barrel, making it rise

effluent overflow

water-filled barrel

heater (power cord goes up and out feeder tube)

HYBRID GENERATOR

As a compromise between simplicity, practicality, and space constraints, I have settled on a hybrid generator system for my own use that takes advantage of the best of both designs and allows for cleaning, tweaking, and upgrading as needed. The hybrid design requires three barrels and piping between them. One barrel is used as the digesting vessel and incorporates the feeding tube, effluent overflow, and gas outlet pipe. The other two barrels are used to make an expandable and isolated gas collection system. I describe how to build this generator in the project on page 256.

Using Biogas

BIOGAS CAN BE used in place of natural gas or liquid propane gas for space and water heating, lighting, and cooking, with some modification to the burner. It can also provide power by fueling an engine or an absorption cooling system, such as a gas refrigerator or chiller.

Purified, or "scrubbed," biogas behaves just like natural gas, but if you choose not to scrub the gas, you will need to deal with the CO_2 and water vapor content by modifying burners or other equipment to ensure good combustion. In other words, you can either modify the gas or modify your equipment. Ideal combustion requires a

---

WHAT TO DO WITH THE EFFLUENT

Effluent is the digested "waste" material from your biogas generator. It's a low-odor blend of compostable solids and nutrient-rich liquid. What you do with the effluent you produce depends upon what you put into the generator, how well the solids are digested, and, perhaps, where you live. You can apply

effluent directly to your garden or fields as a soil amendment, or you can compost it and further refine it for improving your soils.

One of the biggest concerns is the presence of pathogens in animal wastes. Some pathogens will be destroyed in the mesophilic temperature range

over a typical retention period, but some require much hotter temperatures to be completely destroyed. High-temperature composting (greater than 140°F) generally eliminates most pathogens. The only way to be sure is to have a sample tested at a reputable lab.

burner that is designed specifically for a particular gas. Modifying burners will involve compromises in efficiency and performance but should yield a usable flame. But first and foremost: When working with or around biogas, safety is your top priority.

HANDLING BIOGAS SAFELY

Methane is a highly flammable gas that will burn when mixed with air at a ratio of between 5 and 15 percent by volume. Biogas has a flammability range of about 4 to 25 percent concentration in air, and possibly wider, depending on its purity. Always take great care when working with biogas. It is quite possible for a biogas generator to explode, especially if there are leaks in the generator or gas piping, or when the gas pressure at a burner falls too low. It's better to have a small amount of biogas leak to the outdoors (where it will be quickly diluted) than to have air leaking into the generator where it's more likely to reach the critical range of flammable ratios.

Maintain Positive Pressure

Keeping a positive pressure inside the digester and gas lines will help to prevent air from leaking into the generator, as well as prevent the burner flames from traveling through the gas tube back to the generator. The processes within the generator will create positive pressure as gases are produced, but as the gas is consumed, the pressure will drop, possibly allowing flame to burn back through the pipe (flashback) to the generator. Always turn off the burner flame if the gas pressure is too low.

Use a Flame Arrestor

To help reduce the chances of a flashback, use a flame arrestor inside the gas pipe. The idea behind a flame arrestor is to stop gas flow and/or reduce the flame temperature to below the ignition point.

There are two relatively simple (though not foolproof) ways to approach this:

1. Incorporate a water trap into the gas line by creating U-shaped bend, or trap, in the line that

Gas storage system using an inverted barrel inside a larger barrel filled with water. Gas from the digester displaces the water, causing the barrel to rise. If needed, a guide can be built to keep the gas barrel straight as it rises so that weight can be placed on top of the barrel to provide the required pressure. Adding bricks for weight is sometimes called "brickage."

is filled with a few inches of water. The biogas will bubble through the water on its way out of the digester vessel and into the gas storage vessel. In the event of a flashback, the water in the trap will cool and extinguish the flame inside the line. This approach will not work very well on the gas supply side, as the flame will sputter as each gas bubble makes it to the burner. Biogas contains water vapor that will condense in the gas line, so keep an eye on the water level in the trap.

2. The second approach works well on the gas supply line: Pack some fine bronze wool (similar to steel wool) into the gas hose just before the burner. Add enough to fill 3" or 4" of hose length. Gas can still move through the bronze wool, but should it start to flow back through the hose, the bronze wool will effectively extinguish the flame.

Respect the Hazards

Treat biogas with the same respect you would treat any other fuel or flammable product. Keep sparks and flame well away from biogas and biogas generators. Never use biogas indoors or in enclosed spaces. Ethyl mercaptan is the odorant added to both natural and bottled gas so that a leak is readily detectable by smell. When burning biogas — in a burner or other suitable equipment — ensure proper ventilation to prevent a buildup of explosive, toxic, or even deadly gases and combustion byproducts. Burning biogas produces carbon monoxide, along with CO_2, water vapor, and nitrogen oxides.

STORING BIOGAS

Once you've produced biogas in your generator, you can pipe it into a simple holding container. For very small batch generators (of the science-project variety) you can use a latex balloon. Larger amounts of gas can be stored in small barrels inverted into larger barrels that are filled with water. The gas storage container should be airtight, allow for expansion and contraction as gas flows into and out of it, be able to deliver the required pressure to the gas appliance, and not turn into shrapnel in the event of flashback. How much storage capacity you need depends on how much gas you make, along with how much and when you use it.

BURNERS

Compared to natural and propane gas, biogas has less energy per unit due to the dilution by CO_2 and, to a much lesser extent, water vapor. This means that the energy in the flame of unscrubbed biogas will be lower when used with conventional natural gas or propane burning equipment. A good biogas flame requires greater fuel flow, higher pressure, and less combustion air than natural or propane gas.

Nozzle size. The burner nozzle size will need to be larger than what you might use for natural or propane gas to yield a similarly energetic flame. How much larger depends on the purity of the biogas and the fuel gas you are converting from. The Orifice Diameter Multiplier chart, above right, indicates how much larger a fuel nozzle orifice needs to be when converting equipment from natural gas or propane to biogas. If you choose not to enlarge the orifice, you may still get a flame but the energy content of the flame will be reduced.

ORIFICE DIAMETER MULTIPLIER FOR GAS BURNERS

% Methane (CH_4) in Biogas	Orifice Diameter Multiplier	
	Natural Gas	**Propane**
70%	1.32	1.63
65%	1.39	1.72
60%	1.46	1.81
55%	1.54	1.92
50%	1.64	2.04

Here's an example: If your biogas contains 60 percent CH_4, and you want to convert a natural gas appliance with an existing orifice diameter of 0.1", you need to enlarge the orifice to: 0.1 X 1.46 = 0.146" diameter. Some gas appliance nozzles can be removed and a different one installed, but many are pressed in place and cannot be removed. The nozzle orifice can be enlarged with a drill press and the appropriate drill bit.

Airflow. For a good flame, the primary air supply to the burner must be reduced. This is easily accomplished by closing the air shutter on the gas appliance's burner tube until a steady, blue, cone-shaped flame is produced at the burner.

Gas pressure. Pressure in the biogas system can be regulated by applying external weight to the top of the container. Increase the weight until the gas flows to the burner at the required pressure. A gas pressure regulator is used on all gas-burning appliances to prevent too much pressure from being delivered to the burner. However, finding equipment suitable for use with biogas is difficult.

A typical propane gas kitchen range might require a delivery pressure of 11" of water column (wc) pressure. This simply means that where the gas enters the appliance it exerts the pressure required to lift a column of water in a vertical tube 11" tall. Expressed another way, 27.71" of water column is the same pressure as 1 pound per square inch (psi). Delivering biogas to a conventional cook stove burner may require up to 20" wc.

Biogas and Engines

Some gasoline engines are designed — or can be modified — for use with natural gas, propane gas, or biogas. Modifications include installing a special carburetor, adjusting the timing, and regapping the spark plugs. Diesel engines can accept up to 80 percent biogas when the biogas is mixed with the incoming combustion air.

When burning unscrubbed biogas in engines, the oil must be changed more frequently than when using purified biogas. Also, since there is less energy in a volume of biogas compared to natural gas, gasoline, or diesel fuel, the engine will produce less power than its factory rating.

COW POOP AND GAS GENERATION

I realize that most of us don't have a family cow at our disposal, but the following example will illustrate the process of planning a methane generator around available resources and realistic expectations.

Fresh cow manure is often used exclusively for on-farm methane production and is well suited to this application. But cow manure has a less-than-optimal C:N of only 15:1 and relatively low gas production by total volume of raw material. Despite these apparently unfavorable conditions, cow manure works well due to a number of other properties that make it attractive for dairy farmers incorporating anaerobic digestion into their daily operations. These include:

- Large quantities of cow manure produced on a typical farm

- Methanogen-rich quality of cow manure, making it a guaranteed gas producer

- High water content, requiring little additional water

- Co-benefit of capturing manure before it winds up as a pollutant in lakes and streams

- Using the digestate as a high-quality fertilizer

- Potential for producing large amounts of gas to generate electrical power for on-farm use or to sell power back to the utility grid

One cow might produce about 18 gallons (140 pounds) of manure each day. Given the amount of water in cow manure, only about 12 percent of the total weight is TS, and somewhere around 85 percent of TS is the useful VS portion, which may ultimately generate about 60 cubic feet of methane (or around 85 cubic feet of biogas). That amount of gas represents perhaps 3 hours of cooking fuel produced each day.

Compare the output of a dairy cow with the amount of food scraps or human excrement (0.5 lb per person, per day) available, and the logistics of material choice for biogas production becomes clear. That's not to say you can't make useful amounts of biogas with other materials; you just need more of them. But most importantly, you need to use what's available to you.

If you have one cow on your homestead property and want to collect all of its manure to make gas, you would first need to contain the cow so you could collect the poop. If you collect 18 gallons of material, and you add another 18 gallons of water to make a slurry, you quickly realize that you'll need a fairly large, continuous flow generator to handle a daily feeding while allowing for sufficient retention time to make a useful amount of gas.

Alternatively, a batch generator made from a 55-gallon drum loaded with a 50/50 mix of manure and water will begin producing gas within a few days, but the quantity will be limited by the amount of VS you can put in the drum.

To give you an idea of how much weight might be required in practical terms, when my 170-pound body compresses a volume of air under a 12"-diameter barrel, it exerts a pressure of about 20" wc. After enlarging the orifice, my Bunsen burner works well over a wide pressure range, and the flame height varies with pressure. A larger (20,000 Btu) burner delivers a steady, high flame at 15" to 20" wc.

Purifying Biogas

DEPENDING ON THE intended use and quality of the gas required, you may need to remove impurities from the gas before burning it. Biogas contains up to 50 percent carbon dioxide (which won't burn) and possibly traces of hydrogen sulfide (H_2S), which will corrode metals and break down engine oil.

Purifying, or scrubbing, biogas can decrease the detrimental effects of these ingredients, decrease gas storage requirements, and increase available heating energy per unit of gas. Unless you're converting an engine to operate permanently on biogas, there is really no pressing need to purify it. If you choose to scrub biogas, it can be done with costly high-tech purification systems, but our discussion will be limited to simpler methods.

One method for reducing both carbon dioxide and hydrogen sulfide in biogas is by bubbling the gas through a solution of calcium hydroxide (or calcium oxide) and water. Calcium oxide is the chemical known as *lime*, commonly used to fertilize soil. You can further reduce amounts of hydrogen sulfide by passing the gas through iron oxide, or rust. This setup can be as simple as a 4"-diameter x 4-foot-long PVC pipe filled with loosely packed rusty steel wool or coarse iron shavings. The rust will eventually be consumed and need to be replaced at a rate that is dependent upon the H_2S concentration in the biogas. The pipe will need to have gas-tight fittings and cleanouts. If you're making biogas for backyard cooking or tiki torch lighting, there is no need to add the complexity of scrubbing.

ENVIRONMENTAL CARE

Anaerobic digestion is essentially a controlled cow fart. Both cow farts and biogas generators produce methane. Methane is a very potent greenhouse gas and should not be released directly into the atmosphere. If you are going to make methane, you must burn it rather than release it into the atmosphere. Burning methane produces carbon dioxide (also a greenhouse gas but less potent than methane) and water vapor. Carbon dioxide is also produced naturally when organic material decays.

Generator and Storage Materials

Given the potentially corrosive nature of biogas with even small amounts of hydrogen sulfide (which may be produced when some manures are digested), avoid using materials that are susceptible to corrosion. Plastic barrels and PVC pipes will prove more durable over time than metal barrels and galvanized pipe.

Biogas Is More than a Gas

SOME YEARS AGO I wrote the book, and then took some time to do other things, including running a medical device company. But more recently, fate brought me back to this subject of biogas when a dear friend of mine from Sri Lanka came to visit me. He was in my office, which is packed with books, and he noticed the biogas book. He asked me about it, and I explained. He got very excited. "We need this in Sri Lanka!"

DAVID HOUSE is the author of *The Complete Biogas Handbook* (see Resources). During my research for this book, I discovered a handful of people like David who represent an altruistic community of dedicated clean-energy researchers and advocates. Their work goes way beyond the technical and digs deep into how energy choices can affect the human condition. It's too easy for those of us with the convenience of automatic fuel delivery and programmable thermostats to complain about the high cost of energy, while elsewhere the cost of a hot meal is simply unreasonable. In this profile, David comments on the non-energy benefits of renewable energy in the developing world.

This set me on a path to researching and thinking about biogas in the developing world, in a far more detailed and intensive manner than I had before. What I learned (or relearned) was that biogas is associated with strong improvements in health, education, financial well-being, gender equality, and a reduction in deforestation, among other benefits. When you think about it, that's an incredible list for something that for most of us is pretty obscure — biogas — but each item on the list has solid reasons for being mentioned:

- **Health.** The World Health Organization says that nearly two million people, mostly women and children, die every year as a result of indoor smoke from wood cooking fires. That's almost 4,500 people every day, about two a minute. Biogas burns with a smokeless flame.

- **Education.** Gathering firewood is very time-consuming (2 to 6 hours a day), and it is usually the eldest girl who does that work, so she has no time for school. Caring for a biogas digester may take a half-hour a day, and in the evening the family can have the benefit of the very bright light that biogas can provide, with the right lamp, allowing reading and study.

- **Financial well-being.** Improved health and increased time help the family — usually the mother — to start a business, barter work for an animal, or improve their financial status in other ways. Finally, the family has a crucial new increment of time and energy, both personal and household.

- **Gender equality.** The improved health and any improvements in financial strength contributed by the mother imply certain changes.

- **Deforestation.** Less wood is taken from trees cut down to feed the cooking fires. Poor countries cannot afford to replant their forests, and carbon dioxide (CO_2) released into the atmosphere from trees that are burned and not replanted has exactly the same effect as CO_2 from fossil fuels. Some studies indicate that as much as 15 percent of the increase in global CO_2 comes from those disappearing forests.

So naturally one should ask: If biogas is such a powerful tool to help achieve the United Nation's Millennium Development Goals (see Resources), then why isn't it used more often? In fact, it is being used a great deal, but not nearly as much as it *could* be used, because the most commonly built biogas digesters are *expensive*: $350 to $700 for a single-family digester.

Some digester programs offer a subsidy to help purchase one, because funders and governments know what a powerful catalyst for development this technology is. Unfortunately, these programs usually leapfrog right over those who are really poor, because they want

Biogas Is More than a Gas

UN Millennium Development Goals	
	End Poverty and Hunger
	Universal Education
	Gender Equality
	Child Health
	Maternal Health
	Combat HIV/AIDS
	Environmental Sustainability
	Global Partnership

UN Millennium Development Goals

to ensure that these expensive digesters are used, and the poor do not have animals to produce manure to feed the digester. So the subsidies more often go to those who are already doing fairly well.

To address these barriers to using biogas, my work is focusing on the development of a very low-cost biogas digester. The prototypes cost about $10 in parts purchased retail in the United States in low quantity, and these digesters (like all digesters) can be fed grass, leaves, food waste, and similar materials, so they can be used by people who do not have animals. Depending on local costs, a mature and manufactured version may cost from $6 to $8 to produce, and could be subsidized, sold at cost, or sold for two to three times the cost of manufacture. If the carbon market perks up, these digesters could earn carbon credits and be profitable even if given away.

Consider that the wealth of the world is her people, above all. We cannot doubt that people of enormous talent — nascent Mozarts, possible Einsteins, potential geniuses and savants of all stripes and kinds — have lived and died in the world's villages without the benefit of the education which would have unlocked all that was imprisoned in their hearts and minds. These gifts are lost to the whole world.

In addition, poverty begets war, and one of those wars could easily suck our children into its mouth. Swine flu started in a poor village somewhere, where pigs and birds and humans all lived too close together in very difficult circumstances. Will the next global pandemic be born in a mud hut, and travel the world at the speed of a 747? Addressing poverty is not merely of benefit to the world's poor, then: it will benefit everyone on this singular green planet.

Imagine! If we are successful at introducing these very low-cost digesters and showing that modest profits can be made, then the idea could march across the equatorial belt like a breath of hope, helping lift families and whole villages out of poverty.

To draw attention to this project, I am building a solar greenhouse on my Oregon farm which will enclose a 10-cubic-meter digester made from a silage bag (used by farmers). I have an agreement with a local fast-food restaurant to feed it all the food waste it generates. My present estimate is that the digester will provide the equivalent of nearly $4,000 of energy every year.

To learn how you can participate in bringing low-cost biogas projects to places and people in need, visit the website for Bread from Stones (see Resources).

I am building a solar greenhouse on my Oregon farm, which will enclose a 10-cubic-meter digester made from a silage bag (used by farmers). I have an agreement with a local fast food restaurant to feed it all the food waste it generates. My present estimate is that the digester will provide the equivalent of nearly $4,000 of energy every year.

Make a Biogas Generator

Making biogas usually involves smelly, messy ingredients. It will be important to develop systems that work for you to minimize handling while maximizing your potential for success.

Two generator designs are presented here. The first is a very simple "test"-batch processor made from a 5-gallon bucket — good for trying out the process without having to get too involved. The second is a hybrid configuration featuring both batch and plug-flow design elements. Using plastic barrels and common plumbing supply parts, it is fairly simple and inexpensive to make and use, but requires some material processing and waste management.

BATCH VS. CONTINUOUS-FLOW

As discussed earlier, a batch generator breaks down one batch of material at a time: You load the material and seal the vessel. Gas is produced after several days (at a rate and duration that depends on the recipe and the internal conditions). After the gas production stops, you empty the generator, compost the effluent, and repeat the process.

In a continuous-flow generator, material can be loaded at a certain rate and is constantly pushed through the vessel over a period of time that matches the retention rate required by the material and conditions. Effluent flows out the opposite end of the generator from the intake, and gas is produced in a fairly continuous process.

A hybrid system allows you to extend the "batch" into a periodic "plug" so that you don't need to empty the barrel quite so often. At some point, however, you will need to empty out the barrel and remove all the sludge that builds up over time.

5-GALLON BATCH GENERATOR

For a small-scale test or science project, you can start with this simple 5-gallon bucket batch generator. It's a manageable and inexpensive project that will help you get a feel for the process. It's also useful for testing your feedstock recipe before moving on to something more complex.

1. Prepare the lid.

Make sure the bucket's lid gasket is in good condition for an airtight seal. Drill a hole through the top of the lid, using a hole saw properly sized to the bulkhead fitting. To ensure a good seal, be sure the hole is in a flat section of the lid, without curves or raised lettering.

Fit the threaded male end of the bulkhead fitting into the hole and secure it on the underside of the lid with the

MATERIALS

5-gallon bucket with gasketed, airtight lid

One ½" NPT threaded bulkhead fitting

One ½" NPT male threaded x ½" O.D. (outside diameter) hose barb T-fitting

One ½" to ¼" barb hose adapter

Teflon tape

One 9" length ½" I.D. (inside diameter) flexible plastic tubing

One 4-foot length ¼" I.D. (inside diameter) flexible plastic tubing

One PVC ball valve for ½" I.D. tubing

Five hose clamps for ½" tubing

Two hose clamps for ¼" tubing

One large "punching bag" balloon

Fine bronze wool pad

One Bunsen burner designed for "artificial" gas (a burner with an integral "flame stabilizer" works best)

Warning!

Making and using biogas is dangerous — potentially deadly — and can cause fires or explosions. Use common sense and proper safety equipment. Perform all experiments (including these projects) outdoors with plenty of ventilation and appropriate personal safety gear. If you're in doubt about any procedure or have never worked with the materials and equipment described, please work with an experienced helper.

fitting's gasket and nut. Make sure this joint creates an airtight seal through the hole in the lid.

2. Complete the gas outlet and hose assembly.

Wrap the threaded end of the ½" T-fitting with Teflon tape. Screw the taped end into the female threads in the top of the bulkhead fitting, and tighten it snugly by hand.

Cut the 9" length of ½" I.D. plastic tubing into one 1" piece and two 4" pieces. These lengths can be adjusted as needed, depending on your specific requirements.

Fit one 4" piece of tubing onto one end of the T-fitting, and secure it with a hose clamp; this is the gas supply line. Attach the other end of the same tube to one side of the ball valve and secure it with a hose clamp. This valve will act as a shut-off between the digester and burner. Attach one end of the remaining 4" piece of tubing to the other side of the valve and secure it with a hose clamp.

The ½" gas supply line must be reduced to accept a ¼" hose to connect to the Bunsen burner's gas inlet. Install the barbed hose adapter in between the ½" and ¼" hoses and secure each connection with a hose clamp.

Slip the open end of the balloon through the 1" piece of ½" tubing, then stretch the balloon over the remaining port on the T-fitting. Slide the tubing down so the end of the balloon is sandwiched between the fitting port and the tube; secure it with a hose clamp. (The tube merely protects the more fragile balloon material.) The balloon serves as a gas storage vessel as well as a pressurizing system; it can also act as a safety valve by popping if the pressure inside becomes too great.

3. Connect the burner.

Test the system for airtightness before making the final gas connection: Hold your hand over the bottom end of the bulkhead fitting, then blow through the open end of the gas supply tubing, and check for leaks. The balloon should inflate and there should not be any leaks.

Complete the gas line connection by fitting the ¼" tubing to the gas inlet on the burner. Make sure the burner control knob is OFF.

4. Make a simple flashback preventer.

As gas pressure drops, the flame at the burner can burn back through the gas line and into the bucket where some gas remains. The best way to prevent

---

EXPLODING WITH IMPATIENCE

When I made my first biogas, I could not get the burner to light even after bleeding off the first balloonful. In disgust, I pulled the hose off the burner and held a match right up to the hose. A flame popped briefly out of the hose as the balloon quickly deflated, followed by a groaning noise from inside the bucket. I ducked, ran, and turned just in time to see the balloon expand and pop, followed by a dramatic geyser of partially digested slurry rise from the gas outlet pipe where the balloon once was.

The methane was not concentrated enough to light the burner, but the methane-to-air mixture inside the bucket was perfect for combustion. Once the balloon was empty, there was no gas pressure pushing the gas out of the barrel, and the flame flashed back through the gas hose. The mixture inside the bucket ignited, causing a small (but contained) explosion.

Fortunately, all of this happened outdoors, and the amount of gas inside a 5-gallon bucket that is mostly full of noncombustible material does not contain a lot of energy. Had I not been so impatient to burn the gas, and had I bled off another balloonload or two of gas, I would not have been able to relate this cautionary tale — all the better for you, and I am none the worse for wear! Please take care when working with flammable and explosive gases, and do not attempt to repeat this experience!

this is to turn off the gas supply before the pressure drops too low. "Too low" means before the flame starts to sputter. As a secondary precaution, create a flashback preventer by inserting some fine bronze wool (similar to steel wool) into the ½" gas hose. Add enough to fill a few inches of hose length. Gas can still move through the bronze wool, but should it start to flow back through the hose, it will be extinguished by the wool.

NOTE: These prevention techniques are not foolproof, so take all necessary safety precautions!

5. Make and use the biogas.

Fill the bucket about one-third full with ground-up or finely diced food scraps. Add about ½ gallon of cow or pig manure (or a gallon of effluent from another digester) to provide the methanogen inoculant. Add water to fill the bucket no more than three-quarters full, and mix up the slurry. Fit the lid onto the bucket, making sure it seals completely. Check for leaks again to be sure the gasket is well seated. Keep the digester warm, ideally between 80 and 100°F.

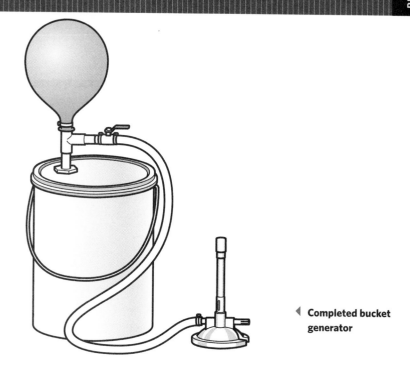

◀ **Completed bucket generator**

After a day or two, the balloon will begin to fill with gas, but this is mostly CO_2 and will not burn. Bleed off this gas by opening the gas valve on the burner, but do not attempt to light the burner. After a few more days (perhaps up to one week) and a

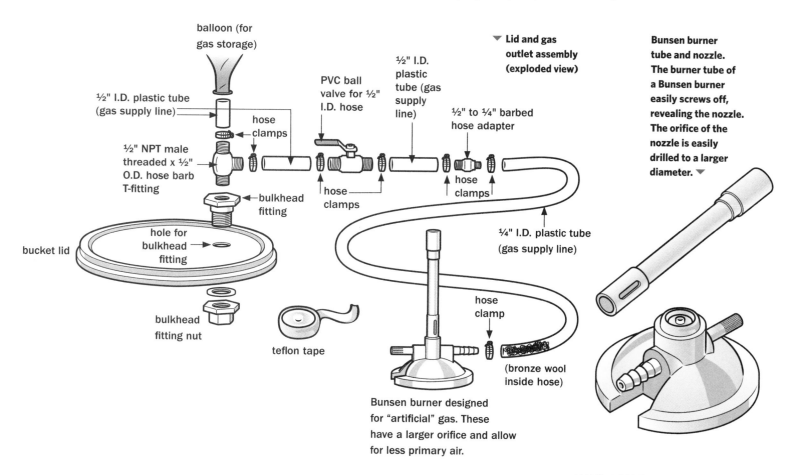

balloon (for gas storage)

½" I.D. plastic tube (gas supply line)

hose clamps

½" NPT male threaded x ½" O.D. hose barb T-fitting

bulkhead fitting

bucket lid

hole for bulkhead fitting

bulkhead fitting nut

teflon tape

PVC ball valve for ½" I.D. hose

hose clamps

½" I.D. plastic tube (gas supply line)

▼ **Lid and gas outlet assembly (exploded view)**

½" to ¼" barbed hose adapter

hose clamps

¼" I.D. plastic tube (gas supply line)

hose clamp

(bronze wool inside hose)

Bunsen burner designed for "artificial" gas. These have a larger orifice and allow for less primary air.

Bunsen burner tube and nozzle. The burner tube of a Bunsen burner easily screws off, revealing the nozzle. The orifice of the nozzle is easily drilled to a larger diameter. ▼

few more balloonfuls, the digester should be producing methane that will burn. When the balloon is full, open the gas valve on the burner and light the gas. If it sputters and doesn't light, the gas is not yet combustible. Bleed off this gas from the balloon and try again later. Keep an eye on the balloon, because as it deflates the pressure will drop. Remember that low pressure in the gas line can cause the flame to roll back into the bucket.

55-GALLON HYBRID GENERATOR

This project shows you the basics of building a hybrid generator using a 55-gallon (or larger) plastic, wide-mouth, screw-top barrel for the digesting vessel, and two additional barrels for gas storage. The essential design elements are to provide an airtight system that lets you put digestible material in, get gas out, and allow for effluent overflow. Once you get the hang of operating this

unit, you'll be able to design your own biogas generator, scaling it up or down, if desired, and adapting the system and materials to suit your specific needs.

If you buy all new parts for this project, you can build the 55-gallon generator for under $300. Optional additional equipment includes a $30 electric heater to keep the slurry warm, a camp stove to burn the biogas, and perhaps a $100 garbage disposer to facilitate grinding up food scraps for feedstock (see Setting Up a Waste Management System on page 240). Most of these parts are available at plumbing supply houses and hardware stores. You can also find tank and barrel connectors online at such places as Tank Depot (see Resources). The thermostatically controlled submersible heater is sold through livestock supply stores or online at Jeffers Livestock (see Resources). Be sure to prepare all materials and test-fit before permanently fastening.

TROUBLESHOOTING YOUR 5-GALLON GENERATOR

If you're making gas but it won't light after a week, try the following:

1. Remove the burner tube from the Bunsen burner to expose the gas nozzle. Open the gas valve and try to light the flame at the nozzle orifice. The blue flame will be very difficult to see in the daylight.

2. If the gas burns at the nozzle but not at the end of the burner tube, either the nozzle orifice is too small or the gas pressure is either too high or too low. Be sure there is sufficient pressure in the balloon. Try squeezing the balloon to

force the gas out, or closing the gas valve to reduce the pressure. Completely close the burner's air inlet.

3. A typical inflated balloon may exert a pressure of about 10" water column (wc), about what is needed to operate a typical gas appliance, but half what you will need with an unmodified biogas burner. Try enlarging the gas nozzle orifice by drilling the hole progressively larger in small increments (0.5 mm or $\frac{1}{64}$").

After experimenting with different burners and pressures, I've found that a 2.5-mm nozzle orifice with very little primary air works well with a Bunsen

burner. A larger portable stove, with a burner rating of 20,000 Btus, requires a 4–6-mm orifice. Both burners work over a wide pressure range, with the flame energy increasing as gas pressure increases. Keep in mind that my biogas is made primarily from food scraps, but the gas you make may be different and require different burner settings.

If the gas still does not burn, you may not be producing methane yet, or there is a leak in the gas line that is diluting the gas with air. A leak anywhere in the system can introduce oxygen that will kill the methanogens, causing your digester to go aerobic and produce only CO_2.

MATERIALS

For Generator

One 55-gallon screw-top barrel (for digester vessel)

One PVC toilet flange to fit over 3" PVC pipe

One 4-foot length 3" PVC pipe

Two 1-foot lengths 2" PVC pipe

Two ½" NPT bulkhead fittings (one is optional for thermometer)

One 2" socket bulkhead fitting (for 2" PVC connection)

Two 2" 90-degree PVC street elbows

One 2" PVC cleanout

One 3" PVC cap

PVC primer and solvent glue

Silicone caulk

Four machine bolts with washers and nuts (sized for toilet flange mounting holes)

One ½" NPT male threaded x ½" O.D. (outside diameter) hose barb fitting

One 1,000-watt submersible heater with thermostat (optional)

Insulation (see step 4)

One thermowell (optional; sized to match ½" bulkhead fitting)

One dial thermometer with stem (optional; sized to match thermowell)

Teflon tape

One 10-foot length ½" I.D. (inside diameter) flexible plastic tubing

One PVC ball valve for ½" I.D. tubing

Four hose clamps for ½" tubing

For Gas Storage and Use

One 55-gallon open-top, wide-mouth barrel (for gas collection)

One 30-gallon open-top barrel (or other size, as needed, to fit inside gas collection barrel)

One ½" NPT bulkhead fitting

One ½" NPT male threaded x ½" O.D. hose barb T-fitting

One 10-foot length ½" I.D. flexible plastic tubing

One PVC ball valve for ½" I.D. tubing

Five hose clamps for ½" tubing

One single-burner propane camp stove

Adapters and hoses as needed to connect gas ½" gas outlet hose to camp stove

Fine bronze wool

1. Install the effluent overflow assembly.

Drill a 2" hole through the side of the barrel, about 6" down from the top. Install the 2" socket (unthreaded) bulkhead fitting into the hole, making sure the joint is airtight. This fitting makes the connection between 2" inch PVC pipes on both the inside and outside of the barrel.

Solvent-glue a 90-degree street elbow to the inside of the bulkhead fitting, pointing the elbow

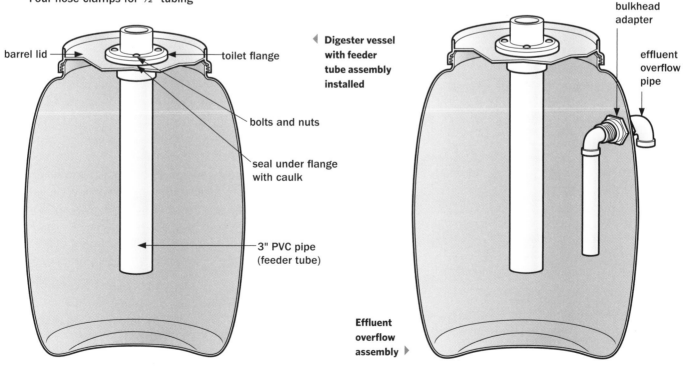

barrel lid — toilet flange

Digester vessel with feeder tube assembly installed

bolts and nuts

seal under flange with caulk

3" PVC pipe (feeder tube)

bulkhead adapter

effluent overflow pipe

Effluent overflow assembly ▸

Alternate Assembly

As an optional feature, you can also install a special thermometer assembly (designed for plumbing applications) that uses a thermowell, a metal probe that projects into the barrel interior and accepts a thermometer on its outside end. Install the thermowell using a ½" threaded bulkhead fitting, as with the gas outlet, locating the fitting about one-third of the way up from the bottom of the barrel. Insert the probe end of the thermometer inside the thermowell to complete the assembly. A long thermowell will provide a better temperature indication.

toward the bottom of the barrel. Cut a piece of 2" pipe to extend from the elbow down to about the middle of the barrel. Glue the pipe to the elbow. The bottom of this pipe must be above the bottom of the feeding tube. When the digester is filled, the pipe must be covered by the slurry at all times to prevent air from entering the barrel.

Glue another 90-degree street elbow to the outside end of the bulkhead, pointing down. Add a pipe to this elbow so it extends down to an effluent collection bucket. Install the cleanout onto the end of this pipe to help keep odors down. As you load the barrel, the slurry will rise in the interior pipe until it reaches the top of the elbows and flows out, keeping the material inside the barrel at a constant level.

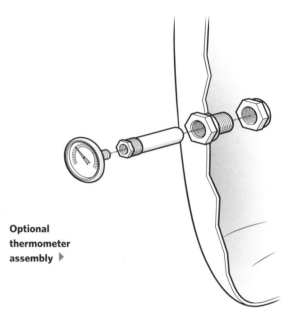

Optional thermometer assembly ▶

2. Install the gas outlet.

Drill a hole into the lid to accept the threaded male end of one of the ½" threaded bulkhead fittings. Install the fitting onto the lid, securing it on the bottom side of the lid with the provided

nut. Make sure this joint creates an airtight seal with the lid.

Wrap the threaded end of the ½" NPT x ½" barb fitting with Teflon tape, and screw the fitting into the top of the bulkhead fitting. Cut a 12" piece from the 10-foot length of ½" plastic tubing, attach it to the barb fitting and secure with a hose clamp. Connect the other end of this hose to a ball valve and secure it with a hose clamp. This valve can be used to isolate the digester from the gas storage.

3. Install the feeder tube.

Cut a hole in the center of the barrel lid, sizing it for a snug fit around the underside of the 3" toilet flange. Test-fit the flange in the hole to confirm a good fit. Mark and drill four holes through the lid for mounting the flange to the lid with machine bolts.

Cut the feeder tube to length from 3" PVC pipe. The length of the tube depends on how tall the barrel is: The pipe runs through the toilet

Gas outlet assembly ▶

flange and should extend several inches above the lid at the top end to about 12" from the bottom of the barrel when the lid assembly is installed. (It's important for the bottom of the tube to be covered by the slurry inside the digester so that no air gets in.)

Solvent-glue the pipe and flange together, following the glue manufacturer's directions. Apply silicone caulk liberally around the hole on the top side of the lid, then insert the pipe and flange through the lid. Secure the flange to the lid with four machine bolts, washers, and nuts. Add more caulk as needed around this joint to ensure that it's airtight. Let the caulk cure completely.

4. Insulate the barrel and add the heater.

If you use a black barrel for the generator, it will absorb solar heat, and some heat will be generated by the digesting process. Depending on your location, there's a good chance you'll want to insulate the barrel and have an additional heat source to keep the slurry at around 100 to 105°F. For my system (which lives in New England), I insulated the barrel with black poly-covered foam wrap designed for insulating beehives (see Resources), but you can make your own similar insulating wrap, if desired.

▲ **Notch detail for submersible pump cord**

Don't forget to insulate underneath the barrel to keep heat from conducting to the ground. One approach is to set the barrel on top of a 2"-thick piece of rigid foam insulation sandwiched in between two pieces of plywood.

A submersible, thermostatically controlled electric stock tank heater will help to ensure the correct temperature. This setup requires about a half kilowatt-hour of heat on a cloudy 70°F day, but on a warm sunny day, the heater does not come on at all.

To install the heater, simply set the unit on the bottom of the barrel, and route the power cord up and out through the feeder pipe. I cut a little notch in the top edge of the pipe to slip the cord through so that the cap can still slide over the top of the 3" feeder pipe.

5. Load the digester.

Start your first batch by adding a 5-gallon bucketful of cow or pig manure (or other source of anaerobic bacteria; see Feeding Your Biogas Generator on page 241). Use more if you have it. To this you can add a mixture of food scraps, grass clippings, or any combination of organic material that results in a C:N ratio of somewhere between 20- and 30-to-1. Be sure the material is chopped up well for best digestion and minimum scum formation.

Fill the barrel about one-third to one-half full with organic material, then cover it with an equal amount of water so that the barrel is about two-thirds to three-quarters full. Don't worry about getting this ratio exact. (You would produce a little bit of gas with only 5 gallons of organic material and 30 gallons of water, but this wouldn't be an efficient use of space inside the generator.)

For subsequent feedings, mix the chopped material 50/50 with water and pour the slurry into the feeder tube through a large funnel. Over time, you can experiment with the ratio of solid and liquid material to find the best mix for your recipe. If necessary, use a stick as a plunger to push the material all the way down and into the barrel. When you're not feeding the generator, fit the 3" PVC cap onto the end of the feeder pipe to help keep odors in and air out.

6. Prepare the gas collection barrels.

When the gas collection system is in use, the 30-gallon barrel (the gas collection barrel) is inverted into the larger, 55-gallon barrel (the water barrel) and receives the gas from the digester by way of the gas outlet hose and a fitting installed on the bottom of the gas collection barrel. As gas fills the collection barrel, the water in the larger barrel creates a seal.

Drill a hole in the bottom of the gas collection barrel and install a ½" NPT bulkhead adapter using the same procedure as before. Wrap Teflon tape around the threaded end of the ½" NPT x hose barb T-fitting and screw it into the bulkhead adapter, hand-tightening it securely.

Begin filling the 55-gallon barrel with water. When it's nearly full, turn the gas collection barrel upside down and push it down so that all of its air escapes through the T-fitting.

7. Make the gas line connections.

Pour a few cups of water into the digester gas outlet hose, and create a dip in the hose so that the water stays at the bottom of this "trap." The water in the hose will act as flashback protection for your digester should the gas in the gas collection barrel ignite. The tubing also lets you gauge the gas production rate by observing bubbles moving through the water.

Connect one end of the hose to the ball valve on the digester's gas outlet. Connect the other end of the hose to one side of the T-fitting on the gas collection barrel, securing it with a hose clamp.

Attach a 12" length of ½" tubing between the other end of the gas outlet T-fitting, then to a PVC ball valve. This valve controls gas flow to the gas burner. Connect the remaining hose to the other end of the ball valve, and finally to the gas valve on the gas burner.

Before making the final connection to the burner, tear off some bronze wool and pack it into the gas hose to fill 3" to 4" of the hose. Pack it as

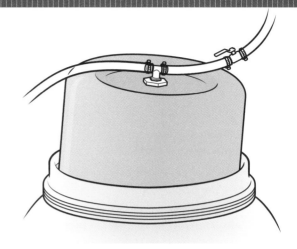

▲ **Gas line connections at collection barrel**

tightly as you can so as to still allow gas to flow through it. Secure all hose connections with hose clamps, keeping in mind that you need a tight seal because the system will be under pressure.

8. Test the gas for flammability.

Over the next few days to a week, gas should start to fill the gas barrel, and it will rise up out of its water bath (be sure the gas inlet valve is open and the gas outlet valve is closed). The first barrel full of gas produced will be primarily carbon dioxide and will not burn. Simply open the gas outlet valve and the weight of the barrel will release the gas. After the oxygen in the digester vessel is depleted, combustible gases will be produced.

To safely test for flammability, sink the gas outlet hose into a bucket of water. Push down on the gas barrel until bubbles rise in the water. Have a helper hold a lighter with a long handle (such as a barbeque lighter with a long shaft) over these bubbles as they pop: If this produces a little flare, you know you're producing biogas. Give it a try in the burner!

ADJUSTING GAS PRESSURE

For best results, use a burner with an adjustable air shutter to help control combustion. Natural and propane gas burners require a pressure of about 10" of water column (wc), which is equivalent to about 0.36 pounds per square inch (psi). It will likely be necessary to increase the gas supply pressure by bricking the inverted barrel until the burner is working satisfactorily (see page 247).

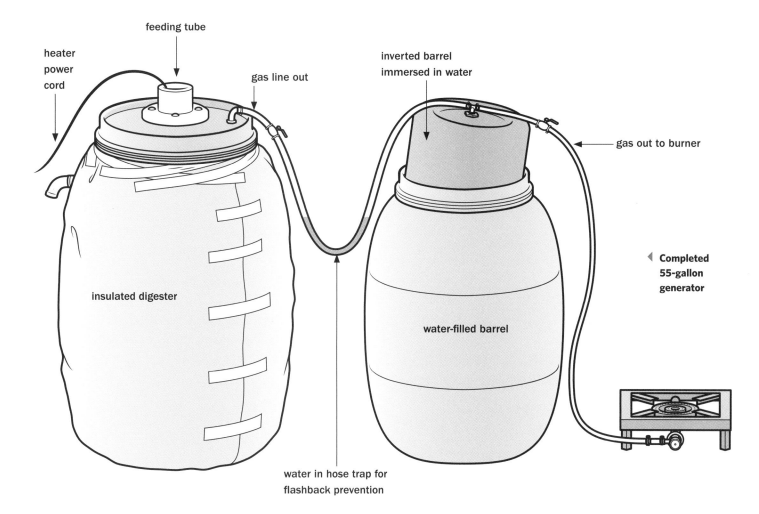

heater power cord

feeding tube

gas line out

inverted barrel immersed in water

gas out to burner

insulated digester

water-filled barrel

▲ **Completed 55-gallon generator**

water in hose trap for flashback prevention

Warning!

Do not hold a flame directly to the outlet of the digester's gas tube. With the right mixture of oxygen and gas, a flashback could occur and the generator could explode. You are working with material that is as flammable, as dangerous, and as useful as natural gas. Use common sense, caution, and the proper safety gear.

Resources Websites by Chapter

INTRODUCTION

Paul Scheckel's Website
www.nrgrev.com

CHAPTER 1 — GETTING READY FOR RENEWABLES

Appropriate Development Solutions
www.approdevsolutions.com
Promoting renewable energy in developing countries

Database of State Incentives for Renewables & Efficiency
U.S. Department of Energy
www.dsireusa.org

Del City
www.delcity.net
Cable, connectors, fuses, and wiring supplies

ENERGY STAR
www.energystar.gov

Global Exchange
www.globalexchange.org

Home Energy Magazine
www.homeenergy.org

Home Power Magazine
www.homepower.com

The Power of Community — How Cuba Survived Peak Oil
Arthur Morgan Institute for Community Solutions
www.powerofcommunity.org

Solar Energy International
www.solarenergy.org

Tax Incentives Assistance Project
www.energytaxincentives.org

CHAPTER 2 — DO YOUR OWN ENERGY AUDIT

Air Conditioning Contractors of America
www.acca.org

Air-Conditioning, Heating, and Refrigeration Institute
www.ahrinet.org

Allergy Control Products
www.allergycontrol.com
Allersearch and De-Mite products

BITS Limited
www.bitsltd.net
Smart Strip Surge Protector

Building Performance Institute, Inc.
www.bpi.org
Technical standards and certified contractors in the *Home Performance with ENERGY STAR* program

CO-Experts
G. E. Kerr Companies, Inc.
www.coexperts.com
Carbon monoxide alarms

Efficiency First
www.efficiencyfirst.org

ENERGY STAR
www.energystar.gov

P3 International Corporation
www.p3international.com
Kill A Watt Power Monitors

Residential Energy Services Network
www.resnet.us
Technical standards and certified contractors in the *ENERGY STAR for New Homes* program

ThinkTank Energy Products Inc.
www.wattsupmeters.com
Watts Up? meters

Additional Resources

Advanced Energy Panels
Windo-Therm, LLC
www.windotherm.com
Removable storm window panels

Battic Door Home Energy Conservation
www.batticdoor.com
Chimney balloon draft stopper

Energy Circle, LLC
www.energycircle.com

Energy Federation Incorporated
www.efi.org

Gordon's Window Decor, Inc.
www.gordonswindowdecor.com

Greenhouse Gas Emissions
U. S. Environmental Protection Agency
www.epa.gov/climatechange/emissions
Information on greenhouse gases

Intergovernmental Panel on Climate Change
www.ipcc.ch

Niagra Conservation
www.niagaraconservation.com

CHAPTER 3 — INSULATING YOUR HOME

Aspen Aerogels, Inc.
www.aerogel.com
Aerogel

BuildingGreen, Inc.
www.buildinggreen.com
Publishers of *Environmental Building News*

Commercial Thermal Solutions, Inc.
www.tigerfoam.com
Tiger Foam Insulation

Cool Roof Rating Council
www.coolroofs.org

Energy Federation Incorporated
www.efi.org

EnviroHomes Ltd.
www.vacuum-panels.co.uk
VacuPor Vacuum Insulation Panels

Green Building Advisor
Taunton Press, Inc.
www.greenbuildingadvisor.com

Guardian Energy Technologies, Inc.
www.sprayfoamdirect.com
Foam It Green Spray Foam Insulation Kits

Louisiana-Pacific Corporation
www.lpcorp.com

NanoPore
www.nanopore.com

Shelter Analytics LLC
www.shelteranalytics.com
Bret Hamilton and Paul Scheckel

Additional Resources

Building Envelopes Program
Oak Ridge National Laboratory
www.ornl.gov/sci/roofs+walls/facts
Handbooks and fact sheets

ENERGY STAR

www.energystar.gov

Provides a Thermal Bypass Checklist Guide (home insulation and air-sealing training material); enter *Thermal Bypass Checklist Guide* into the search bar tool

The Home Energy Diet

Written by Paul Scheckel. New Society Publishers, 2005.
www.nrgrev.com

Open Sash

www.opensash.com
Window retrofits

University of Bath

www.bath.ac.uk
United Kingdom study of embodied energy in building materials: *Inventory of Carbon & Energy* (ICE) database

U.S. Department of Energy

http://energy.gov
Various topics on energy and efficiency

CHAPTER 4 — DEEP ENERGY RETROFITS

Alliance for Low-E Storm Windows

www.low-estormwindows.com

Benjamin Obdyke, Inc.

www.benjaminobdyke.com
HydroGap

Davis Energy Group Inc.

www.davisenergy.com
NightBreeze

Dow Chemical Company

http://building.dow.com
Perimate insulation

Efficient Windows Collaborative

www.efficientwindows.org

Grace Construction Products

www.graceconstruction.com
Perm-a-Barrier

National Fenestration Rating Council

www.nfrc.org

New York State Energy Research and Development Authority

www.nyserda.ny.gov

Passive House Institute US

www.passivehouse.us

RenewABILITY Energy, Inc.

www.renewability.com
Power-Pipe, drain water heat recovery system

Swing-Green, LLC

www.swing-green.com
Green Fox, drain water heat recovery system

Tremco Commercial Sealants & Waterproofing

www.tremcosealants.com
ExoAir

Additional Resources

Building America, Building Technologies Program

U.S. Department of Energy
www.buildingamerica.gov

Building Science Corporation

www.buildingscience.com
Performs high-level work on high-efficiency buildings; site includes publications for download

BuildingGreen, Inc.

www.buildinggreen.com
Publishers of *Environmental Building News*

Fine Homebuilding Magazine

Taunton Press
www.finehomebuilding.com

Green Building Advisor

www.greenbuildingadvisor.com
Information on building energy and green building

Health House

American Lung Association
www.healthhouse.org

Home Ventilating Institute

www.hvi.org

Ventilation products, information, and ratings

Passive House Institute US

www.passivehouse.us

Source for new homes; exemplary standards for efficiency and durability

CHAPTER 5 —
HOME ENERGY MONITORING

APRS World, Inc.

www.aprsworld.com

Arduino

www.arduino.cc

Microcontroller platform

Bidgely

www.bidgely.com

BizEE Software Limited

www.degreedays.net

Degree Days

Blue Line Innovations

www.bluelineinnovations.com

Power Cost Monitor

Campbell Scientific, Inc.

www.campbellsci.com

Efergy Technologies Limited

www.efergy.com

The Energy Detective

www.theenergydetective.com

Horizon Fuel Cell Technologies

www.horizonfuelcell.com

Hotwatt, Inc.

www.hotwatt.com

Intellergy, Inc.

www.intellergy.net

Itron

www.itron.com

Pulse-enabled gas meter

Navien America, Inc.

www.navienamerica.com

On-demand water heaters

NRG Systems, Inc.

www.nrgsystems.com

Data logger

Omega Engineering, Inc.

www.omega.com

In-line water meter

Onset Computer Corporation

www.onsetcomp.com

OpenEnergyMonitor

www.openenergymonitor.org

OutBack Power Technologies, Inc.

www.outbackpower.com

P3 International Corporation

www.p3international.com

Plot Watt

www.plotwatt.com

Powerhouse Dynamics

www.powerhousedynamics.com

eMonitor

PowerWise Systems

www.powerwisesystems.com

RainWise, Inc.

www.rainwise.com

Serious Energy, Inc.

www.seriousenergy.com

Serious Energy Manager software

SimpleHomeNet

www.simplehomenet.com

Smarthome

www.smarthome.com

SmartLabs, Inc.

www.insteon.net

Insteon

ThinkTank Energy Products Inc.

www.wattsupmeters.com

Watts Up? meters

Water Heater Innovations, Inc.
Rheem Manufacturing Company
www.marathonheaters.com
Marathon water heaters

Weather Underground, Inc.
www.wunderground.com

Web Energy Logger
www.welserver.com

ZigBee Alliance
www.zigbee.org
Wireless communication standard

CHAPTER 6 — SOLAR HOT WATER

American Solar Energy Society
http://ases.org

Armacell Enterprises GmbH
www.armacell.com
Armaflex

Build-It-Solar
www.builditsolar.com

Community Hydro, LLC
www.communityhydro.biz

Florida Solar Energy Center
www.fsec.ucf.edu

McMaster-Carr
www.mcmaster.com
Industrial supply retailer

National Renewable Energy Laboratory
www.nrel.gov
Home of the Renewable Resource Data Center,
as well as analysis tools and maps

Solar Radiation Data Manual for Flat-Plate and Concentrating Collectors.
Written by William Marion and Stephen Wilcox.
National Renewable Energy Laboratory, 1994.
http://rredc.nrel.gov/solar/pubs/redbook
Solar radiation data

Solar Rating & Certification Corporation
www.solar-rating.org

W.W. Grainger, Inc.
www.grainger.com

CHAPTER 7 — SOLAR ELECTRIC GENERATION

Array Technologies, Inc.
http://arraytechinc.com
Solar tracking racks

Dow Chemical Company
www.dowpowerhouse.com
Solar shingles

DPW Solar
Preformed Line Products
www.dpwsolar.com
PV array mounting systems and information

In My Backyard Tool
National Renewable Energy Laboratory
www.nrel.gov/eis/imby
Solar power estimator

National Climate Data Center
National Oceanic and Atmospheric Administration
www.ncdc.noaa.gov

National Geophysical Data Center
National Oceanic and Atmospheric Administration
www.ngdc.noaa.gov
Geomagnetic map

National Renewable Energy Laboratory
www.nrel.gov
Home of the Renewable Resource Data Center
and the In My Backyard solar and wind power
estimator

Shelter Analytics
www.shelteranalytics.com
Energy improvement analysis tool

Solar Pathfinder
www.solarpathfinder.com

Solmetric, Inc.
www.solmetric.com
SunEye

Zomeworks Corporation
www.zomeworks.com
Solar tracking racks

Additional Resource

Solar Energy International
www.solarenergy.org
Classes on solar and wind energy

CHAPTER 8 — WIND ELECTRIC GENERATION

American Wind Energy Association
www.awea.org

Bergey WindPower Co.
www.bergey.com
Wind generators

Blow by Blow: Our Best-ever Guide to Home-scale Wind Turbines
Home Power Magazine
www.homepower.com
June/July 2010 issue, #137

Cable and Wire Shop
www.cableandwireshop.com

Chance Civil Construction
www.abchance.com
Anchors and tools

Distributed Wind Site Analysis Tool
The Cadmus Group, Inc.
http://dsat.cadmusgroup.com

Hubbell Power Systems
www.hubbellpowersystems.com
Anchors and design guide resources

Inspeed.com, LLC
www.inspeed.com
Wind measurement and data loggers

Kestrel Wind Turbines
www.kestrelwind.co.za

Kingspan Renewables Ltd.
www.kingspanwind.com
Proven wind generators

Loos & Company, Inc.
www.loosnaples.com

National Climatic Data Center
National Oceanic and Atmospheric Administration
www.ncdc.noaa.gov

NRG Systems, Inc.
www.nrgsystems.com
Wind towers, measurement, and data loggers

ROHN Products, LLC
www.rohnnet.com
Rohn towers, parts, engineering, and design

Sabre Industries, Inc.
www.sabresitesolutions.com
Tower parts

Small Wind Certification Council
www.smallwindcertification.org

Talco Electronics
www.talco.com

TESSCO
www.tessco.com
Tower parts

Wind Energy Resource Atlas of the United States
Written by D. L. Elliott, C. G. Holladay,
W. R. Barchet, H. P. Foote, and W. F. Sandusky,
1986.
National Renewable Energy Laboratory
www.nrel.gov/rredc/wind_resource.html

Wind Powering America
U.S. Department of Energy
www.windpoweringamerica.gov

Ian Woofenden
Home Power Magazine
https://homepower.com

Antenna Systems
www.guywire.net
Guy wire for towers

Solar Energy International
www.solarenergy.org
Classes on wind energy

Wind-Works.org
www.wind-works.org
Paul Gipe's wind website

CHAPTER 9 — HYDRO ELECTRIC GENERATION

Alternative Power & Machine
www.apmhydro.com
Hydroelectric systems

Ampair Energy Ltd.
www.ampair.com

Canyon Hydro
www.canyonhydro.com
Hydroelectric systems

Energy Systems & Design
www.microhydropower.com
Hydroelectric systems

Harris Hydroelectric
www.thesolar.biz/Harris_Hydro.htm

Hydroscreen, LLC
www.hydroscreen.com
Debris screens

RockyHydro
www.rockyhydro.com
Hydroelectric systems

Small Hydropower & Micro Hydropower
www.smallhydro.com
Small hydro resource site

Clean Water Act, Section 401 Certification
United States Environmental Protection Agency
http://water.epa.gov/lawsregs/guidance/
wetlands/sec401.cfm
Water Quality Certificate

Federal Hydro Licensing
Federal Energy Regulatory Commission
http://ferc.gov/industries/hydropower.asp

StreamStats
U.S. Geological Survey
http://streamstats.usgs.gov

U.S. Army Corps of Engineers
www.usace.army.mil

***Water Measurement Manual*, revised ed.**
www.usbr.gov/pmts/hydraulics_lab/pubs/wmm
United States Department of the Interior, Water
Resources Research Laboratory, 2001.

CHAPTER 10 — RENEWABLE ELECTRICITY MANAGEMENT

Del City
www.delcity.net

National Fire Protection Association
www.nfpa.org
NFPA 70: National Electrical Code

Inverters, chargers, controllers, and monitors

Apollo Solar, Inc.
www.apollosolar.com

Blue Sky Energy, Inc.
www.blueskyenergyinc.com

Bogart Engineering
www.bogartengineering.com
Tri-Metric power system monitor

EXCELTECH

www.exeltech.com

Fronius International GmbH

www.fronius.com

MidNite Solar

www.midnitesolar.com

Morningstar Corporation

www.morningstarcorp.com

OutBack Power Technologies, Inc.

www.outbackpower.com

SMA Solar Technology

www.sma-america.com

Solectria Renewables

www.solren.com

Specialty Concepts, Inc. (SCI)

www.specialtyconcepts.com

Xantrex Technology, Inc.

www.xantrex.com

Generators

Kohler Co.

www.kohler.com

Northern Lights, Inc.

www.northern-lights.com

Batteries

East Penn Manufacturing Company, Inc.

www.dekabatteries.com
Deka batteries

Interstate Battery System of America, Inc.

www.interstatebatteries.com

Rolls Battery Engineering

Surrette Battery Manufacturing
www.rollsbattery.com

Trojan Battery Company

www.trojanbattery.com

Additional Resources

Delta Lightning Arrestors

www.deltala.com

ETI

www.etigroup.eu

Mapawatt Blog

www.mapawatt.com

Zephyr Industries, Inc.

http://zephyrvent.com
Battery box vent

CHAPTER 11 —
BIODIESEL

Biodiesel: Do-it-yourself Production Basics
National Sustainable Agriculture Information
Service
www.attra.ncat.org/attra-pub/biodiesel.html
Written by Rich Dana. NCAT, rev. ed. 2012.

Biodiesel Fuel Education Program

University of Idaho
www.uiweb.uidaho.edu/bioenergy

Conney Safety Products

www.conney.com

Frey Scientific

www.freyscientific.com

Jeffers, Inc.

www.jefferspet.com

Journey to Forever

www.journeytoforever.org

Kitchen Biodiesel

www.kitchen-biodiesel.com
Biodiesel books, resources, how-to

make-biodiesel.org

www.make-biodiesel.org
Biodiesel books, resources, how-to

McMaster-Carr

www.mcmaster.com

Neptune
Pump Solutions Group
www.psgdover.com/neptune/mixers/drum-mixers
Drum mixers

Northern Tool + Equipment Catalog Co.
www.northerntool.com

Oilseed Processing for Small-Scale Producers
National Sustainable Agriculture Information Service
www.attra.ncat.org/attra-pub/oilseed.html
Written by Matt Rudolf. NCAT, 2008.

PolyDome
www.polydome.com

United States Plastic Corporation
www.usplastic.com

W. W. Grainger, Inc.
www.grainger.com

SVO Conversion Kits

Alternative Technology Group GmbH
www.diesel-therm.com
Diesel-Therm

Frybrid Diesel/Vegetable Oil
www.frybrid.com

Greasecar Vegetable Fuel Systems
www.greasecar.com

Heaters

Alternative Technology Group GmbH
www.diesel-therm.com

Five Star Manufacturing
http://fivestarmanufacturing.com

Biodiesel Kits and Supplies

B100 Supply
www.b100supply.com
Biodiesel parts and supplies

Diesel Toys, LLC
http://dieseltoys.com

Ever Green Recovered Energy Distribution, Inc.
http://evergreenred.com/

Utah Biodiesel Supply
www.utahbiodieselsupply.com

Additional Resources

Alternative Fuels Data Center
U.S. Department of Energy
www.afdc.energy.gov

ASTM International
www.astm.org

Biodiesel Emissions Analyses Program
United States Environmental Protection Agency
www.epa.gov/OMS/models/biodsl.htm

Biodiesel Handling and Use Guide, 4th ed.
www.nrel.gov/vehiclesandfuels/npbf/feature_guidelines.html
Prepared by the National Renewable Energy Laboratory, 2009.

Fuel Economy Guide
U.S. Department of Energy
www.fueleconomy.gov
Learn about auto fuel economy and tax incentives for all vehicles

National Biodiesel Board
www.biodiesel.org

Vehicle Technologies Program
U.S. Department of Energy
www1.eere.energy.gov/vehiclesandfuels

CHAPTER 12 — WOOD GAS

All Power Labs
www.gekgasifier.com
Gasifier Experimenters Kit

Aprovecho Research Center
www.aprovecho.org

Biochar Activity Kit
Greater Democracy
www.greaterdemocracy.org/archives/1316

biochar.org
www.biochar.org
Biochar information

BioEnergy Discussion Lists
www.bioenergylists.org

CO-Experts
G. E. Kerr Companies, Inc.
www.coexperts.com

Global Alliance for Clean Cook stoves
www.cleancook stoves.org

ProTech Safety
www.protechsafety.com

Slower Traffic Keep Right
www.tweecer.com
TwEECer

Spenton LLC
www.spenton.com

Victory Gasworks
www.victorygasworks.com

Additional Resources

Biomass Energy Foundation
http://biomassenergyfndn.org

Food and Agriculture Organization of the United Nations
www.fao.org
"Wood Gas as Engine Fuel," 1986;
enter "woodgas" in the search bar

Low-tech Magazine
www.lowtechmagazine.com

Mother Earth News
www.motherearthnews.com

CHAPTER 13 — BIOGAS

B and B Honey Farm
www.bbhoneyfarms.com
Colony Quilt, an insulating beehive wrap

Bread from Stones
www.breadfromstones.org

Jeffers, Inc.
www.jefferslivestock.com
Thermostatically controlled submersible heater
(product #W-449)

On-Farm Composting Handbook
Cornell Waste Management Institute
http://compost.css.cornell.edu/OnFarmHandbook/onfarm_TOC.html
Published by NRAES, 1992.

Tank Depot
www.tank-depot.com
Tank and barrel connectors

United Nations Development Programme
www.undp.org

Biogas Books and Articles

"3-Cubic Meter Biogas Plant: A Construction Manual"
http://pdf.usaid.gov/pdf_docs/PNAAP417.pdf
Published by Volunteers in Technical Assistance.
Available for download from various sources
online.

Biogas Books and Articles *continued*

A Chinese Biogas Manual
Journey to Forever
www.journeytoforever.org/biofuel_library.html
Edited by Arian van Buren. Intermediate Technology Publications, 1979.

The Complete Biogas Handbook
www.completebiogas.com
Written by David House. Alternative House Information, 2010.

Handbook of Biogas Utilization, 2nd ed.
General Bioenergy
www.bioenergyupdate.com
Written by Charles C. Ross and T. J. Drake III. Environmental Treatment Systems, 1996.

"Methane Recovery from Animal Manures: The Current Opportunities Casebook"
ManureNet, Conservation Ontario
http://agrienvarchive.ca/bioenergy/download/methane.pdf
Written by P. Lusk. National Renewable Energy Laboratory, 1998.

Additional Resources

AgSTAR Program
U.S. Environmental Protection Agency
www.epa.gov/agstar

Beginners Guide to Biogas
University of Adelaide
www.adelaide.edu.au/biogas

"Methanol Recovery: Dickinson College Biodiesel"
Dickinson College
http://dickinson.edu/about/sustainability/biodiesel/content/Presentations
Methanol recovery presentation by Matt Steiman.

Metric

Conversion Charts

Unless you have finely calibrated measuring equipment, conversions between U.S. and metric measurements will be somewhat inexact. It's important to convert the measurements for all of the ingredients in a recipe to maintain the same proportions as the original.

GENERAL FORMULA FOR METRIC CONVERSION

Ounces to grams	multiply ounces by 28.35
Grams to ounces	multiply grams by 0.035
Pounds to grams	multiply pounds by 453.5
Pounds to kilograms	multiply pounds by 0.45
Cups to liters	multiply cups by 0.24
Fahrenheit to Celsius	subtract 32 from Fahrenheit temperature, multiply by 5, then divide by 9
Celsius to Fahrenheit	multiply Celsius temperature by 9, divide by 5, then add 32

APPROXIMATE EQUIVALENTS BY VOLUME

U.S.	Metric
1 teaspoon	5 millileters
1 tablespoon	15 millileters
$1/4$ cup	60 milliliters
$1/2$ cup	120 milliliters
1 cup	230 milliliters
$1^1/4$ cups	300 milliliters
$1^1/2$ cups	360 milliliters
2 cups	460 milliliters
$2^1/2$ cups	600 milliliters
3 cups	700 milliliters
4 cups (1 quart)	0.95 liter
1.06 quarts	1 liter
4 quarts (1 gallon)	3.8 liters

APPROXIMATE EQUIVALENTS BY WEIGHT

U.S.	Metric
$1/4$ ounce	7 grams
$1/2$ ounce	14 grams
1 ounce	28 grams
$1^1/4$ ounces	35 grams
$1^1/2$ ounces	40 grams
$2^1/2$ ounces	70 grams
4 ounces	112 grams
5 ounces	140 grams
8 ounces	228 grams
10 ounces	280 grams
15 ounces	425 grams
16 ounces	454 grams
(1 pound)	

Metric	U.S.
1 gram	0.035 ounce
50 grams	1.75 ounces
100 grams	3.5 ounces
250 grams	8.75 ounces
500 grams	1.1 pounds
1 kilogram	2.2 pounds

Index

Other Storey Titles You Will Enjoy

The Backyard Homestead, edited by Carleen Madigan.
A complete guide to growing and raising the most local food available anywhere —
from one's own backyard.
368 pages. Paper. ISBN 978-1-60342-138-6.

The Backyard Lumberjack, by Frank Philbrick & Stephen Philbrick.
Practical instruction and first-hand advice on the thrill of felling, bucking, splitting, and stacking wood.
176 pages. Paper. ISBN 978-1-58017-634-7.

Just In Case, by Kathy Harrison.
An empowering guide that takes readers through the process of setting up a sensible home system
that takes over when outside services are disrupted.
240 pages. Paper. ISBN 978-1-60342-035-8.

A Landowner's Guide to Managing Your Woods, by Ann Larkin Hansen, Mike Severson &
Dennis L. Waterman.
How to maintain a small acreage for long-term health, biodiversity, and high-quality timber production.
304 pages. Paper. ISBN 978-1-60342-800-2.

Storey's Basic Country Skills, by John and Martha Storey.
A treasure chest of information on building, gardening, animal raising, and homesteading —
perfect for anyone who wants to become more self-reliant.
576 pages. Paper. ISBN 978-1-58017-202-8.

The Year-Round Vegetable Gardener, by Niki Jabbour.
How to grow your own food 365 days a year, no matter where you live!
256 pages. Paper. ISBN 978-1-60342-568-1.
Hardcover. ISBN 978-1-60342-992-4.

**These and other books from Storey Publishing are available
wherever quality books are sold or by calling 1-800-441-5700.
Visit us at *www.storey.com*.**